1Ø 120v, 20A, 2wire circuit

Hint. Diagram.

Modern Commercial Wiring

Based on the 2005 NEC®

by

Harvey N. Holzman

Master Electrician

Member, International Association of Electrical Inspectors

Based on the 2005 National Electrical Code®

Publisher

The Goodheart-Willcox Company, Inc.

Tinley Park, Illinois

www.g-w.com

Manufactured in the United States of America.

Library of Congress Catalog Card Number 2004059918

International Standard Book Number 1-59070-438-X

1 2 3 4 5 6 7 8 9—05—09 08 06 07 06 05

Library of Congress Cataloging in Publication Data

Holzman, Harvey N.
Modern commercial wiring / by Harvey N. Holzman.

p. cm.
"Based on the 2005 National electrical code."
Includes index.
ISBN 1-59070-438-X
1. Electrical wiring, Interior.
2. Commercial buildings–Electric equipment. I. Title.
TK3284.H65 2005 2004059918
621.319'24–dc22

Contents

Preface

The purpose of this text is to help the electrical practitioner—student, apprentice, journeyman, designer, engineer, contractor, and maintenance person—working in the commercial electrical trade. It is written in clear and simple language from one electrician to others.

Basic electrical concepts and theory are reviewed, but it is assumed the reader has some prior knowledge and experience with electrical principles and electrical wiring. Every chapter is geared toward providing the practitioner with state-of-the-art design criteria, methods, and requirements of the *National Electrical Code*®. A concentrated effort has been made to be accurate in order to give the reader useful instruction and guidance that can be implemented in the real, day-to-day practices of commercial wiring. This includes (but is not restricted to) installing, designing, troubleshooting, and maintaining electrical systems, devices, and equipment.

The chapters are arranged in a logical sequence, but are not dependent on one another. Each could be studied as a stand-alone entity. The chapters begin with clear objectives and close with review questions to reinforce what has been learned.

Hopefully, this volume will provide many instructive and useful bits of information to the reader. Your comments and suggestions for additions and improvements are most welcome and will serve to enhance future editions.

Harvey N. Holzman

Acknowledgments

Few technical texts are produced by the efforts of one person. This text is no exception. Many individuals contributed their time and energy to provide much appreciated assistance and materials that enhanced the project. The following individuals, companies, and organizations deserve credit for helping to create this quality text. I thank each and every one of you for your unselfish contributions.

Information and photos

Allen-Bradley
Appleton Electric Company
Arlington Industries, Inc.
Automatic Switch Company
AVO International and Biddle Instruments
Bussmann Division of Cooper Industries
Calculated Industries, Inc.
Channellock, Inc.
Chloride Systems
Construction Specification Institute
Cooper Industries, Cooper Power Systems
Crouse-Hinds Division, Cooper Industries
Eaton Corporation, Cutler-Hammer Products
Electrical Testing Laboratories
Fluke Corporation
Furnas Electric Company
General Electric Company
Greenlee Textron, Inc.
Hipotronics, Inc.
Ideal Industries, Inc.
International Association of Electrical Inspectors
Marc Kaducak
Jack Klasey
Klein Tools, Inc.

Kohler Company
The Langford Group
L. S. Starrett Company
O/Z Gedney Electric
National Fire Protection Association
PW Industries, Inc.
RACO, Inc.
Reliance Electric
Ridge Tool Company
S&C Electric Company
Southwire Company
Square D Company
Star Products Division, U.S. Trolley Corp.
Underwriters Laboratories, Inc.
Vermont American Tool Company
Wavetek Corporation
Westinghouse
The Wiremold Company

Illustration assistance

Marylin Agee, Square D Company
Patsy Campbell, Wavetek Corporation
Gary R. Carr , Greenlee Textron, Inc.
Sandra L. Davis, Ideal Industries
Mike Harrington
Peter Hayward, Westinghouse
Diane M. Khar, Square D Company
W.H. Korb, PW Industries, Inc.
Gale Langford, The Langford Group
Claudia K. Lester, Calculated Industries, Inc.
Ron Mabry
Marilyn Muscenti, Reliance Electric
Michael D. Myer, Star Products Division
Peg O'Neil, AVO International
Randy Polito
Kenneth R. Reinehr, Appleton Electric Company
Sheryl L. Schaffer, Electrical Testing Laboratories
Holly Jo Schubert, Underwriters Laboratories, Inc.
Amy Scoggin, Square D Company
Franny Singleton, Square D Company
Vicki Snapp, Square D Company
Beverly A. Summers, Fluke Corporation
Amy E. Takas, Chloride Systems
Anne Tighe, Hipotronics
Candace Tyndall, Square D Company
Lynn Watkins, Bussmann Div., Cooper Industries
Steve Wilcox, Furnas Electric Company
Janice A. Zimmerman, Klein Tools, Inc.
Richard J. Zuccaro, General Electric Company

Manuscript typing

Dahlia Tracey
Leslie Tracey
Melissa Rodriquez

Text Features

Many features have been incorporated into the design of this text. These features are designed to aid in your understanding of commercial wiring and to improve your familiarity with the *National Electrical Code*®.

WARNING

Warnings indicate situations that may be harmful or dangerous. Read Warnings carefully—what you learn may prevent an accident someday.

CAUTION

Cautions warn of situations that may result in damage to equipment or materials.

NOTE

Notes provide additional information related to a topic.

NEC NOTE 90.3

NEC Notes provide a direct reference to applicable material in the *National Electrical Code.* The *Code* section from which the note is taken is referenced in the heading. For instance, this note is labeled as being contained in *Section 90.3.* NEC Notes contain only a portion of the material in the referenced section—they do *not* serve as a substitute for the *National Electrical Code.* Always refer to the *Code* section for the appropriate information.

Pedagogical Features

The following features have been included to improve comprehension:

- **Technical Terms**—Each chapter opens with a list of Technical Terms. Read the list before studying the chapter, pay special attention when a term is defined in the text, and then review the list after completing the chapter to be sure you understand all the terms.
- **Learning Objectives**—Included in each chapter, the Learning Objectives define the knowledge and skills that will be gained when the chapter is completed.

- **Test Your Knowledge**—Located at the end of each chapter, these questions test comprehension of the material presented in this text.
- **Using the NEC**—These questions are designed to improve your comfort and familiarity with the *National Electrical Code.* You must use the *Code* to answer these questions because the answers are not contained in this text.

Note on Material from the National Electrical Code

Tables and excerpts from the *National Electrical Code* are reprinted with permission from NFPA 70-2005, the *National Electrical Code®,* Copyright© 2005, National Fire Protection Association, Inc., Quincy, Massachusetts, 02169. This reprinted material is not the complete and official position of the National Fire Protection Association, which is represented only by the standard in its entirety.

Changes in 2005 NEC

Every edition of the *Code* includes many improvements. The following are three of the biggest changes you will find in the 2005 *NEC:*

- **Administration and Enforcement**—*Article 80* has been relocated to Annex G.
- **Chapter 3, Wiring Methods and Materials**—Most articles, *300–392*, have had some important changes affecting the way various wiring methods are to be executed in the field.
- **New Articles**—There are many new articles appearing in the 2005 *Code*, as well as some that have been removed or relocated.

Trademarks

National Electrical Code® and *NEC®* are registered trademarks of the National Fire Protection Association, Inc., Quincy, Massachusetts.

E-Z Check™ GFI Circuit Tester is a registered trademark of Ideal Industries, Inc., Sycamore, Illinois.

CUBEFuse™ is a trademark of Cooper Bussmann, Inc. St. Louis, MO.

MasterFormat™ is a trademark of The Construction Specification Institute, Alexandria, Virginia.

Chapter 1

Electrical Fundamentals Review

Technical Terms

Alternating current (ac)
Alternator
Branched circuit
Capacitance
Conductors
Current
Direct current (dc)
Electromagnetic induction
Electromotive force (emf)
Equivalent circuit
Frequency
Impedance
Inductance
Instantaneous voltage
Insulator
Nominal voltage
Ohm's law
Parallel circuit
Power
Power factor
Rated voltage
Resistance
Root-mean-square (RMS) voltage
Series circuit
Voltage
Work

Objectives

After completing this chapter, you will be able to:

- Define the basic units for electrical voltage, current, resistance, energy, and power.
- Calculate electrical power.
- Discuss the characteristics of series, parallel, and complex circuits.
- Calculate voltage, current, and resistance in series, parallel, and complex circuits.
- Describe and compare alternating current and direct current.
- Explain the functions of basic components of an alternator.
- Explain the effects of inductance and capacitance in ac circuits.
- Calculate power factor and impedance in ac circuits.
- Describe single-phase and three-phase electrical systems.
- Recognize wye and delta configurations.
- Distinguish between *nominal voltage* and *rated voltage.*

This chapter reviews the fundamental concepts of electricity. This review will help you master the material in this book. The material covered in this chapter is intended as a refresher for those already familiar with the basic concepts and fundamental electrical relationships. If you have no prior training or experience with electrical theory, this chapter will provide the basics and encourage you to pursue further study.

Basic Electrical Circuit Units

The flow of electricity in a circuit can be compared to the flow of water through pipes. There are several measurable quantities associated with water flow. There is the rate of flow, the force pushing the water through the pipe, and the opposition to the flow caused by friction with the pipe. An electrical circuit has similar characteristics. The rate of flow of electricity is called *current,* the force pushing the electricity is called *voltage,* and an opposition to the flow of electricity is called *resistance.* These are the basic units of electricity.

Current

All matter is made up of tiny particles called atoms. Each atom has a central nucleus around which electrons orbit in much the same way the planets orbit around the sun, **Figure 1-1.** Depending on the type of atom and its structure, some electrons are free to move to other nearby atoms. Under normal conditions, this occurs randomly without any specific direction or net gain or loss within the substance.

If an electron force is applied to the substance, the electrons can be "pushed" in one direction. This directional movement is ***current.*** Thus, current is to electricity as flow is to water, **Figure 1-2.**

Certain substances, particularly metals, are composed of atoms that have many easily moved electrons. These

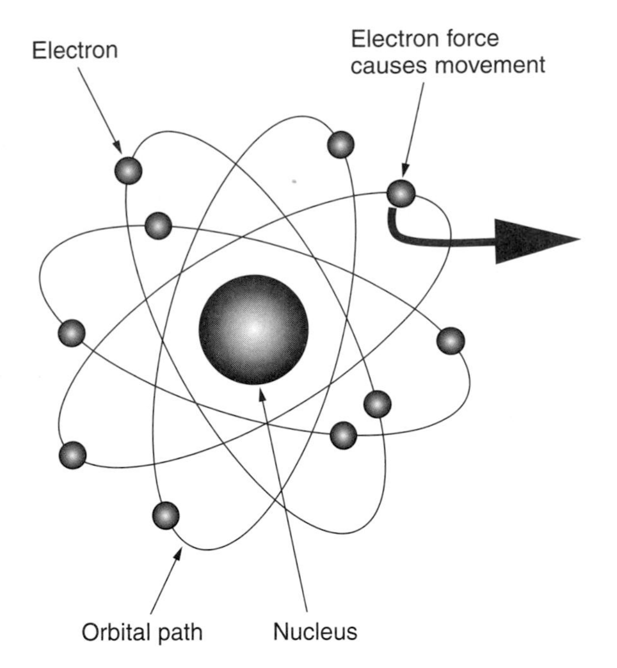

Figure 1-1. In an atom, electrons orbit the nucleus the same way planets in the solar system orbit the sun.

substances—such as silver, copper, and aluminum—are called ***conductors.*** Other substances—such as rubber, plastic, and wood—are composed of atoms with few easily moved electrons. These substances, which resist electron flow, are called ***insulators.*** They are unable to carry a current.

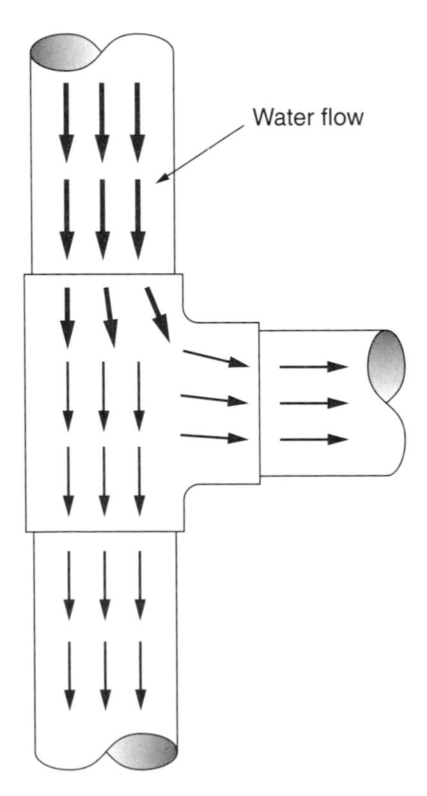

Figure 1-2. Current flow through a wire is similar to water flow through a pipe.

Current is a measurable quantity. The number of electrons that flow through a conductor in a given amount of time is measured in units called ***amperes.*** One ampere is equivalent to 6.25×10^{18} electrons passing a given point in one second. Amperes are usually abbreviated as *amps* or *A*. The symbol used in formulas to represent current is *I*.

Voltage

For a current to flow in a conductor, a force must be applied at one end. This electron-pushing force is called the ***electromotive force (emf)***. This force can be produced in various ways—chemically, mechanically, or by other forms of energy (such as sunlight). Specific examples are batteries (chemical), generators (mechanical), and photovoltaic cells (sunlight). Voltage is a measurement of the emf produced by a source. The ***volt (V)*** is the basic unit. It is defined as the potential difference between two points in an electric circuit when the energy needed to move one ampere between the points is one *joule*.

NOTE

In formulas, voltage is represented by either *E* or *V*. *E* will be used in this text.

Resistance

The opposition to the flow of electrons through a circuit is called ***resistance*** and is measured in units called ***ohms.*** The symbol for ohms is the Greek letter omega (Ω). One ohm is defined as the electrical resistance that allows one ampere to flow when one volt is applied.

For an electric circuit having a constant resistance, the current and voltage are directly related. That is, if the voltage is doubled, the current will double; if the voltage is increased ten times, the current will increase ten times. Thus, for any given circuit (with a constant resistance), the ratio of voltage to current is constant. This relationship is called ***Ohm's law*** and can be expressed mathematically as

$$R = \frac{E}{I}$$

where

R = Resistance (ohms, Ω)
E = Voltage (volts, V)
I = Current (amps, A)

As we go on, we will see how Ohm's law is applicable to many types of problems involving electrical circuits.

Power, Energy, and Work

The basic purpose of an electric circuit is to perform work. ***Work*** is the process by which energy is transformed from one type to another. For example, an electric motor performs work by changing electrical energy into mechanical energy. *Any form of energy has the ability to do work.* Both work and energy are practically identical and can be thought of interchangeably.

Power is the rate at which energy is transformed; it is the rate of doing work. Electric power is measured in ***watts (W),*** and is the product of voltage times current. Since a watt and a volt-ampere are the same, power is often expressed in ***volt-amperes (VA).*** Power is expressed mathematically as

$$P = E \times I$$

or

$$P = I^2 \times R$$

where

P = Power
E = Voltage
I = Current
R = Resistance

Figure 1-3 shows the mathematical relationships of power, voltage, current, and resistance.

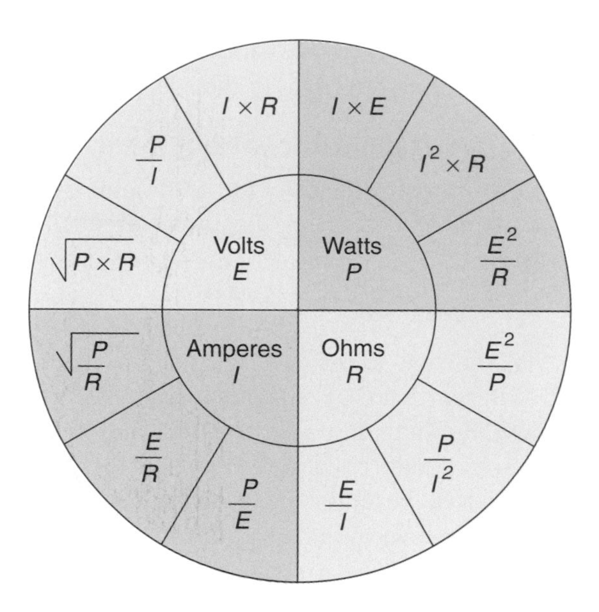

Figure 1-3. Mathematical relationship of power, voltage, current, and resistance. Within each quadrant, the three outer expressions are equal to the inner quantity.

Basic Circuits

Simple and complex circuits are composed of the same components. All circuits must have a source, a device, and a conductor. The source provides the electromotive force. The device, resistance, or element (these terms are equivalent) uses the energy to perform some desired function. The conductor (most often a wire) carries the current from the source to the device and back to the source. Other components found in a circuit are controllers (such as switches) and circuit protection (such as a fuse or circuit breaker). A very basic circuit is shown in **Figure 1-4.**

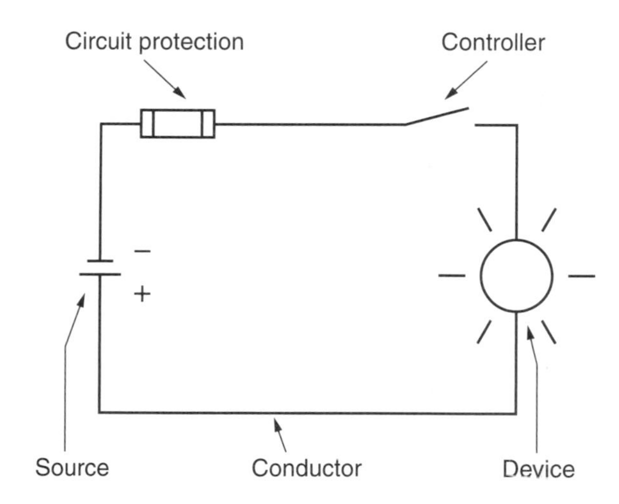

Figure 1-4. A simple electric circuit. The source supplies electromotive force, which moves current through the conductor to the other components.

There are several types of circuits. The various types are differentiated by the arrangement of the resistors and other elements. There are series circuits, parallel circuits, and complex circuits.

Series Circuits

A ***series circuit,*** **Figure 1-5,** is a circuit in which the loads are connected in such a way that there is only one current path. As shown in the illustration, the current

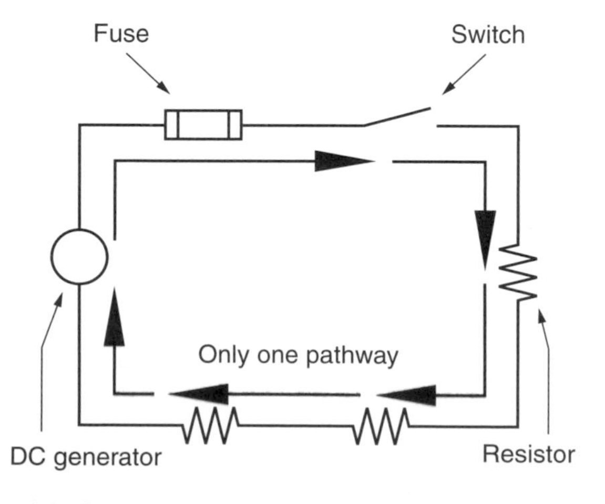

Figure 1-5. Series circuits have only one pathway for current.

must pass through the first resistor, then the second, and then the third before returning to the source.

Characteristics of Series Circuits

- The current *(I)* is the same through every element and every resistor in the circuit.

$$I_{TOTAL} = I_1 = I_2 = I_3 = I_N$$

- The voltage across each resistor is a product of the current times the individual resistance. Recall Ohm's law ($E = I \times R$). The voltage across each resistance can be different. However, the total voltage must equal the sum of the individual voltages across the resistors.

$$E_{TOTAL} = E_1 + E_2 + E_3 + \ldots + E_N = E_T$$

- The total resistance of the circuit is the sum of the individual resistances.

$$R_{TOTAL} = R_1 + R_2 + R_3 + \ldots + R_N = R_T$$

- Ohm's law is applicable to any part of the circuit or to the circuit as a whole. This is extremely important to remember. It is the key to understanding and solving circuit problems.

The following example illustrates the characteristics and relationships of a series circuit and how Ohm's law is used to solve circuit problems.

Sample Problem 1-1

Problem: Using the following figure, calculate the current flowing through the circuit and the voltage at each resistor.

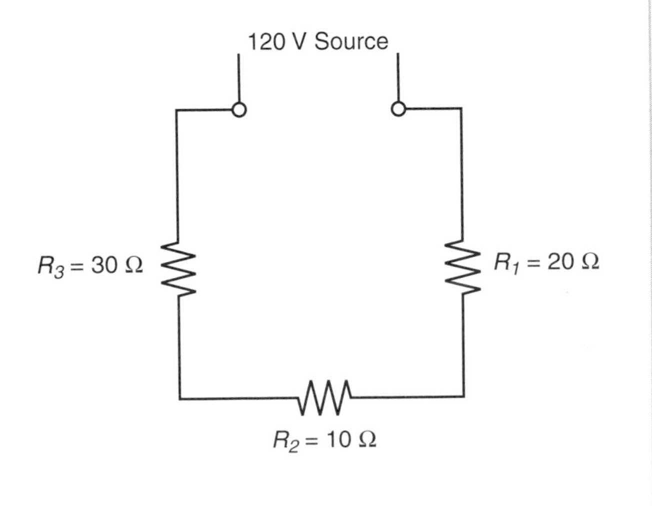

Formulas:

$$I_T = \frac{E_T}{R_T}$$

$$R_T = R_1 + R_2 + R_3$$

$$I_T = I_1 = I_2 = I_3$$

$$E_N = I_N R_N = I_T R_N$$

Solution: The current can be calculated using the formula

$$I_T = \frac{E_T}{R_T}$$

To find the total resistance, add the individual resistances together:

$$\begin{aligned} R_T &= R_1 + R_2 + R_3 \\ &= 20\ \Omega + 10\ \Omega + 30\ \Omega \\ &= 60\ \Omega \end{aligned}$$

The voltage is 120 volts, so the current is

$$\begin{aligned} I_T &= \frac{120\text{ V}}{60\ \Omega} \\ &= 2\text{ A} \end{aligned}$$

The voltage at each resistor can be calculated using the fact that current is constant throughout the circuit:

$$I_1 = I_2 = I_3 = I_T$$

Therefore,

$$\begin{aligned} E_1 &= I_1 R_1 \\ &= I_T R_1 \\ &= 2\text{ A} \times 20\ \Omega \\ &= 40\text{ V} \end{aligned}$$

$$\begin{aligned} E_2 &= I_2 R_2 \\ &= I_T R_2 \\ &= 2\text{ A} \times 10\ \Omega \\ &= 20\text{ V} \end{aligned}$$

$$\begin{aligned} E_3 &= I_3 R_3 \\ &= I_T R_3 \\ &= 2\text{ A} \times 30\ \Omega \\ &= 60\text{ V} \end{aligned}$$

The current flowing through the circuit is 2 amps, with resistor voltages of 40 volts (E_1), 20 volts (E_2), and 60 volts (E_3).

Series circuits are rarely encountered in practical commercial, residential, or industrial wiring applications. These circuits have several inherent problems:

- Any change of a resistance in the circuit affects the current throughout the circuit. This, in turn, changes the voltage across the other resistors in the circuit.
- If any part of the path is broken (opened), the circuit is dead. If one element fails, the entire circuit is shut off.
- Switches or other controlling devices cannot be used for individual loads within the circuit. If one load is turned off, all loads lose power.
- Voltage across any load depends on the overall circuit arrangement. It is almost impossible to design circuit devices that can operate at so many different voltages.

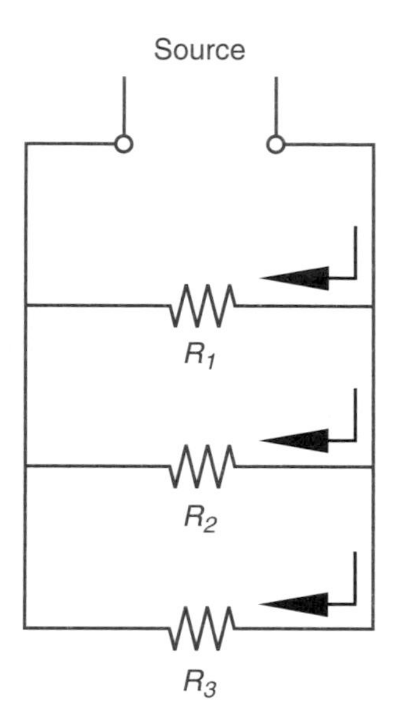

Figure 1-6. Current moves through parallel circuits using more than one pathway.

Parallel Circuits

A ***parallel circuit*** is one in which the elements are arranged in such a manner that there are several paths for the current. **Figure 1-6** illustrates a parallel circuit. In electrical circuits, the term *parallel* does not necessarily mean physically or geometrically parallel, but merely signifies alternate routes or branches. Parallel circuits are often called ***branched circuits.***

Characteristics of Parallel Circuits

- The voltage across all branches of the circuit is the same as the total voltage at the source.

$$E_{TOTAL} = E_A = E_B = E_C = E_N$$

- The total current is equal to the sum of the currents flowing through each of the branches. Thus,

$$I_{TOTAL} = I_A + I_B + I_C + \ldots + I_N$$

- The total resistance in a parallel circuit is the reciprocal of the sum of the reciprocals of the separate branch resistance. To express this in a much clearer manner, let us look at it mathematically:

$$\frac{1}{R_{TOTAL}} = \frac{1}{R_1} + \frac{1}{R_2} + \frac{1}{R_3} + \ldots + \frac{1}{R_N}$$

- Ohm's law applies to the entire circuit and any of the branches.
- Any opening or break of a branch does not stop the current flow through other branches. For this reason, parallel circuits have a distinct advantage over series circuits.

Notice that the characteristics of a parallel circuit are in many ways opposite those of a series circuit. The following example illustrates the relationships at work in a parallel circuit.

Sample Problem 1-2

Problem: Using the following figure, find (a) the current through each resistor, (b) the total current of the circuit, and (c) the total circuit resistance.

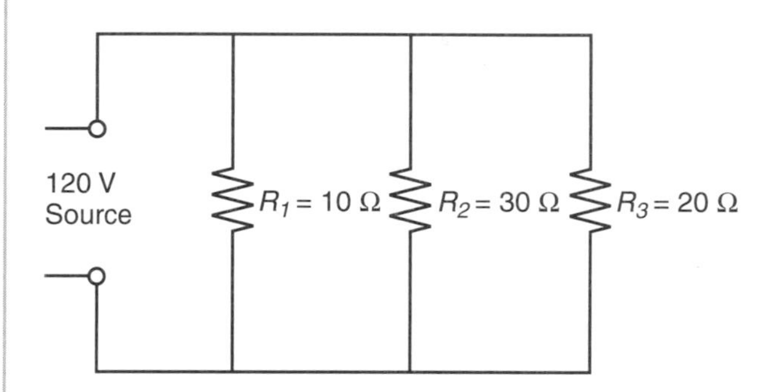

(a) Formulas:

$$E_T = E_1 = E_2 = E_3$$

$$I_1 = \frac{E_1}{R_1}$$

$$I_2 = \frac{E_2}{R_2}$$

(Continued on the following page.)

Sample Problem 1-2 *Continued*

(a) Solution:

$$E_T = E_1 = E_2 = E_3 = 120\ \text{V}$$
$$I_1 = \frac{120\ \text{V}}{10\ \Omega} = 12\ \text{A}$$
$$I_2 = \frac{120\ \text{V}}{30\ \Omega} = 4\ \text{A}$$
$$I_3 = \frac{120\ \text{V}}{20\ \Omega} = 6\ \text{A}$$

(b) Formula:

$$I_T = I_1 + I_2 + I_3$$

(b) Solution:

$$I_T = 12\ \text{A} + 4\ \text{A} + 6\ \text{A} = 22\ \text{A}$$

(c) Formula:

$$\frac{1}{R_T} = \frac{1}{R_1} + \frac{1}{R_2} + \frac{1}{R_3}$$

(c) Solution:

$$\frac{1}{R_T} = \frac{1}{10\ \Omega} + \frac{1}{30\ \Omega} + \frac{1}{20\ \Omega} = \frac{6}{60} + \frac{2}{60} + \frac{3}{60} = \frac{11}{60} = 0.18333$$
$$R_T = \frac{60}{11} = \frac{1}{0.18333} = 5.45\ \Omega$$

Arranging elements in a parallel circuit is far more practical than arranging them in a series circuit. Devices are subject to almost no adverse effects from other elements, as is the case with series circuits.

Complex Circuits

Most circuits are not simply series or parallel, but a complex arrangement consisting of series and parallel portions. A series-parallel combination circuit can be broken down into an ***equivalent circuit*** by applying the concepts mentioned previously.

Sample Problem 1-3

Problem: Refer to the following complex circuit. The source voltage is 12 volts. What is the (a) total resistance, (b) total current, and (c) voltage across each resistor?

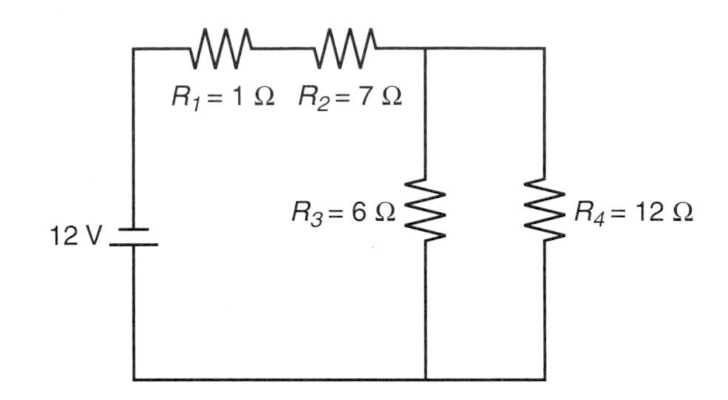

(a) Formulas:

$$\frac{1}{R_{EQ}} = \frac{1}{R_3} + \frac{1}{R_4}$$
$$R_T = R_1 + R_2 + R_{EQ}$$

(a) Solution: Think of the circuit as a series circuit where R_1 is the first resistor, R_2 is the second, and R_3 and R_4 combine to make the third resistor.

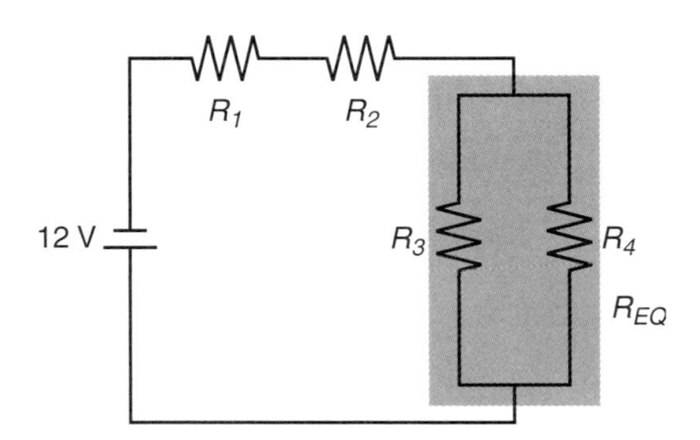

The total resistance of the two parallel branches (R_3 and R_4) is the equivalent resistance. The equivalent voltage is the total voltage in the parallel branches.

$$\frac{1}{R_{EQ}} = \frac{1}{6}\ \Omega + \frac{1}{12}\ \Omega = \frac{2}{12} + \frac{1}{12} = \frac{3}{12}\ \Omega = \frac{1}{4}\ \Omega$$
$$R_{EQ} = 4\ \Omega$$
$$R_T = 1\ \Omega + 7\ \Omega + 4\ \Omega = 12\ \Omega$$

(Continued on the following page.)

Sample Problem 1-3 *Continued*

(b) Formulas:

$$I_T = \frac{E_T}{R_T}$$

(b) Solution:

$$I_T = \frac{12\text{ V}}{12\ \Omega}$$
$$= 1\text{ A}$$

(c) Formulas:

$$I_1 = I_2 = I_{EQ} = I_T$$
$$E_1 = I_1R_1$$
$$E_2 = I_2R_2$$
$$E_{EQ} = I_{EQ}R_{EQ}$$
$$E_3 = E_4 = E_{EQ}$$

(c) Solution:

$$I_1 = I_2 = I_{EQ} = 1\text{ A}$$
$$E_1 = 1\text{ A} \times 1\ \Omega$$
$$= 1\text{ V}$$
$$E_2 = 1\text{ A} \times 7\ \Omega$$
$$= 7\text{ V}$$
$$E_{EQ} = 1\text{ A} \times 4\ \Omega$$
$$= 4\text{ V}$$
$$E_3 = 4\text{ V}$$
$$E_4 = 4\text{ V}$$

Notice in Sample Problem 1-3 how the summation of the voltage across each of the resistors within the equivalent series circuit is equal to the source voltage. The power is also conserved:

$$P_T = E_TI_T$$
$$= 12\text{ V} \times 1\text{ A}$$
$$= 12\text{ W}$$

$$P_1 + P_2 + P_{EQ} = E_1I_1 + E_2I_2 + E_{EQ}I_{EQ}$$
$$= (1\text{ V} \times 1\text{ A}) + (7\text{ V} \times 1\text{ A}) + (4\text{ V} \times 1\text{ A})$$
$$= 1\text{ W} + 7\text{ W} + 4\text{ W}$$
$$= 12\text{ W}$$

The two power values are equal. The various characteristics of electric circuits are summarized in **Figure 1-7.**

Alternating Current

Alternating current (ac) has current constantly changing direction and voltage continuously changing value and polarity. Therefore, just understanding voltage, current, and resistance is not enough (as it is with direct current). We must be familiar with how alternating current is generated and its inherent characteristics and relationships.

Alternating current is produced by generators and then distributed by a vast transmission system to plants, commercial buildings, and homes. Alternating current is easier and less costly to generate than ***direct current (dc)*** and is conveniently distributed. In addition, the voltage and current can be altered using transformers to suit specific requirements.

Circuit Characteristics

	Series Circuits	Parallel Circuits	Complex Circuits
Resistance Symbol: R Unit: Ohm (Ω)	Sum of individual resistances $R_T = R_1 + R_2 + R_3 + \ldots + R_N$	$\frac{1}{R_T} = \frac{1}{R_1} + \frac{1}{R_2} + \frac{1}{R_3} + \ldots + \frac{1}{R_N}$	Total resistance equals resistance of parallel portion and sum of series resistors.
Current Symbol: I Unit: Amperes (A)	The same throughout the entire circuit $I_{TOTAL} = I_1 = I_2 = I_3 = \ldots = I_N$	Sum of individual currents $I_T = I_A + I_B + I_C + \ldots + I_N$	Series rules apply to series portion of the circuit. Parallel rules apply to parallel part of the circuit.
Voltage Symbol: E Unit: Volts (V)	Sum of individual voltages $E_T = E_1 + E_2 + E_3 + \ldots + E_N$	Total voltage and branch voltage are the same $E_T = E_A = E_B = E_C = \ldots = E_N$	Total voltage is sum of voltage drops across each resistor and each of the branches of the parallel portion.

Figure 1-7. Summary of the characteristics of electrical circuits.

Electromagnetic Induction

When a conductor is moved across a magnetic field, as shown in **Figure 1-8,** an electromagnetic force is produced in the conductor. If the conductor forms a closed loop (circuit), the emf will cause a current to flow around this loop. This is referred to as ***electromagnetic induction.***

In a generator, conductors form loops or circuits that are moved through a magnetic field. This creates a source of emf. See **Figure 1-9.**

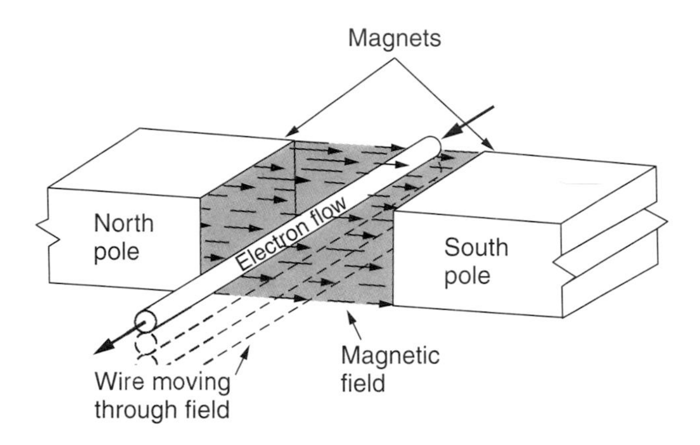

Figure 1-8. Moving a conductor up and down between the poles of a magnet will displace electrons in the conducting wire and cause them to move through the conductor. This is called electromagnetic induction.

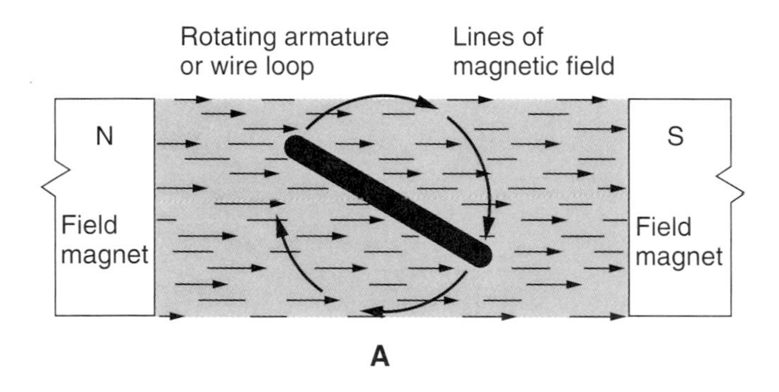

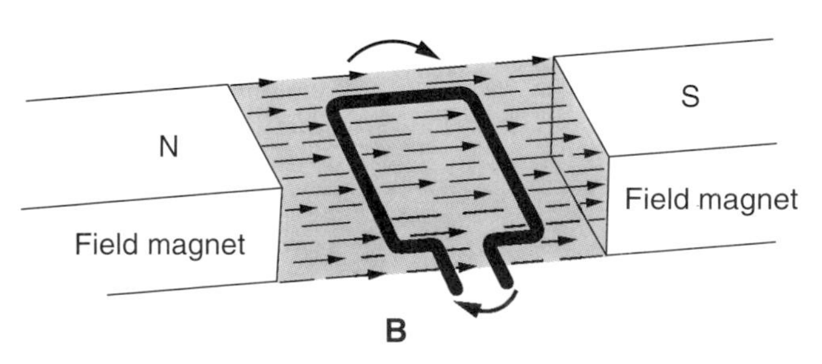

Figure 1-9. A simple generator of electricity. A rotating coil of wire has current induced in it as it cuts a magnetic field created between the poles of a magnet. A—Straight-on view of the generator. Voltage is low with loop in this position. The wire is cutting the field at a sloping angle. B—Perspective view showing loop cutting the magnetic field at a right angle. Voltage is highest in this position.

Alternators

An ac generator, or ***alternator,*** consists of several parts, **Figure 1-10.** The following is a brief description of the four basic parts:

- A ***coil*** or ***armature*** rotates. Its conducting wire cuts across a magnetic field. The electron flow begins in this part.
- Stationary ***poles*** create the necessary magnetic field. In some generators, the poles are magnetized by a portion of the electrical current generated.
- Metal ***slip rings*** are always in contact with one of the terminals of the coil or armature. They transfer the current to the brushes.
- The ***brushes*** transfer current from slip rings to the external circuit. One brush is in contact with each slip ring.

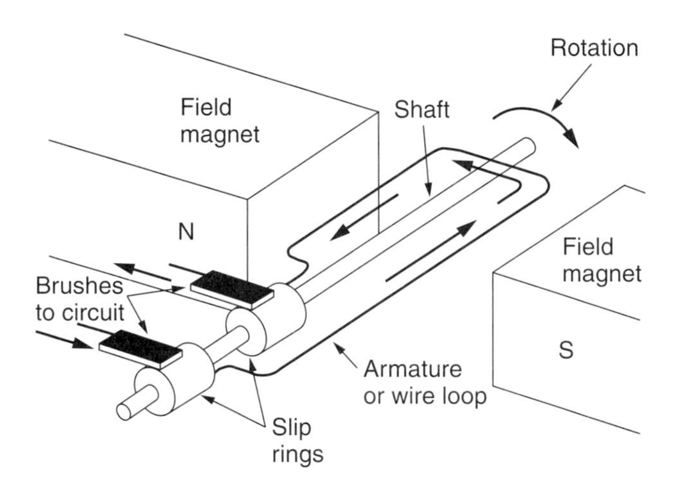

Figure 1-10. Basic elements of an alternator. The wire loop carries induced current. Electrons flow out of one brush, through the circuit, and back in through the other brush.

Alternator operation

As the armature is turned, it cuts through the magnetic field created by the field magnets. Electrons flow through the wires of the armature. The electrons move into one of the slip rings, then to one of the brushes and, finally, into the external circuit. See **Figure 1-11.**

Every half-turn, the electron flow reverses. Then the electrons flow out to the external circuit, through the opposite slip ring and brush. This creates the reversal of electron flow that accounts for alternating current.

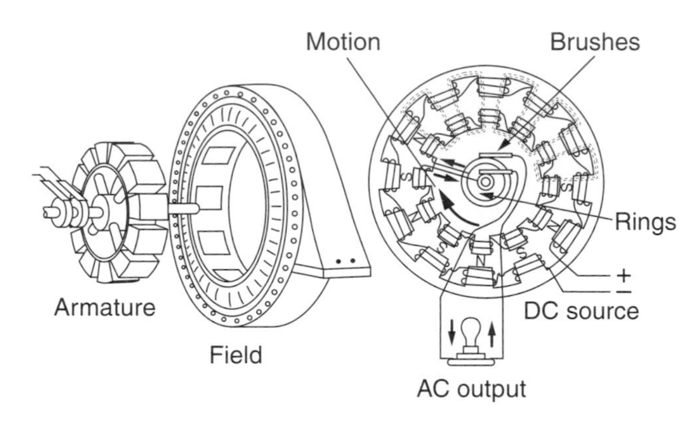

Figure 1-11. A commercial alternator looks much different from the simple sketches you have seen. This unit uses some of the current it produces to magnetize the field.

The speed at which the generator is turned determines the ***frequency*** of the current. One complete turn is called a ***cycle,*** **Figure 1-12.** Frequency is measured with units of ***hertz (Hz),*** which represent cycles per second.

NOTE

In the United States, the frequency of alternating current is 60 Hz. An alternator produces 60 complete turns and 120 changes of current direction per second. The frequency is so high that the reversals cannot be detected as so much as a flicker in an electric light.

Instantaneous, Peak, and Effective Voltage

Instantaneous voltage is the voltage at a particular instant. For example, looking again at our sine curve, the maximum positive voltage occurs at 90° and the maximum negative voltage occurs at 270°. These instantaneous voltage values are particularly important since they also represent the peak positive and negative voltages. Another important instantaneous voltage occurs at 45° and 135° in the positive part of the cycle, and again at 225° and 315° during the negative part. The voltage at these points is referred to as ***effective voltage,*** or ***root-mean-square (RMS) voltage,*** and is equal to 70.7% of the peak voltage. ***Nominal voltage,*** such as 120 volts or 240 volts, refers to the effective voltage of a circuit.

NOTE

Voltmeters read the effective voltage of a circuit. Equipment is rated for the effective voltage; calculations and specifications are made using the effective voltage values.

Phase Relationship

In ac circuits, the voltage and current vary continuously with time. This relationship is often called the ***phase relationship.*** If alternating current is run through a purely resistive circuit, both voltage and current, although changing, will vary together and are said to be *in phase*. See **Figure 1-13.** However, most circuits contain components that cause the voltage and current to be out of phase with each other. See **Figure 1-14,** for instance, where the current lags behind the voltage. That is, the current reaches its peak *after* the voltage has reached its peak. This is ***lagging current.*** Other circuit conditions might cause the current to peak ahead of the voltage, as illustrated in **Figure 1-15.**

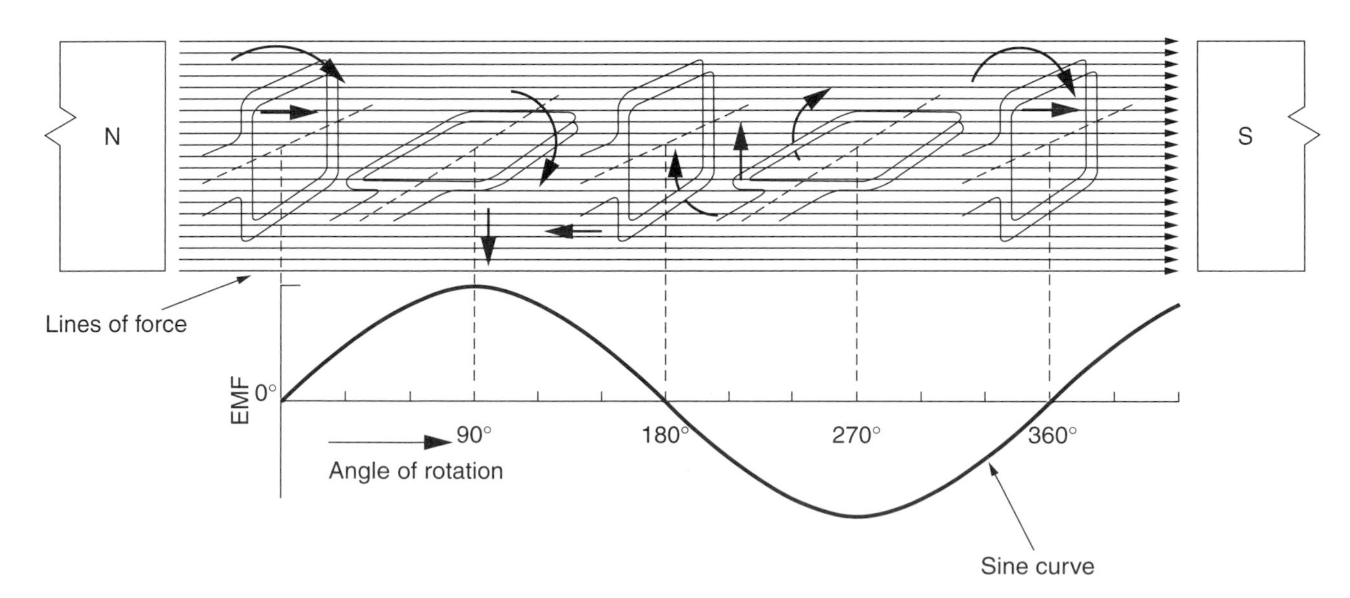

Figure 1-12. Voltage produced by an alternator varies. Compare the position of the coil above with the sine curve below. Peaks in the wave represent points of highest voltage during one complete cycle.

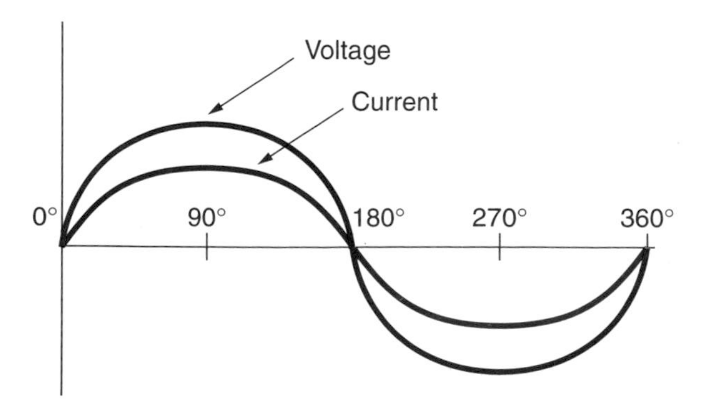

Figure 1-13. Voltage and current in phase.

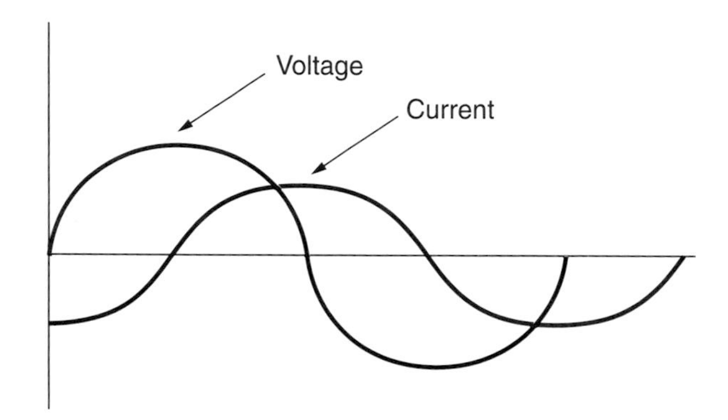

Figure 1-14. Voltage and current out of phase. The current lags behind the voltage.

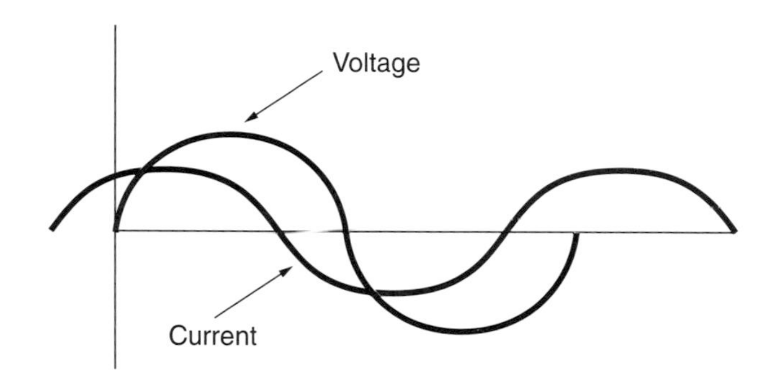

Figure 1-15. Current leading the voltage.

Inductance

Inductive devices are another component of ac circuits. These devices create electromagnetic fields that decrease the response speed of the circuit.

The most basic and useful type of inductive device is a coil. A ***coil*** is a conductor formed into a series of loops that may have a solid, movable core, **Figure 1-16.** Electricians working with lighting controls are familiar with the variable inductor (another name for a coil). As illustrated in **Figure 1-17,** the lighting brightness is controlled by inserting or withdrawing the iron core. The coil has the least inductance when the core is withdrawn and the greatest inductance when the core is fully inserted. Current is greatest when inductance is least.

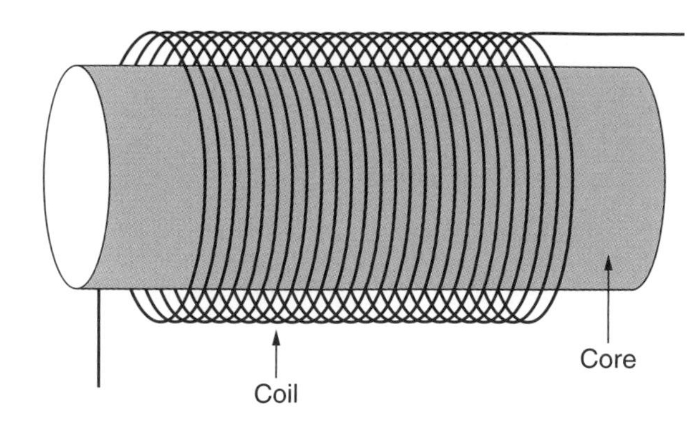

Figure 1-16. A coil with a solid core is one of the simplest types of inductive devices.

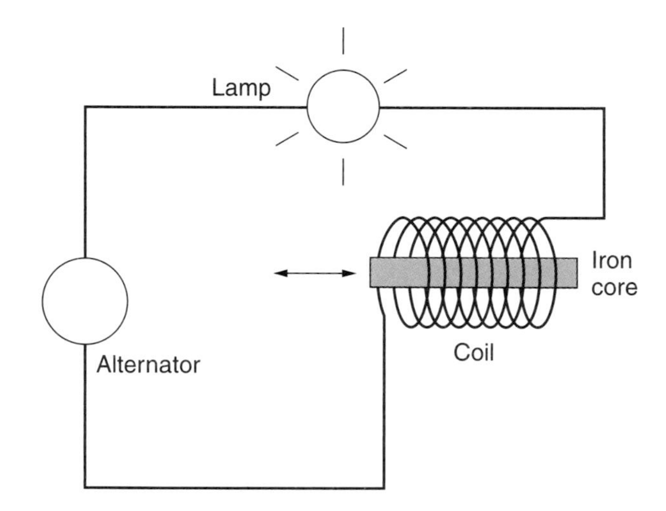

Figure 1-17. The lamp brightness is controlled by insertion of the iron core. When the core is fully inserted, inductance is greatest, current is least, and lamp is dimmest.

Inductance opposes any change in current flow. If the voltage across a coil increases, the current flow increases slowly due to the self-induced voltage (which opposes current change). Conversely, if the voltage across a coil decreases, the current value decreases gradually due to the self-induced voltage.

The major difference between an ac circuit with resistance alone and one with induction is response speed. The current flow in a circuit having only resistance will change instantly if the voltage changes. Current flow in an inductive circuit will respond slowly if a voltage change occurs.

This opposition due to inductance is called ***inductive reactance.*** It is measured in ohms, as is resistance. The symbol for inductive reactance is X_L. ***Inductance (L)*** is measured in units called *Henrys (H).* An inductance of one Henry is present if one volt is applied to an inductor and the current flow increases at a rate of one ampere per second.

Ohm's law, as applied to a circuit with pure inductance, can be rewritten as follows:

$$I = \frac{E}{2\pi fL} = \frac{E}{X_L}$$

where

I = Current (A)

E = Voltage (V)

π = Pi (3.14)

f = Frequency (Hz)

L = Inductance (H)

X_L = Reactance (Ω)

In other words, resistance in a purely inductive circuit is due to inductive reactance, and is numerically equal to $2\pi fL$. One key idea to remember about purely inductive circuits is that the current lags the voltage by 90°. Also, since inductive reactance increases or decreases with the frequency of a circuit, it has the same effect as resistance and impedes the current flow.

Capacitance

Capacitance (C) is the condition whereby a circuit stores electrical energy. This occurs whenever there are two conducting materials, such as metal plates, separated by an insulating material (such as air or paper). A device specifically constructed in such a manner is called a ***capacitor*** or a ***condenser,*** **Figure 1-18.**

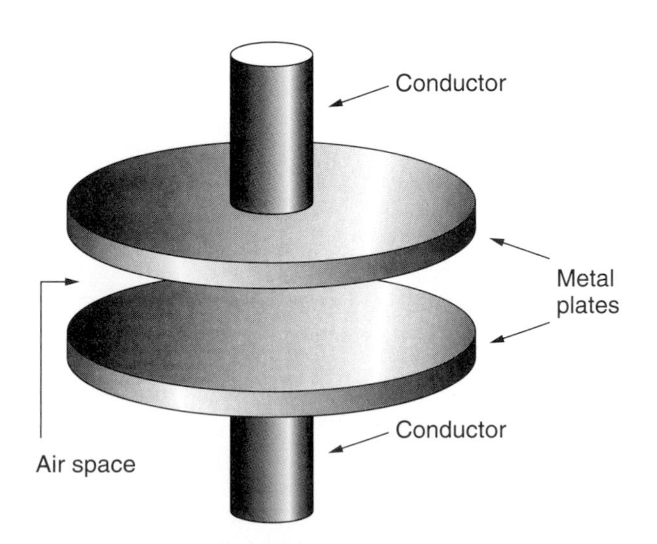

Figure 1-18. A simple capacitor is composed of two layers of conducting material (metal plates) separated by a layer of insulating material (air).

If a source of direct current is connected to the two plates of a capacitor, the current will flow initially, but then gradually decrease to zero. The electrical energy will be stored in the capacitor. The capacitor is charged. If the source is not disconnected and the two plates are connected by a wire or resistor, current will flow through that connection for a limited time. When the current reaches zero, the capacitor is discharged. However, a capacitor in a dc circuit cannot maintain a charged condition because the current flows in one direction only.

In an ac circuit, the voltage is constantly changing direction. The current is initially in one direction, charging the capacitor, and then in the opposite direction, discharging the capacitor. Therefore, current continues to flow, alternating back and forth, the capacitor plates charging and discharging.

The unit of measure for capacitance is the ***farad (F).*** ***Capacitive reactance (X_C)*** is the opposition to the flow of current due to capacitance in the circuit. It is measured in ohms, just as is inductance and resistance.

In a purely capacitive ac circuit (without any inductance or resistance), the current will lead the voltage by 90°, or 1/4 of a cycle. Mathematically, the value of the capacitive reactance in a circuit is

$$X_C = \frac{1}{2\pi fC}$$

where

X_C = Capacitive reactance (Ω)

π = Pi (3.14)

f = Frequency (Hz)

C = Capacitance (F)

Ohm's law is applicable to a purely capacitive circuit and is expressed as

$$I = \frac{E}{X_C} = 2\pi fCE$$

Impedance

In ac circuits, the total resistance is the result of a combination of opposing factors. Some circuits have inductance and resistance; some have resistance and capacitance, and others have all three factors. ***Impedance (Z)*** is the total opposition to the flow of alternating current. Impedance in a circuit can be calculated using the following formulas:

- For a circuit having resistance, inductance, and capacitance

$$Z = \sqrt{R^2 + (X_L - X_C)^2}$$

- For a circuit having resistance and capacitance

$$Z = \sqrt{R^2 + (X_C)^2}$$

- For a circuit with resistance and inductance

$$Z = \sqrt{R^2 + (X_L)^2}$$

- For a purely resistive circuit

$$Z = R$$

- For a purely capacitive circuit

$$Z = X_C$$

- For a purely inductive circuit

$$Z = X_L$$

Impedance represents the total resistance in a circuit, so Ohm's law for ac circuits becomes

$$Z = \frac{E}{I}$$

Power Factor

As noted earlier, power in a dc circuit is the product of voltage and current. In most ac circuits, determining the power is a bit more involved, requiring that the power factor be considered. The ***power factor*** is the ratio of the true power to the apparent power. It is expressed as a decimal or a percentage. Power factor may be anywhere from 0 to 1.0 (0–100%).

Recall, as was noted earlier, that power is calculated by multiplying the current and voltage. This still holds true for dc circuits as well as ac circuits in which the current and voltage are in phase and all energy from the source is fed into the external circuits. However, when the current and voltage are out of phase (which is quite often the case with practical ac circuits), some energy is fed back to the source from the external circuit. In this case, simply multiplying voltage and current would not represent the actual power. In fact, such a value is called ***apparent power.***

The apparent power is determined by measuring the voltage with a voltmeter, the current with an ammeter, and then multiplying the actual readings. True power, on the other hand, is measured with a wattmeter. Then, to find the power factor, the following formula is applied:

$$\text{Power Factor} = \frac{\text{True Power}}{\text{Apparent Power}}$$

Sample Problem 1-4

Problem: In the circuit shown, true power as indicated by the wattmeter (W) is 1900 watts, the ammeter (A) reads 10 amps, and the voltmeter (V) reads 240 volts. What is the power factor?

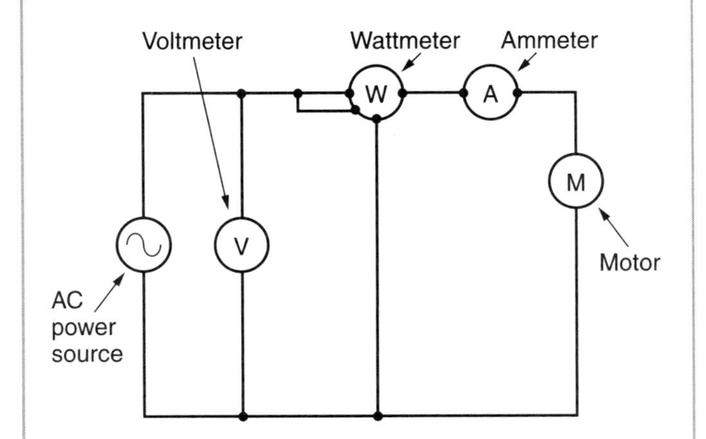

Formulas:

$$\text{Apparent Power} = V \times I$$

$$\text{Power Factor} = \frac{\text{True Power}}{\text{Apparent Power}}$$

Solution:

$$\text{Apparent Power} = 240\text{ V} \times 10\text{ A} = 2400\text{ VA}$$

$$\text{Power Factor} = \frac{1900\text{ W}}{2400\text{ VA}} = .79$$

A low power factor, as in our example, can be improved by certain corrective measures, such as changing various systems components and installing capacitors.

Utilization Voltages

There are five basic voltages derived from utility-supplied distribution. The commercial building application depends on load requirements and economic considerations.

- **Single-phase, three-wire system**, although most commonly used in residences, is often applicable to small commercial buildings and apartment complexes. See **Figure 1-19.** The 240-volt portion can be used for the larger power loads, while the 120-volt portion serves the lighting and receptacles.

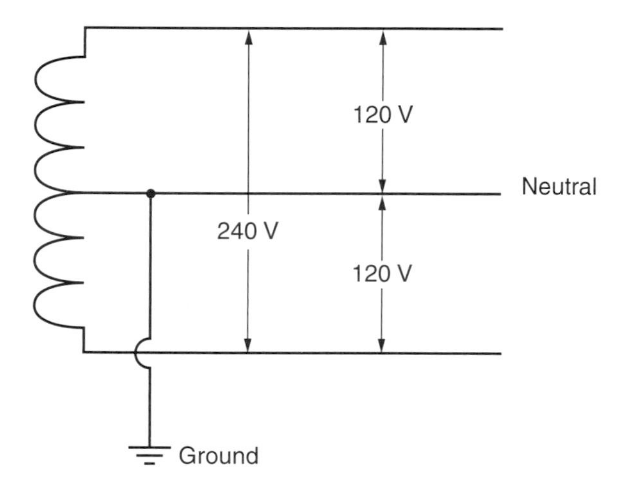

Figure 1-19. Single-phase, three-wire system. By connecting a "hot" wire to the grounded wire, 120-volt service is obtained. Connecting the two hot wires supplies 240 volts.

- **Three-phase, three-wire system** is a common secondary system used in commercial buildings. This is a delta-connected secondary with voltages of 240 volts, 480 volts, or 600 volts between phase conductors. The system is used when motor loads make up the majority of the overall load. Step-down transformers can be used to provide 120-volt service for lighting and receptacles. See **Figure 1-20.**
- **Three-phase, four-wire wye** is the most common three-phase system, particularly the type providing 120/208-volt service. This system can serve a variety of power and lighting load conditions: four-wire, 120/208-volt loads; three-wire, 120/208-volt loads; three-wire, 208-volt loads; two-wire, 208-volt loads; and two-wire, 120-volt loads. **Figure 1-21** illustrates this highly adaptable and flexible system.
- **Three-phase, 480/277-volt, four-wire system** is another extremely flexible variation of the four-wire wye. It is a very economic commercial system and has advantages over the 120/208-volt system. This system provides for three types of loads: 277-volt, single-phase; 480-volt, three-phase; and 120/208-volt or 120/240-volt loads with the addition of a step-down transformer. **Figure 1-22** shows the 480/277-volt, four-wire wye configuration.
- **Three-phase, four-wire, delta system** is basically a modification of the three-phase, three-wire delta. In this arrangement, the grounded neutral

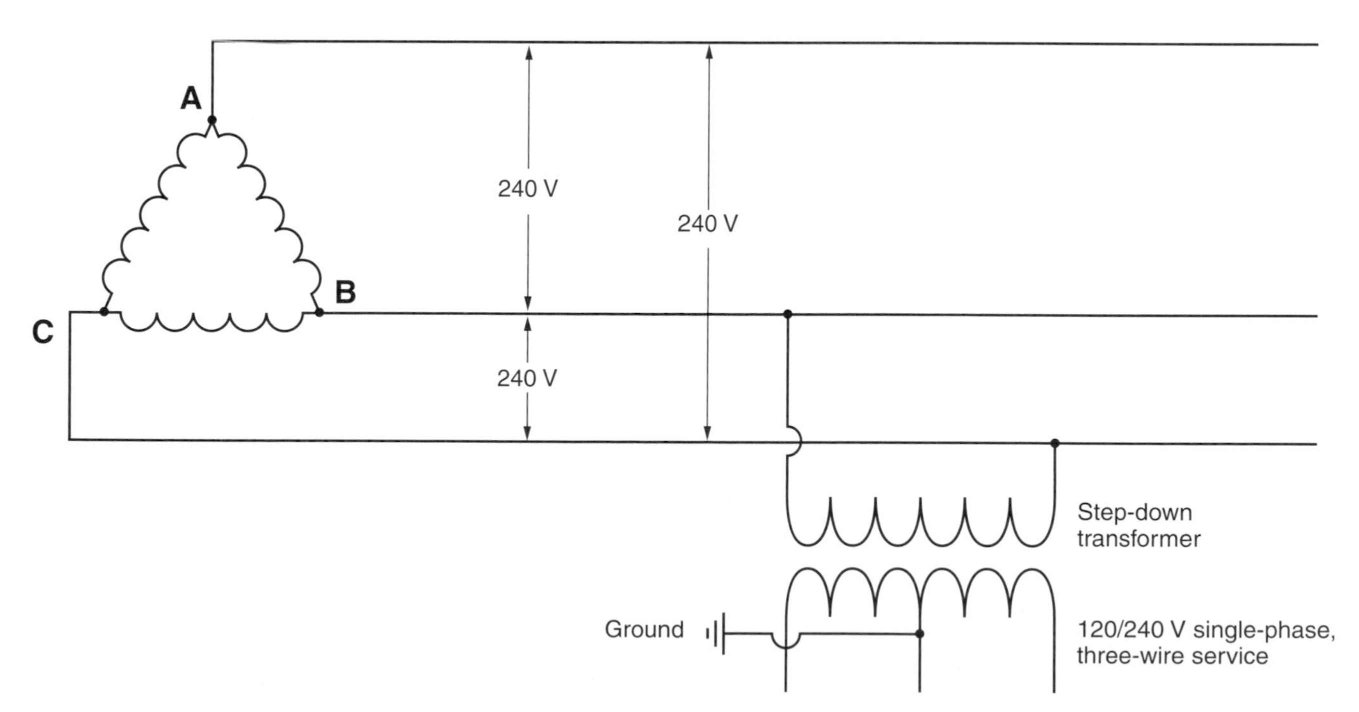

Figure 1-20. Three-phase, three-wire delta system. This system provides 240 volts and 480 volts. A step-down transformer is used to supply 120-volt service.

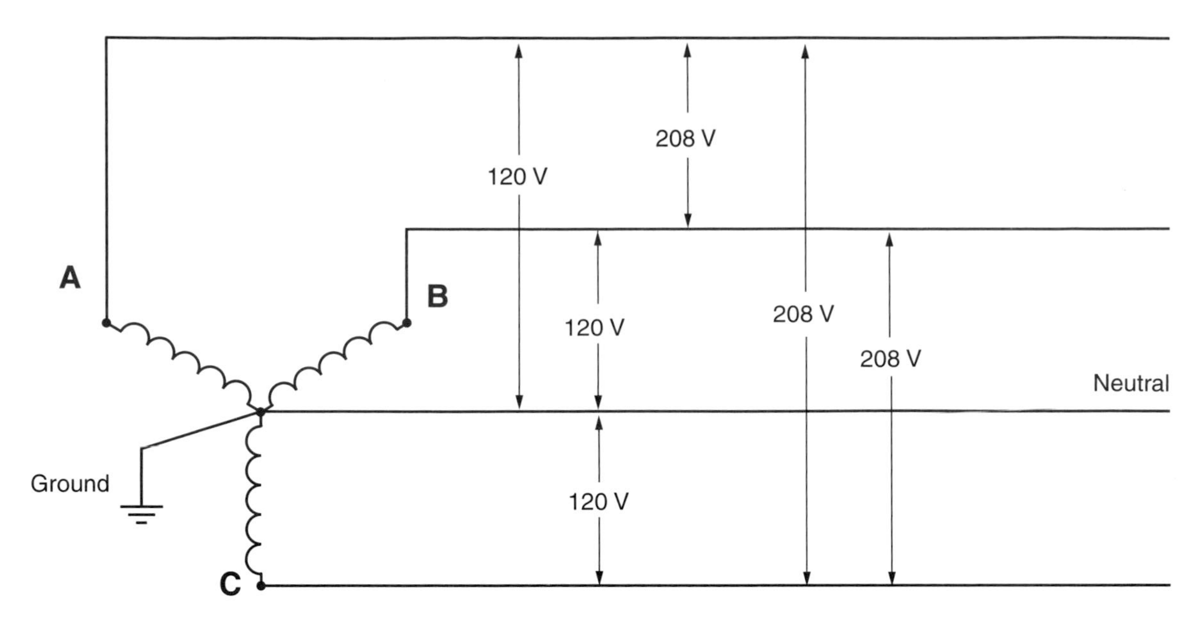

Figure 1-21. The three-phase, four-wire wye system is the most common three-phase system.

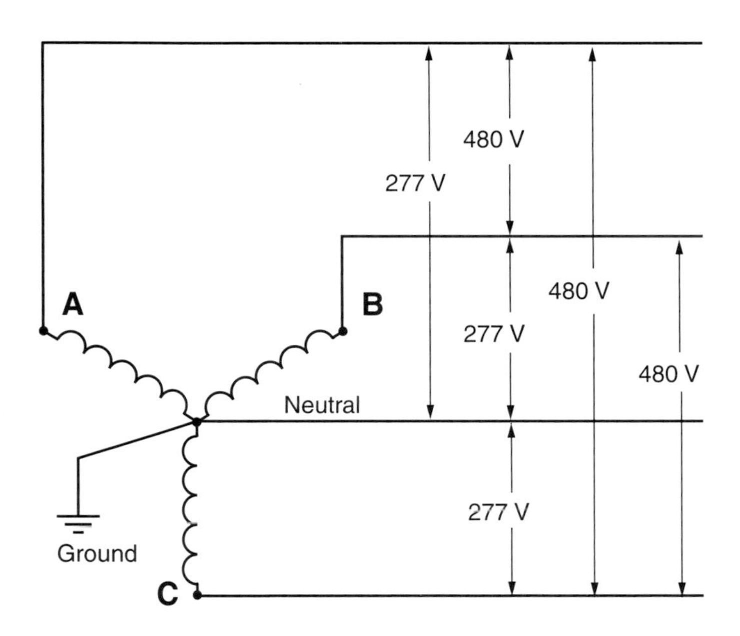

Figure 1-22. A three-phase, four-wire, 480/277-volt wye system is similar to the 120/208-volt, four-wire wye system.

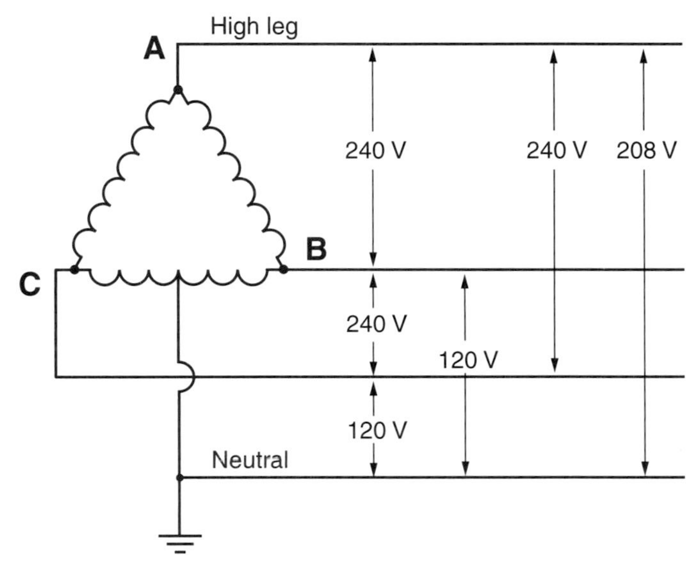

Figure 1-23. The three-phase, four-wire delta system is a modified three-wire delta system. A grounded neutral wire is connected to one of the secondary transformer windings.

is added as a center tap to one of the secondary transformer windings, as shown in **Figure 1-23.**

CAUTION

In a four-wire delta system, the phase that is *not* connected to the tapped winding is called the *high leg.* The voltage difference between the high leg and the neutral is greater than the difference between the other phases and the neutral. The high leg conductor must be identified with orange marking to prevent an electrician from mistaking it for one of the other phase conductors.

There are other system configurations used for both commercial and industrial applications. However, they are essentially derived from these five basic types.

Phase is represented by the Greek letter ϕ, wire is abbreviated with a *w,* and a wye system is designated with a Y. For example, a three-phase, four-wire 120/208-volt wye system is abbreviated as 3ϕ, 208 Y/120 V, 4 w.

Voltage Standards

Nominal voltage indicates a desired level of the supply voltage for the electrical system. ***Rated voltage*** is the preferred voltage for equipment, particularly motors.

Commercial electrical systems are within the category of low and medium voltage; high voltage is reserved for power company distribution systems. The table in **Figure 1-24** lists the nominal and rated voltages for low and medium voltage categories as issued by the American National Standards Institute (ANSI).

Review Questions

Answer the following questions. Do not write in this book.

1. Name the three basic characteristics of electrical circuits and indicate their symbols.
2. Explain the difference between an insulator and a conductor.
3. What is the force that moves electrons and how is it measured?
4. List three characteristics of a series circuit.
5. List three characteristics of a parallel circuit.
6. How does inductance affect phase of an ac circuit?
7. How does capacitance affect phase of an ac circuit?
8. Name the parts of an ac generator.

Nominal Voltages for Commercial Electrical Systems

Nominal Voltage	Rated Voltage (Motor Rating)	General Voltage Class
Low Voltage		
1ϕ, 120/240 V, 3 w	115/230 V	125/250 V
3ϕ Electrical Sytems		
208 Y/120 V, 4 w	200 V	125/250 V
240 V, 3 w	230 V	250 V
480 Y/277 V, 4 w	460 V	600 V
480 V, 3 w	460 V	600 V
600 V, 3 w	575 V	600 V
Medium Voltage		
2400 V, 3 w	2300 V	5 kV
4160 Y/2400 V, 4 w	4000 V	5 kV
4160 V, 3 w	4000 V	5 kV
4800 V, 3 w	4600 V	5 kV
6900 V, 3 w	6600 V	15 kV
13800 Y/7970 V, 4 w	13200 V	15 kV
13800 V, 3 w	13200 V	15 kV

Figure 1-24. This table lists nominal voltage, rated voltage, and general voltage class of standard supply systems.

9. Briefly explain instantaneous, peak, and effective voltages.
10. What is impedance?
11. What is nominal voltage?
12. What is the phase relationship of a purely resistive ac circuit?
13. List three common secondary supply voltages used in commercial buildings.
14. What is RMS voltage and how is it related to peak voltage?
15. Sketch a single-phase, three-wire system and show the voltage present between the conductors.
16. In the circuit shown below, determine the total current and the voltage at each resistor.

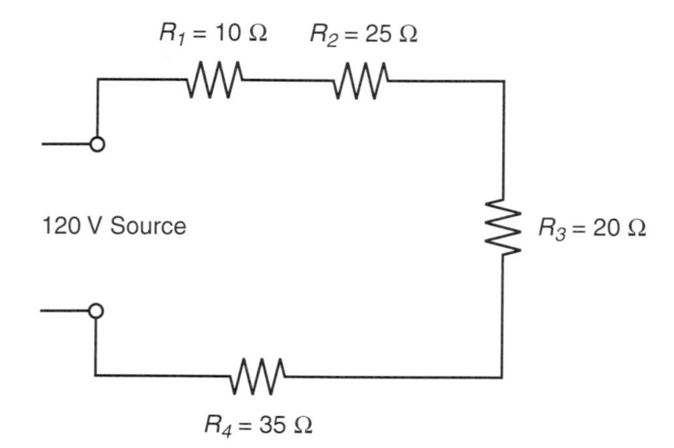

17. If the current in this circuit is 2 amps, what is the resistance of R_2?

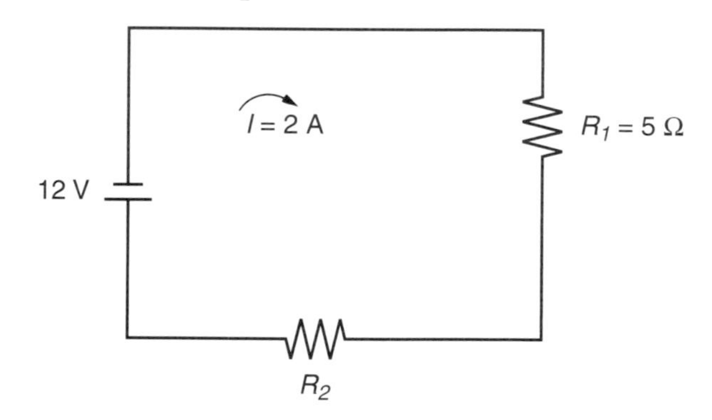

18. In the following circuit, determine the (a) current through each resistor, (b) total current in the circuit, and (c) total circuit resistance.

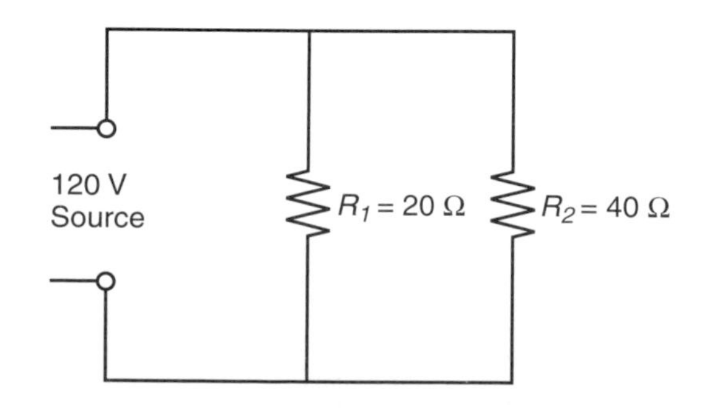

19. The total current in this circuit is 4 amps. Determine the resistance in R_3.

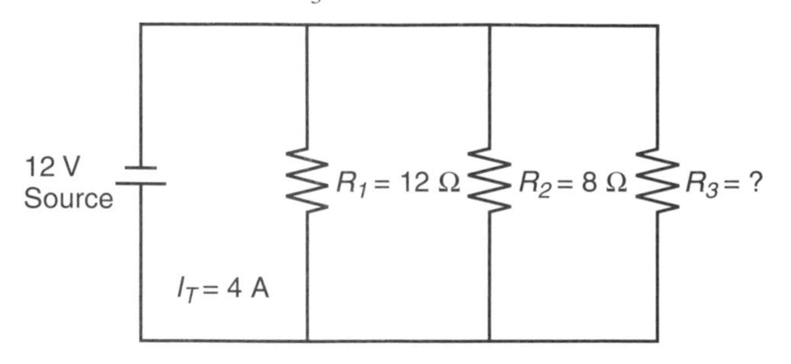

20. In the following circuit, determine the (a) total resistance, (b) total circuit current, and (c) voltage across each resistor.

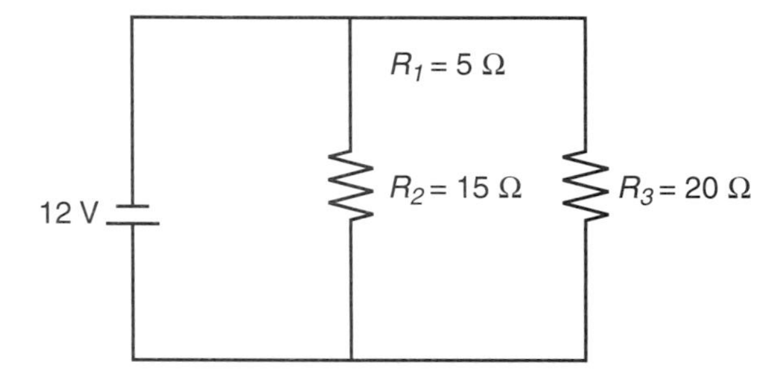

21. In the following circuit, determine the (a) total resistance, (b) total circuit current, and (c) voltage across each resistor.

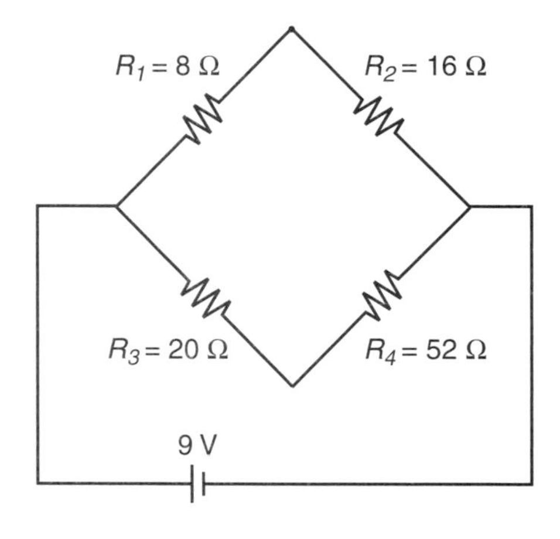

22. The total current in this circuit is 4 amps. Determine the resistance of R_5.

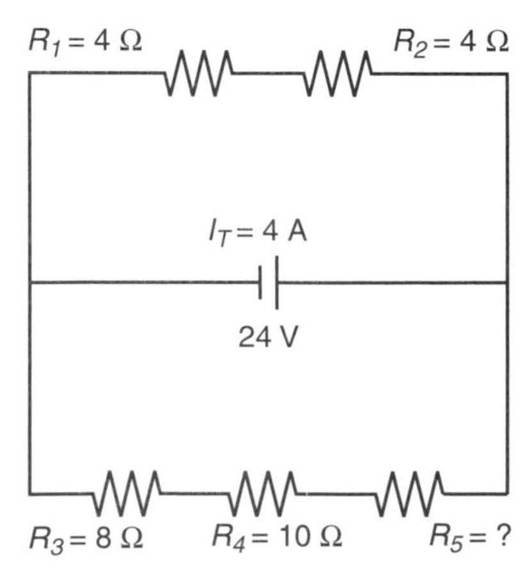

23. In the following circuit, determine the (a) total resistance, (b) total circuit current, and (c) voltage across each resistor.

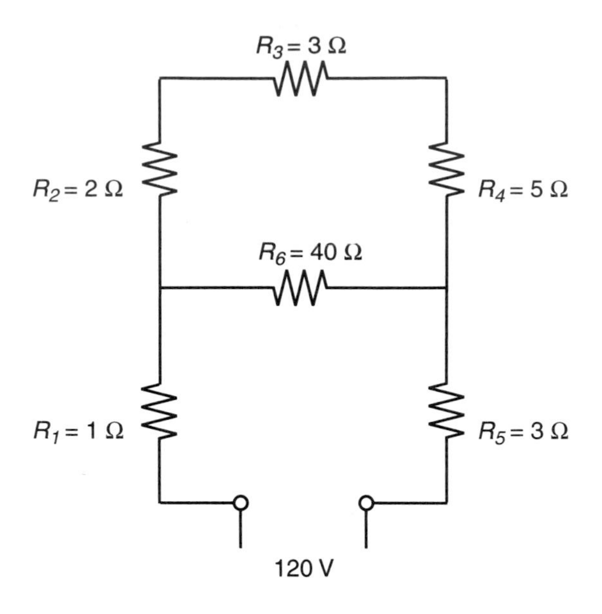

24. Determine the power factor in the following circuit.

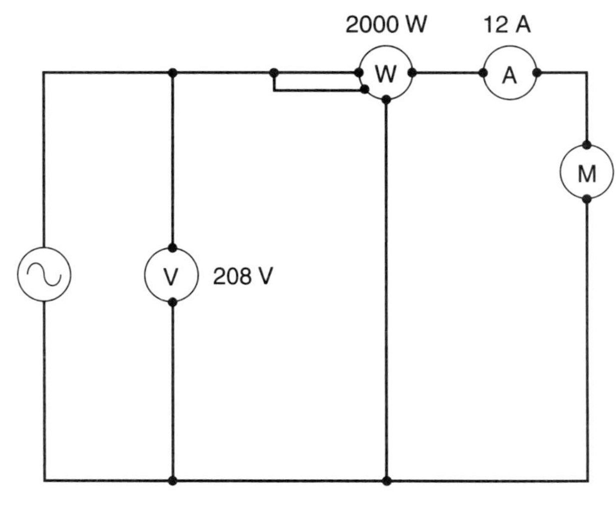

25. If the true power in the following circuit is 0.875, what is the reading on the wattmeter?

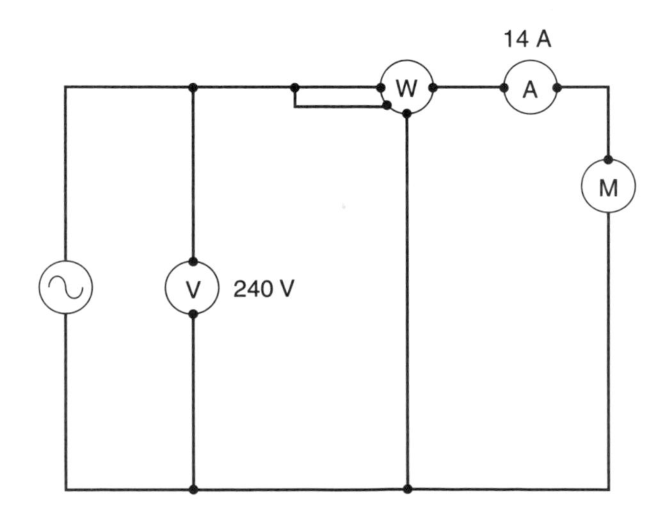

Chapter 2

Safety

Technical Terms

Confined spaces	Fibrillation
Current path	Lockout/tagout procedures

Objectives

After completing this chapter, you will be able to:

- ❍ Cite examples of hazardous situations.
- ❍ Identify hazards associated with electrical work.
- ❍ Explain basic safety rules.
- ❍ Describe safety equipment and protective clothing.
- ❍ Follow basic procedures designed to aid an injured worker.

Safety is very important in every aspect of electrical work. Extreme caution must be exercised around electrical equipment (particularly when using power tools and test equipment).

Electricians are not immune to electrical shock hazards. In fact, persons working on electrical equipment are those who are most often injured by electrical shock. In many cases, accidents occur when an experienced electrician becomes careless or lazy and doesn't follow the proper safe working procedures.

Always be completely aware of what you are doing. If you are not sure of something, ask questions before performing any work on electrical equipment.

Cause of Accidents

Accidents are normally caused by a combination of events. First, there must be a potentially hazardous situation. Some examples of these situations include the following:

- Performing work on energized equipment.
- Performing work near an energized circuit.
- Performing work on a ladder, scaffolding, or in an area where a fall hazard exists.
- Performing work in a confined space.
- Performing work below other workers.

The first step in preventing accidents is to identify the potential hazards before beginning work. Once these hazards are identified, you must take the appropriate steps to protect yourself from them.

Working on energized, or "live," equipment is very dangerous. You must go to additional lengths to ensure your safety. These additional safety precautions are discussed later in the chapter.

Disconnecting the circuit being worked on does not guarantee your safety. After disconnecting the circuit, always test the circuit, using an appropriate testing instrument, to make sure there is no current. The most dangerous situation is working on energized equipment you think is de-energized. Also, identify any nearby circuits that you have not disconnected.

When a fall hazard exists, wear an appropriate safety harness. Always use nonconducting fiberglass ladders. An aluminum ladder can become energized if a live conductor comes in contact with it.

Confined spaces, such as trenches, tunnels, and manholes, can provide many hazards. The tight working conditions may make it difficult for you to work carefully. An inadequate oxygen supply and the presence of explosive or poisonous gases are additional hazards. When work is performed in a confined space, a worker should be posted outside the confined space to get help if an accident occurs.

Work should never be performed below other workers without taking proper precautions. A physical barrier should be present to protect the lower workers from any falling debris created by the higher workers.

WARNING

The most important rule for preventing electrical accidents is to disconnect the circuit or equipment before doing work on it.

Electrical Hazards

Shocks and explosions caused by electrical faults are the two most dangerous electrical hazards. These hazards can only occur when working on or near energized equipment.

Electrical shock can be painful and sometimes fatal. The severity of a shock depends on several conditions:

- **Moisture**—If the floor, equipment, or clothing is wet, the shock will be more severe than under dry conditions.
- **Flooring**—The better the floor conducts electricity, the more severe the shock potential.
- **Grounding**—Ungrounded equipment and circuits are much more dangerous than grounded ones.
- **Duration of contact**—An electrical shock may cause muscle paralysis, preventing a worker from releasing the equipment supplying the shock.
- **Current path**—As electric current passes through the body, internal tissue may be burned. Current passing through the heart may cause an erratic heartbeat and ***fibrillation*** (a potentially fatal condition where the heartbeat and pulse are not synchronized). Therefore, a ***current path*** through the heart (such as from hand to hand) is more dangerous than a current path that avoids the heart (such as right hand to right foot). See **Figure 2-1.**

Voltage does not have as much to do with shock severity as current does. Remember, 10,000 volts is no more dangerous than 120 volts. A current of 10 milliamperes is painful and can have severe effects. A current of 100 milliamperes is almost always fatal. The table in **Figure 2-2** shows the effects of various current levels.

On high-voltage equipment, a short circuit or fault may produce an electric arc and explosion. The arc is hazardous for three reasons:

- **Heat**—A tremendous amount of heat can be generated by an electric arc. The heat may cause burns, and hot metal components may be propelled onto the worker.
- **Noise**—The sudden expansion of air may cause an explosion, which could result in hearing damage.
- **Explosive force**—The explosion may throw a worker back from the equipment. This helps the worker escape the heat, but the worker may suffer physical injury if thrown into other equipment.

Accident Prevention

The best way to prevent electrical accidents is to always de-energize the circuit before you work on it.

Figure 2-1. The current path of an electric shock will affect the extent of injury. If current passes through the heart, fibrillation may occur. A hand-to-hand current path is the most dangerous path.

Average Effects of Electrical Current on the Body

Current in Amperes	Effect on Body
1 milliampere (mA)	Barely felt as a tingling sensation
1 mA to 10 mA	Causes muscles to “freeze” or contract, preventing release of the energized object
10 mA to 100 mA	Fatal after several seconds
over 100 mA	Almost always fatal

Figure 2-2. Current determines the severity of shock. This table lists various currents and the shock results.

When you disconnect the circuit, tag or mark the circuit so that it is not mistakenly reconnected before you complete your work. ***Lockout/tagout procedures*** involve placing a lock on the disconnect to prevent the circuit from being connected. See **Figure 2-3.** When the work is completed, the lock is removed and the circuit is

Figure 2-3. These safety lockouts allow up to six workers to lock the circuit off. (Ideal Industries, Inc.)

re-energized. Be aware of any specific lockout/tagout procedures and follow them.

Basic Safety Rules

Certain safety rules should be followed during the installation and maintenance of electrical equipment. Obeying the following rules will help to prevent accidents and injury:

- Disconnect the circuit from the source, if possible. Work on live circuits only when absolutely necessary. Never assume the circuit is de-energized—check it with a test light or meter before starting work.
- Keep the work area dry and clear of debris. Cover wet floors with wooden planks.
- Make sure all equipment is properly grounded.
- Use tools safely, as outlined in Chapter 3 of this text.
- Remain alert at all times. Always think about what you are going to do before you do it. Avoid quick movements. Work slowly and deliberately. Plan each step of the work.

Remember, safety in electrical work may save your life and the lives of those working with you. Never compromise safety in order to complete a job quickly.

Working on Energized Equipment

Normally electricians de-energize equipment before performing work. However, there are rare times when the equipment must be "hot" or "live" while you work on it. When this occurs, you must obey the following additional safety procedures:

- Be sure to use insulated tools.
- Wear rubber gloves and eye or face protection.
- Place a rubber blanket (or other suitable insulated shield) over exposed live parts adjacent to the work area.

WARNING

Only work on energized equipment when it is absolutely necessary.

Safety Equipment

Construction work of any type can be dangerous. Electrical workers must perform their tasks as safely as possible. The following are some of the items included as basic safety equipment for an electrician:

- **Hard hat**—A molded, impact-resistant helmet protects the worker from accidental blows to the head. The hard hat should have an inner webbing that fits snugly against the head, leaving an air cushion between the helmet and the head to prevent direct impact from falling objects. A proper hard hat will also protect a worker from electric shock. Hard hats are required on all construction sites.
- **Eye protection**—Safety glasses, goggles, and face masks are several types of eye protection required, depending on the task to be performed. They protect your eyes from flying debris, dust, liquids, and sparks. Safety glasses are minimum eye protection required on all construction sites. They are made of shatter-proof plastic or glass and have side shields. Goggles fit snug to the skin and are used mostly where liquids or mists may present a potential danger. **Figure 2-4** illustrates safety glasses and goggles. The full face mask is required where sparks or splashing liquids could be present.
- **Ear protection**—Earplugs or earmuffs are used where noises (particularly loud and high-frequency noises) are generated. In certain severe conditions, both plugs and earmuffs are needed. Working in areas where drilling, hammering, or sawing are continuous throughout the day warrants the use of ear protection. A constant exposure to high-decibel noise can cause permanent damage to the middle

Figure 2-4. Goggles and safety glasses protect your eyes from airborne debris.

and inner ear. Various types of hearing protection are shown in **Figure 2-5.**

- **Safety belt and harness**—Most fatal construction accidents are caused by falls. A fall of 6′ can be very serious or even deadly. Fall protection in the form of safety belts and body harnesses can prevent these types of accidents. As with the other types of personal protection, safety belts are required on most construction sites, particularly when working at heights of 6′ and up.
- **Rubber gloves**—Always wear rubber gloves when working on energized equipment. The insulating material provides extra protection against electrical shocks. Rubber gloves should be tested every six months. See **Figure 2-6.**
- **Insulated hand tools**—If a metal screwdriver or other tool contacts a live wire or energized piece of equipment while in contact with a grounding path, an electrical fault can occur. The occurrence of these types of accidents is greatly reduced when insulated hand tools are used. The insulated material around the tool serves as a barrier, preventing the flow of electricity. See **Figure 2-7.**
- **Nonconductive ladders**—Fiberglass ladders should be used for electrical work. Aluminum ladders should be avoided because they can become energized if they come into contact with an electric source.

Perhaps the most important item needed by the electrician is protective clothing. Protect yourself by minimizing the amount of exposed skin. Clothing that is resistant to flash flame helps to prevent burns if a fault causes an explosion.

Wear rubber boots or shoes with rubber soles. Use rubber gloves whenever possible. Do not wear clothing with exposed zippers, buttons, or other metal fasteners.

Never wear metal objects (such as rings, wristwatches, or bracelets) when working around electrical equipment. These conducting items increase the risk of electrical shock.

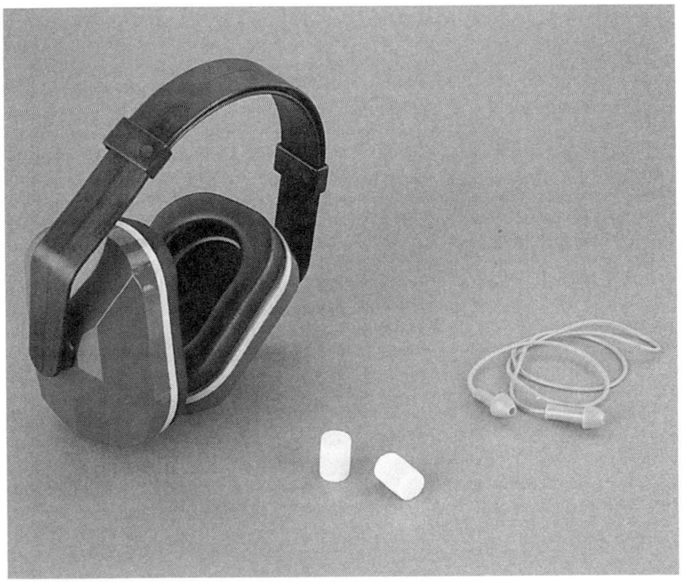

Figure 2-5. Earmuffs are used to protect against very loud noises. Earplugs connected by a string can be wrapped around your hard hat band so hearing protection is always handy.

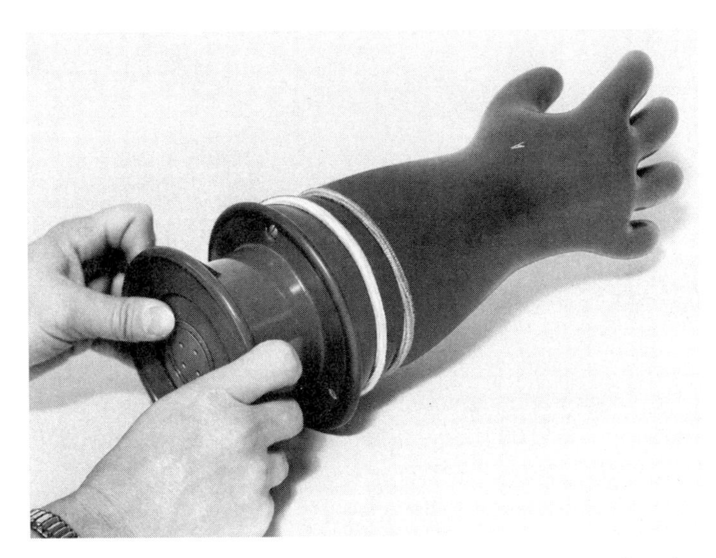

Figure 2-6. Using a pneumatic glove tester ensures that the glove is still sound. Rubber gloves should be tested every six months. (AVO International)

Figure 2-7. This insulated screwdriver helps to protect the electrician from electrical shocks. (Greenlee Textron)

WARNING

Always wear a hard hat and safety glasses around construction areas.

Helping an Injured Worker

When an injury occurs to another worker, you must remain calm. Do not panic. Consider your own safety first. If the accident was caused by an unsafe condition, does the condition still pose a danger? If so, *do not* place yourself in danger trying to help the injured worker. Remember, you're not helping anyone by getting injured yourself—you're only making matters worse.

If the injury may be serious, call for help as soon as possible. The survival of the injured worker may depend on how quickly trained medical personnel arrive or by how quickly the victim can be transported to a trauma center. Remain calm on the telephone, provide as much information as possible, and stay on the line until the dispatcher tells you to hang up.

Someone should wait by the work site entrance to help the emergency medical personnel find the location. It is best for several workers to wait in case multiple vehicles arrive. Remember, the medical personnel are unfamiliar with work sites. Be sure that the area is clear and safe. Lead them from their vehicles to the injured worker.

NOTE

If you are not aiding an injured worker, stay out of the way of people who are helping.

As a general rule, if the injured worker is in a life-threatening situation and you can do something to remove the worker from the situation, do it before you call for help. Some examples of this include the following:

- If a shock victim's muscles are paralyzed and they are still in contact with the electric source, stop the flow of electricity through the victim, and then call for help.
- If an accident victim is not breathing and you have cardiopulmonary resuscitation (CPR) training, perform CPR immediately and send someone else to call for help.

WARNING

When another worker is injured, think of your own safety first. Do not put yourself in danger trying to help—this will only make matters worse.

First Aid Preparation

The best way to prevent accidents from happening is to identify unsafe conditions and correct them. You should also be as prepared as possible when an accident occurs. By preparing for an accident before it occurs, you will be able to react more quickly and more usefully if an accident does occur.

There are many ways to prepare for accidents. Try to do as many of the following as possible:

- Always know where the work site first aid kit is located.
- Always know where a phone is located on the work site. If there is no number for an ambulance or the fire department posted, dial 911 and ask the operator to connect you with the emergency medical personnel. See note below.
- Take a first aid and CPR class. Organizations such as the YMCA, American Red Cross, or your local park district may offer a one-day first aid and CPR class. The first aid training will provide guidance in dealing with many different types of injuries. Investing one day of time and taking the course may save a life someday.

NOTE

Most areas have 911 emergency service. If an injured worker may be in a life-threatening situation, use the 911 service. However, for less serious injuries, dial 0 and have the operator connect you with the ambulance company.

WARNING

When helping an injured worker, avoid direct contact with body fluids. The work site first aid kit should contain rubber gloves and a face shield for performing mouth-to-mouth procedures.

A first aid class will generally cover methods of treating various types of injuries. The following lists some injuries and the generally-accepted methods of treatment.

- **Shock victim**—Immediate action must be taken to help a shock victim, but take every precaution not to place yourself in jeopardy. If the individual is still in contact with the source of current, use an insulated or nonconductive object to move either the victim or the source. If the disconnect means is available, open the circuit. Then call medical personnel to the scene.
- **Bleeding**—If the victim is bleeding, first find the source of the blood. You may need to cut clothing in order to do this. Apply pressure to the wound

with a clean rag or cloth to reduce the bleeding. Use the materials in the first aid kit to apply a pressure dressing to the wound. If an appendage is severed, place the severed part in a dry plastic bag and place it in a cooler. Many appendages can be reattached if the victim receives treatment quickly.

- **Puncture wounds**—When an object is lodged in the victim, do not remove it. The object may do additional damage if you remove it. Try to stop the bleeding and prevent the object from moving. If the victim is impaled on an object that cannot be moved, wait for the medical personnel to arrive before cutting the lodged item.
- **Burns**—The severity of burns varies. Most burns should be cooled with water to lessen the injured worker's pain. Burns can be overcooled, so do not use ice. If the skin is black and charred, the nerves are burned and there will be little pain, so it is not necessary to cool with water.
- **Heat**—When working on a hot and humid day, be sure to drink plenty of water to prevent dehydration. Lack of water may produce heat cramps, heat exhaustion, or heat stroke. Move a worker suffering from any heat-related condition out of direct sunlight immediately. A worker with heat cramps or heat exhaustion should rest and drink water. Heat stroke causes a person to turn bright red and lose consciousness. If this occurs, the person must be cooled as quickly as possible, normally using water.

NOTE

This brief first aid information is not sufficient to prepare you to treat an injured worker. It is only presented to illustrate how treatment varies among injuries and to encourage you to enroll in a first aid class.

Review Questions

Answer the following questions. Do not write in this book.

1. Why are fiberglass and wooden ladders preferred over aluminum ladders for electrical work?
2. Name the most important rule for preventing electrical accidents.
3. Describe the three dangerous results of an electrical explosion.
4. Why is a current path running from right hand to left hand more dangerous than a current path from right hand to right foot?
5. Explain the purpose of lockout/tagout procedures.
6. What safety precautions are needed when working on energized equipment?
7. Describe the differences in protection offered by safety glasses, goggles, and face masks.
8. In case a worker is injured, you should know the locations of two items on the work site. Name these two items.
9. Explain why an object lodged inside a victim should not be removed immediately.
10. Describe the symptoms and treatment for heat stroke.

Chapter 3

Tools

Technical Terms

Cable bender
Fish tape
Hole saw
Hot box
Insulated pliers
Meggers
One-shot bend method
Progression bend method

Objectives

After completing this chapter, you will be able to:

- Recognize the basic construction tools used by electricians.
- Identify the tools used specifically for electrical installation.
- List basic safety rules for using tools.
- Identify various types of bending tools and pulling equipment.
- Explain the functions of various electrical testing devices.

This chapter describes basic and complex tools used by commercial electricians to perform various wiring installations. The principles of tool use and proper care are emphasized. The use of both hand and power tools is covered.

An electrician must have the correct tools for the job. Tools must be kept in good working order and properly repaired as needed. If they are in poor condition, tools are not conducive to good work and should be replaced.

Tools used for commercial or industrial installations can be quite expensive and often complicated to operate. They require a combination of skill, training, and experience to be used correctly, efficiently, and safely.

Standard Construction Tools

Many tools are used in several areas of construction. For example, carpenters, electricians, and plumbers all use hammers, saws, and drills in the course of their work. Anyone working in the building trades should be familiar with these standard construction tools. The following sections describe some of these tools and their specific use for electrical work.

WARNING

Most injuries occurring when using hand tools can be prevented by obeying the following safety rules:

- Always wear eye protection when using a tool.
- Use the correct tool for the job and use tools properly.
- Keep tools in good working condition. Replace worn and damaged tools.

Striking Tools

Always select a type and size of hammer appropriate for the task. For example, you would not use a sledgehammer to drive nails for small cable brackets, nor would you use a small tack hammer to drive grounding rods into the ground. If the hammer is too large and heavy, the struck materials may be damaged. If the hammer is too small, it might not produce enough force to drive the object. A small hammer may also require its user to try to overcompensate by swinging harder and with less control, resulting in a "miss" and damage to materials.

Never strike tools and objects that are not intended to be struck. Do not use the butt of a screwdriver or a flashlight to pound a nail. Strike only objects designed to be struck, such as nails, chisels, and punches. Be sure the hammer is appropriate for the nail size.

Always wear safety glasses when working with a hammer. Materials can be chipped accidentally, sending small objects into the air. Safety rules for hammers are summarized in **Figure 3-1.**

Hammer Safety

- ➤ Always choose the right size and weight hammer for the intended purpose.
- ➤ Never strike tools together unless they are designed for such use.
- ➤ Striking tools should strike a surface squarely—never at an angle.
- ➤ Always wear safety glasses when working with striking tools.

Figure 3-1. Basic safety rules for using hammers.

The following are some of the hammers used for an electrical installation:

- **Claw hammer**—Used for driving and pulling nails. The two most common types of claw hammers are the curved-claw hammer, used for general carpentry, and the straight-claw hammer, or framing hammer, which is favored by electricians. The straight claw hammer is better suited for breaking out plaster and drywall during remodeling.
- **Club hammer**—Often used to demolish masonry and drive steel chisels as well as masonry nails. The electrician may find this type of hammer useful to drive ground rods and bust through masonry.
- **Sledgehammer**—Used for heavy work, particularly for driving stakes and splitting stone. This tool is very useful when driving ground rods.

WARNING

Never use a hammer that has a loose or damaged head.

Sawing Tools

Sawing tools are needed to cut materials and remove obstructions. The requirements of the specific task determines whether a handsaw or power saw is used. Be sure the selected saw is appropriate for the task.

Keep your saws well maintained. Follow the tool care instructions in the owner's manual. Always work with a sharp blade. At the very least, a dull blade forces you to work more slowly. A dull blade can also bind in the material, making cutting more dangerous.

Force is required to propel the saw through the cut, but be careful not to use too much force. Let the blade do the cutting and do not rush it. Trying to rush through the cut by forcing the saw puts added pressure on the blade and could cause the blade to break.

Always wear gloves and safety glasses when sawing. This protects your hands and eyes from any loose particles. **Figure 3-2** summarizes these safety rules for cutting tools.

Many different types of saws are used by electricians. These include both handsaws and power saws. The following are some of the more commonly used saws and some of their uses:

- **Handsaw**—Used to trim studding or modify joists to accommodate panels, boxes, and fixtures. A fine-tooth handsaw can be used to cut PVC conduit. See **Figure 3-3.**
- **Keyhole saw**—Handy for making neat, accurate openings in drywall for receptacle boxes.
- **Reciprocating saw**—Used for heavy-duty sawing into wooden and metal structural members.
- **Band saw**—Needed to cut through heavy-gage metal framing and structural steel. They are also used to cut conduit and metal supports.
- **Hacksaw**—This saw can be used to cut conduit and metal obstructions.
- **Hole saw**—This cylindrical saw is mounted on an auger with a pilot drill. The auger is then fitted into an electric drill. The saw is used to cut holes in wood and other materials. See **Figure 3-4.**

Cutting Tool Safety

- ➤ Wear gloves and safety glasses when using cutting tools.
- ➤ Use the correct size and type of cutting tool for its designed purpose.
- ➤ Never force a cutting tool past reasonable pressure.
- ➤ Keep cutting edges sharp.

Figure 3-2. Summary of basic cutting tool safety rules.

Figure 3-3. Nonmetallic rigid conduit (PVC) can be cut with a fine-tooth handsaw. This saw is made especially for cutting PVC, but it can also cut nail-embedded wood, plasterboard, and plywood. (L. S. Starrett Company)

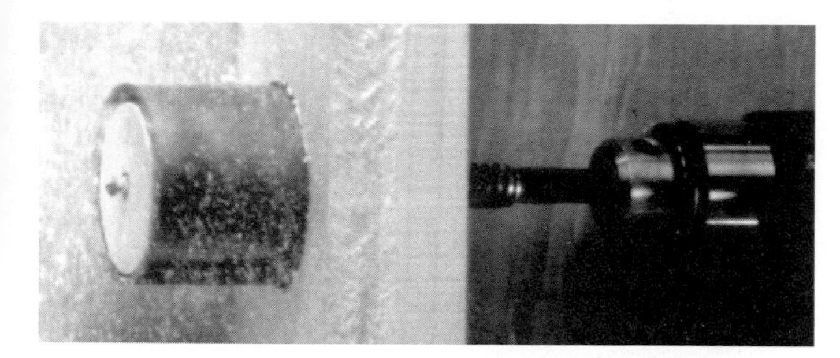

Figure 3-4. A hole saw is used to cut holes in studs and joists. Conduit or cable can then be installed. (Vermont American Tool Company)

CAUTION

Hole saws are operated at different speeds for different materials. Be sure to check the manufacturer's information for the correct speed.

Fastening Tools

The most popular fastening tools are screwdrivers and wrenches. Screwdrivers are used to install and tighten screws in terminal blocks and devices such as receptacles, fixtures, boxes, box covers, and panels. They should be insulated when used for electrical work.

Screwdrivers are designed to install and remove screws. Do not use a screwdriver for chiseling or prying. Never strike a screwdriver with a hammer.

CAUTION

The size of the screwdriver should be appropriate for the screw. Using the wrong size screwdriver can cause damage to the screwhead and the screwdriver.

Wrenches are another versatile tool used by electricians to fasten items. There are numerous types and styles available. Most wrenches are forged from carbon steel or chrome vanadium, giving them superior strength and light weight. They are available in both English and Metric standard sizes and lengths.

When working with fastening tools, be sure to avoid overtightening or over-torquing. This can damage the threads or the part being tightened.

Drilling Tools

Augers, gimlets, reamers, brace and bits, hand drills, and power drills are all members of a family of tools used for drilling or enlarging holes. The electrician makes use of several of these when installing electrical systems, particularly the power drill. Drill bits specially designed for wood and masonry are found in many sizes and lengths in the electrician's toolbox—they are absolutely essential components.

Miscellaneous Construction Tools

Electricians use many standard construction tools when installing or repairing an electrical system. The following is a partial list of tools that will be needed:

- **Toolboxes and cabinets**—These products are designed to hold tools. Do not stand or work on them. Always push wheeled cabinets, do not pull them.
- **Utility light**—A utility light is needed when the work area does not have proper lighting. This is often the case with electrical work, where the lighting circuit may be de-energized or lighting fixtures may not be installed. Other types of portable lights are also useful.
- **Tape measure and folding rule**—These instruments are used to measure distances. A measuring tape with a wide blade will not bend as easily as a thin-blade tape. The stiffer tape makes measuring easier.
- **Ladders**—Use wood and fiberglass ladders only. Do not use ladders made of conductive material (aluminum, for example).
- **Chalk line**—This tool marks straight lines. It can be used as an aid when cutting, measuring, and aligning items.
- **Level**—A level is used when placing conduit and equipment.
- **Plumb bob**—Used to establish vertical lines, this tool is basically a string with a pointed weight.
- **Gas generator**—Generators are used to supply power for tools and equipment requiring an electrical source.
- **Extension cords**—Extension cords are needed to supply power to areas without available receptacles. Consider both the current load and the cord length when selecting an extension cord.
- **Space heaters**—When working in cold climates, heaters are needed to create a safe, warm working environment.
- **Portable fans**—In hot climates, fans are used to help keep workers cool. Fans also circulate air and transport dust and small particles away from the work area.

Electrical Tools

In addition to standard construction tools, electricians use a variety of specialized tools. These tools are designed to be used when working with electrical conductors, cables, and conduit. Electronic equipment is used to test the condition of circuits and equipment.

Electrician's Hand Tools

Normally, electricians use hand tools to cut, strip, and crimp cables and conductors. The size of the cable or conductor determines the type of tool needed. For small conductors, pliers and combination tools are all that is needed. For large conductors and cables, specialized tools are used.

Pliers

When working with wires, electricians use a variety of pliers to shape, bend, cut, and strip wire. There are many types of pliers, each suited for a particular set of tasks. Those types used by electrical workers are discussed below and illustrated in **Figure 3-5.**

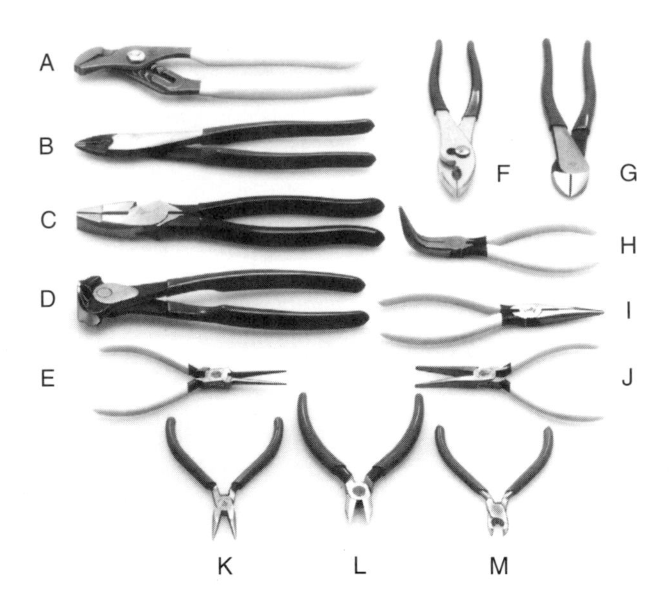

Figure 3-5. There are many types of pliers. A—Tongue-and-groove pliers. B—Long-reach pliers. C—Side-cutting pliers. D—End-cutting pliers. E—Needle-nosed pliers. F—Slip-joint pliers. G—Diagonal-cutting pliers. H—Curved needle-nosed pliers. I—Long-nosed pliers. J—Flat-nosed pliers. K—Long-nosed cutter pliers. L—Diagonal-cutting pliers. M—End-cutting pliers. (Klein Tools, Inc.)

- **Linesman's pliers**—These pliers are also known as combination pliers or engineer's pliers. They have flat, serrated jaws with side cutters located close to the pivot.
- **Flat-nosed pliers**—These pliers are also called square-nosed pliers or duckbill pliers. They are used to bend and shape wires. They range in size (length) from 4 1/2″ to 7 1/2″. Some have side cutters near the pivot.
- **Long-nosed pliers**—More commonly called needle-nosed pliers, these are extremely handy for shaping, bending, and cutting small-gage wires used in instrument and panel wiring. They have a variety of shapes—curved nose, bent nose, and straight nose. They have side cutters located just ahead of the pivot.
- **Diagonal-cutting pliers**—Referred to as side-cutting pliers, these pliers are primarily used by electricians to cut wires flush with a surface. They are occasionally provided with a coil spring between the handles to automatically open. The handles are fairly curved and the overall length is rarely more than 8″.
- **End-cutting pliers**—These pliers are also used to cut wire close to a surface, while keeping hands (particularly knuckles) free from injury. Since the jaws are just ahead of the pivot, these pliers can exert great force and are capable of cutting moderate size wire. They generally range from 5″ to 9″ in overall length.
- **Slip-joint pliers**—The function of slip-joint pliers is the same as linesman's pliers. However, slip-joint pliers can open at two different jaw widths due to the pivot design. Slip-joint pliers are mostly used in plumbing, jewelry applications, and general work. Many slip-joint pliers have jaws that are partially flat and partially curved so they can grip either flat or round workpieces.
- **Tongue-and-groove pliers**—These pliers have unique slots in the pivot area so they can be adjusted to several different widths. Originally designed for working on plumbing fittings, they are also called pump pliers. The tool has long handles to provide excellent leverage. Although they are adjusted like slip-joint pliers, the grooves secure the jaw very well so these pliers won't slip when working with substantial force. They are often used by electricians as substitutes for adjustable wrenches and fixed wrenches.
- **Vise-grip pliers**—Vise-grip jaws are adjusted by turning an adjuster knob. When closed, the jaws exert a tremendous force on the workpiece and stay locked until the release lever is actuated. This makes the tool similar to a vise. Vise-grip pliers are often used to temporarily clamp items together. There are several different jaw styles and shapes: flat jaw, curved serrated jaw, curved smooth jaw, and C-clamp jaw.

Use pliers for their intended purpose. Do not use pliers to strike other objects or tools. Using pliers to tighten nuts and bolts can damage the fastener. Use a wrench instead.

Insulated pliers should be used for electrical work, **Figure 3-6.** The insulation helps protect the electrician

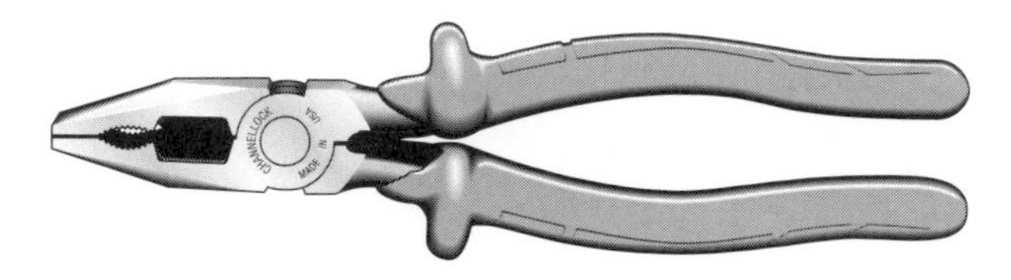

Figure 3-6. The insulated handles of these linesman's pliers provide added protection against shocks. Always use insulated tools when working near energized circuits. (Channellock, Inc.)

from shock and may prevent the pliers from accidentally connecting two energized objects.

Do not extend handles for greater leverage. Doing so may cause the handles to break or bend because they are not designed to support the larger force. In situations where the pliers are not providing enough clamping force, use a larger set of pliers. **Figure 3-7** summarizes the safety rules for pliers.

Pliers Safety

- Never use pliers to strike other objects or tools.
- Generally, pliers should not be used to tighten nuts on bolts. Wrenches are far better for that purpose.
- Keep pivot oiled.
- For electrical work, be sure handles are insulated.
- Do not use pliers to cut hardened steel wire, unless they are specifically made for that purpose.
- When cutting electrical wire, always cut at right angles.
- Never extend handles for greater leverage. Use longer handled pliers or cutters designed for greater leverage.

Figure 3-7. Basic safety rules for using pliers.

Cutting tools

Cables are cut on the work site using a wide variety of tools. Thick cables and large conductors are cut with long-handled, heavy-duty cable cutters, ratchet cable cutters, or hydraulic cable cutters, **Figure 3-8.** For most small conductors, wire cutters or cutting pliers are used.

CAUTION

Most cable cutters are designed to cut copper and aluminum. Do not use them to cut steel or other materials unless they are approved for such use.

Stripping and crimping tools

A wide variety of stripping and crimping tools are also available. For conductor sizes up to and including 10 AWG, a small combination tool can be used for stripping, crimping, and cutting. For larger conductors and cables, more specialized strippers and crimpers are needed. **Figure 3-9** shows several types of small strippers.

Compression tools are used to attach lugs and connectors to the end of cables and conductors. Manual and hydraulic compression tools are available. See **Figure 3-10.**

Bending Tools

Conduit is normally purchased in straight lengths. In order to route conductors around corners, the conduit must be bent. Several tools are used to bend conduit. The type of conduit bender used depends on the type and size of conduit being bent.

Conduit hickeys and hand benders can be used to bend small sizes (up to and including 1″) of rigid steel and rigid aluminum conduit. These tools are used to bend offsets in 1 1/4″ rigid conduits. Manual benders can also be used to bend EMT in sizes up to and including 1 1/4″. See **Figure 3-11.**

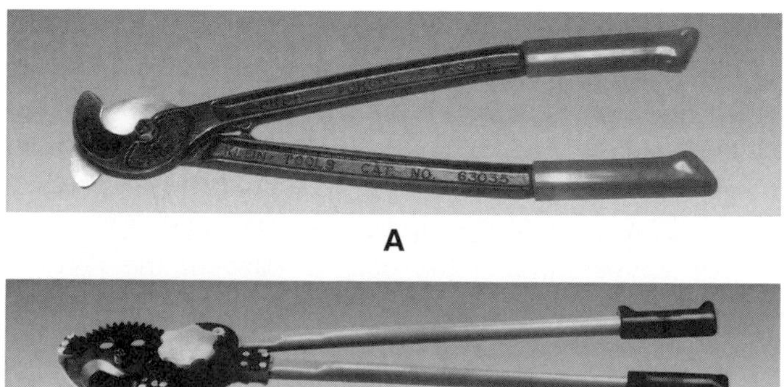

Figure 3-8. Cable cutters. A—Long-handled manual cable cutter. B—Heavy-duty ratchet cable cutter. C—Compact ratchet cable cutter. (Klein Tools, Inc. and Greenlee Textron)

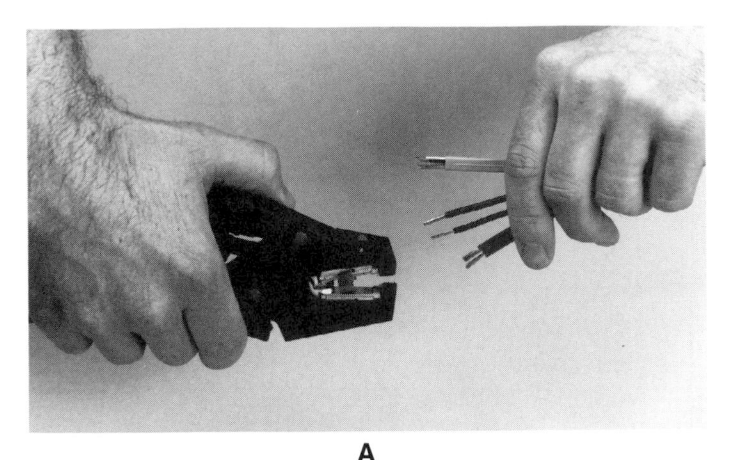

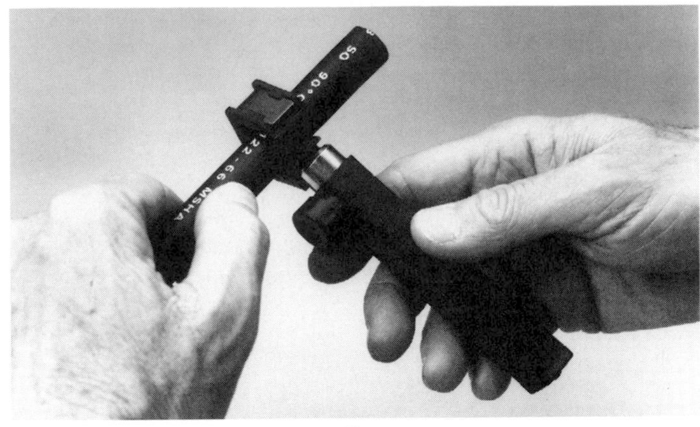

A B

Figure 3-9. Small stripping tools are satisfactory for most jobs. A—This tool strips and cuts conductors as large as No. 12 AWG. B—A pocket cable stripper is used for larger conductors. (Greenlee Textron)

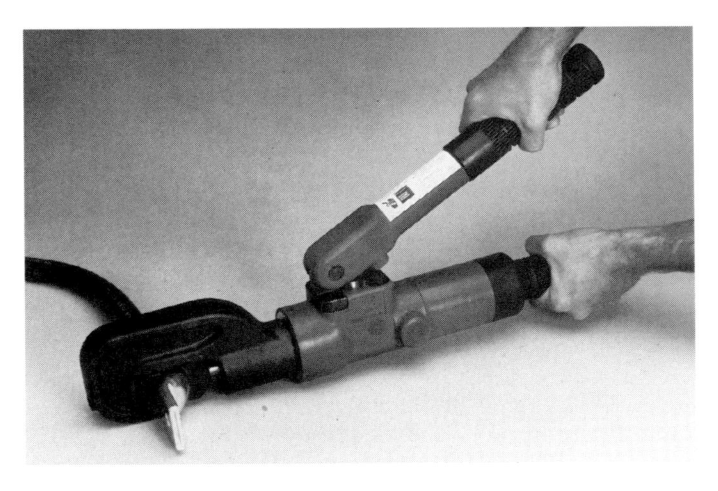

Figure 3-10. Using a hydraulic crimper to attach a lug to the end of a large conductor. (Greenlee Textron)

Power conduit benders are used to form bends in larger conduit sizes. These benders are powered hydraulically or electrically. Several conduits can be bent at the same time.

Electricians use two methods of conduit bending. The method is selected based on the type of conduit being bent, the size of the conduit, the size of the bend, and the tools available.

In the ***progression bend method,*** the final bend is the result of many smaller bends. First the conduit is marked. The first small bend is made, then the conduit is repositioned. Another small bend is made, and the conduit is again repositioned. This continues until the desired bend is attained.

The ***one-shot bend method*** bends the conductor in a single step. There is no need to adjust the conduit in the bender several times. Compared to the progression bend method, one-shot bending is faster and easier. However, this method requires larger and more expensive bending equipment.

Figure 3-11. An electrician bending EMT with a hand bender. (Greenlee Textron)

Rigid nonmetallic conduit (PVC) is normally too brittle to bend using hand or hydraulic benders. However, if the conduit is heated, it becomes more ductile (able to be bent). A PVC heater/bender, or electric ***hot box,*** is used to prepare the piece for bending. The hot box is connected to a 120-volt power source and preheated. Once the proper temperature in the box is reached, the PVC is inserted and rotated until it becomes pliable. The conduit is removed and then bent into the proper shape.

PVC conduit can be overheated. Therefore, never heat conduit with a concentrated heat source, such as a flame from a torch. The heat source should be uniform and constant. Use proper tools for cutting PVC conduit, **Figure 3-12.**

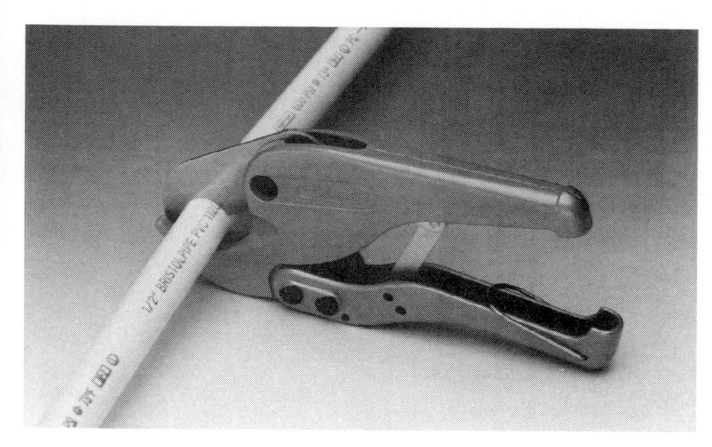

Figure 3-12. This PVC cutter is able to cut conduit sizes up to 1 5/8″. (General Tools Mfg. Co., Inc.)

Large conductors and cables can be bent with a ***cable bender.*** The leverage provided by the long handles allows thick cables to be bent by hand.

Conduit can go through many procedures from the time it arrives at the work site until it is installed. In addition to benders, the following tools may be needed:

- **Pipe cutter**—This tool can be used to cut conduit. A hacksaw or a power saw may also be used.
- **Pipe reamer**—Cutting conduit can produce sharp metal burrs that can damage conduit insulation. A pipe reamer is used to remove these burrs. All conduit should be reamed before it is installed. If a pipe reamer is not available, a file can be used to remove burrs and sharp edges from conduit.
- **Conduit threader**—Threads must be cut into rigid conduit so couplings and fittings can be screwed on.

Pulling Tools

After the conduit and boxes are installed, conductors must be pulled through the system. If the conductors are relatively small, ***fish tape*** can be used. The fish tape is first pushed through the conduit between the boxes where the ends of the conductor will be located. See **Figure 3-13.** The conductor is then wrapped securely to the fish tape, and both are pulled back through the conduit and to the other box.

Wire pulling lubricant is used to ease cable pulling. The conductors are coated with the lubricant and then pulled through the conduit. After pulling, the lubricant dries to a thin film or powder.

A power vacuum/blower fishing system can also be used to move fish tape through the conduit. A piston with the line attached to it is placed in the conduit. The vacuum/blower is attached to the conduit and blows the piston to the next outlet box. As the piston is blown through the conduit, it drags the tape behind it.

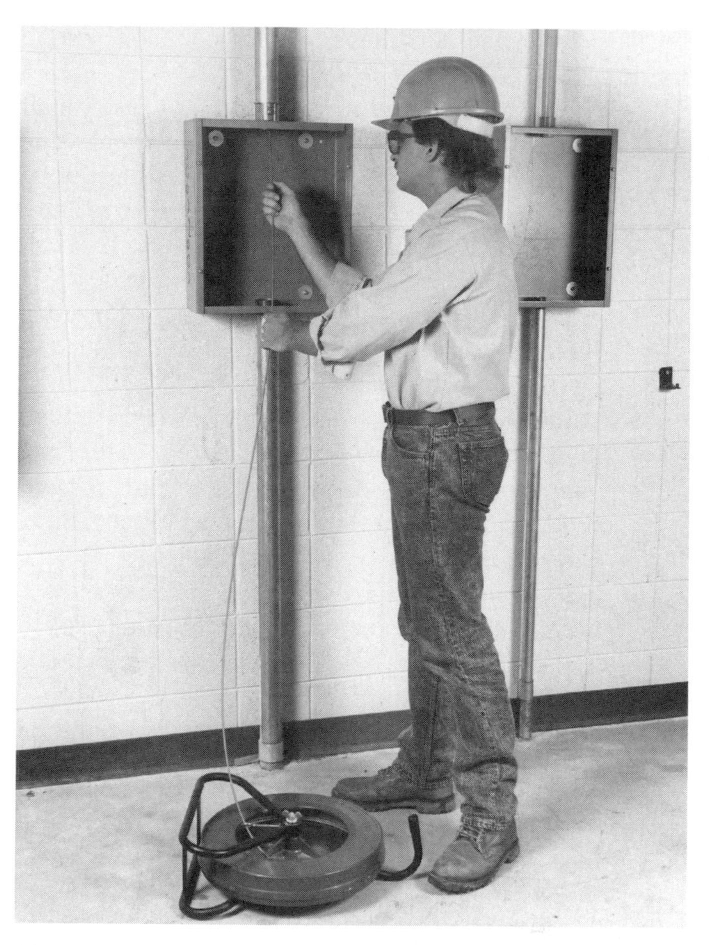

Figure 3-13. Feeding fish tape into conduit. When the fish tape reaches the other junction box, the conductor will be attached and pulled back through. (Greenlee Textron)

Cable puller systems are used to pull large cables through conduit. Cable pullers can be manual or may include an electric motor.

Electrical Testing Tools

In order to determine what is wrong with an electrical system, testing equipment is needed. The better the testing equipment, the easier the job and the more accurate the readings.

There are numerous tools available for testing electrical components and circuitry. From the simple test light to the most sophisticated circuit analysis instruments, electricians have a wide variety of units to choose from for testing, analyzing, and troubleshooting electrical systems. Great care should be taken while using any test equipment to protect the equipment and the operator.

Be sure to read all the instructions provided with the equipment before attempting to use it. Keep the instruction booklet and refer to it when necessary to ensure the equipment is being used correctly and operating accurately.

The following are some of the common testing instruments:

- **Voltage and continuity testers**—These small battery-operated testers are used to determine the circuit voltage and polarity. Most testers can be used over a wide voltage range. See **Figure 3-14.**
- **Receptacle circuit testers**—These testers check for various problems in a circuit. The features vary among models, but many test for open ground, correct wiring, open hot or neutral, and reverse polarity. See **Figure 3-15.**
- **Clamp-on meters**—These meters clamp around a conductor and measure the current in the line. Be sure clamp-on meter jaws are closed completely and are on one conductor at a time. See **Figure 3-16.**
- **Megohmmeters**—Also called ***meggers,*** these instruments measure resistance. They are used to detect insulation deterioration and the presence of dirt and moisture.
- **Frequency meters**—These meters measure the frequency of an alternating current. A frequency meter can be used to verify that a generator is working properly.
- **Multimeters**—Generally, multimeters are used to measure the voltage, current, and resistance in a circuit. A multimeter is shown in **Figure 3-17.**

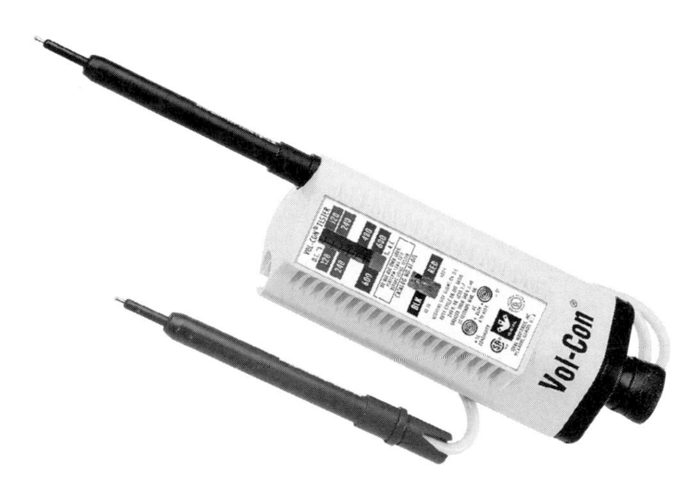

Figure 3-14. This heavy-duty voltage and continuity tester can test circuits up to 600 volts. (Ideal Industries, Inc.)

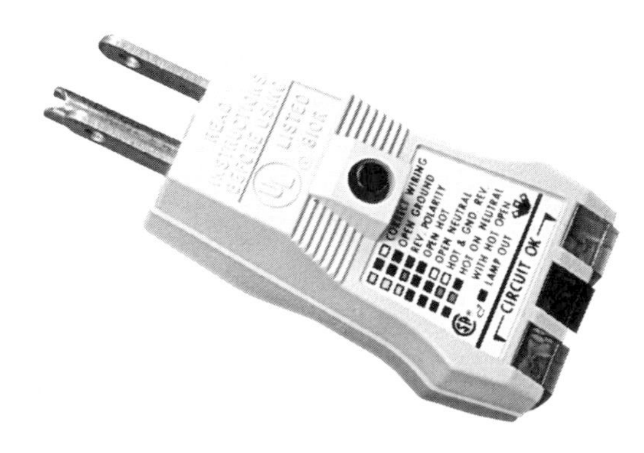

Figure 3-15. This E-Z Check® GFCI circuit tester identifies the condition of the circuit. (Ideal Industries, Inc.)

Figure 3-16. This clamp-on meter measures the current in a conductor. It is constructed to survive a six-foot drop. (Fluke)

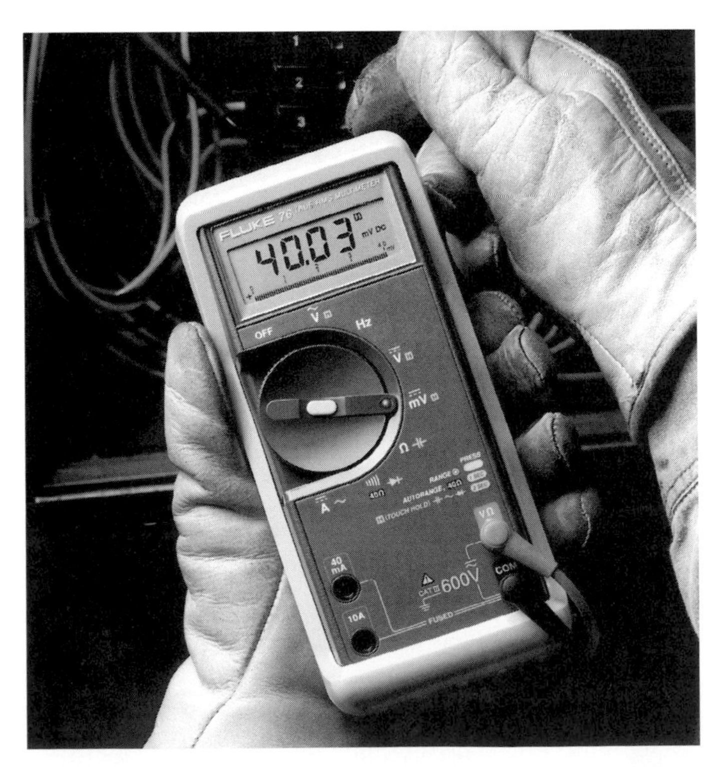

Figure 3-17. An electrician uses multimeters to measure voltage, current, and resistance in circuits. (Fluke)

Inspect the insulation on the test leads before and after testing. Look carefully for any cuts, abrasions, or nicks. Do not use the tester if there is any damage.

Use the test equipment for its intended purpose. Do not exceed the limitations of the equipment.

Keep testing equipment in good, clean condition. Make sure all components are in proper condition and are well maintained. Internal fuses, power supply cords, and batteries should be in good condition.

When using instruments with several scales, always begin testing with the highest scale to prevent overloading. Work your way down through the scale until the appropriate scale is reached. This protects the instrument from damage.

When making resistance measurements, never work on an energized circuit.

Miscellaneous Electrical Tools

Electricians use many tools besides those discussed in this chapter. The following are several additional tools an electrician must be comfortable using:

- **Calculator**—Electricians perform calculations to size conductors, conduit, and overcurrent protective devices. Calculations are also needed to determine material lengths and to set up conduit bends. Calculators designed specifically for electrical work are also available, **Figure 3-18.**
- **Soldering iron**—For joining wires, where soldering is necessary.
- **Fuse puller**—A fuse puller is used to remove fuses from panelboards.
- **Wire gage**—This instrument is used to check wire sizes.

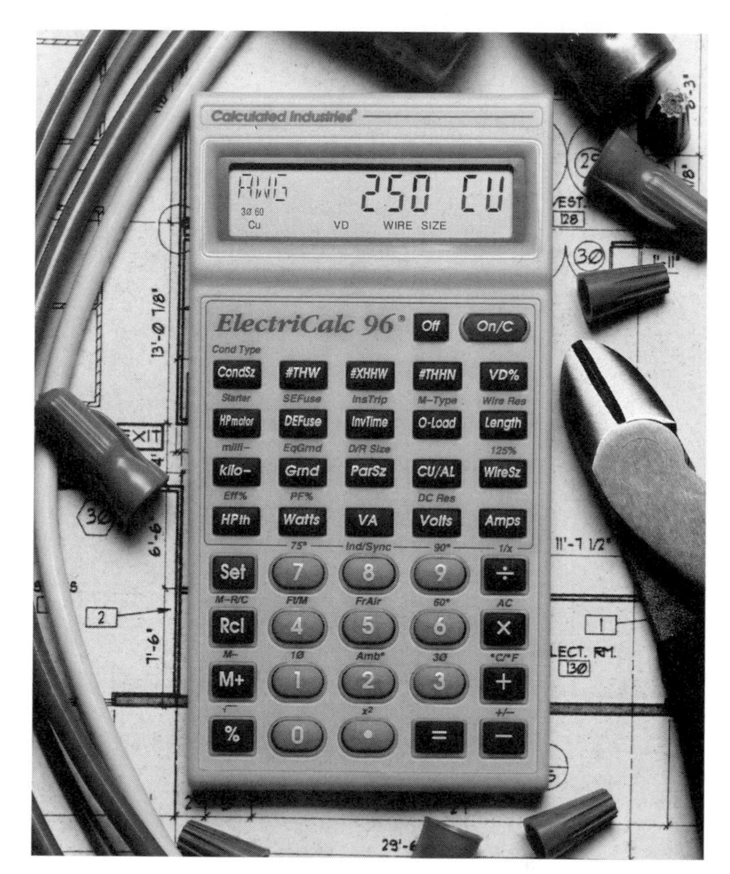

Figure 3-18. This calculator is designed for electrical work. It can calculate wire and conduit sizes, determine motor full-load amps, and compute fuse and breaker sizes. (Calculated Industries, Inc.)

Review Questions

Answer the following questions. Do not write in this book.

1. List the three safety rules that, if obeyed, will prevent most hand tool accidents.
2. Describe the hazards involved in using a hammer that is too big for the job and a hammer that is too small for the job.
3. Why should you wear gloves when cutting material with a saw?
4. In what type of situation are end-cutting pliers useful?
5. Why is it dangerous to extend handles on pliers to gain leverage?
6. Explain the differences between the progression bend method and the one-shot bend method.
7. Describe the process and equipment used to bend rigid nonmetallic (PVC) conduit.
8. Explain why conduit must be reamed or filed after it is cut.
9. How does a power vacuum/blower fishing system work?
10. When using an electrical testing device with multiple scales, why should you begin testing on the highest scale?

This cutter is used to cut BX cable. (Greenlee Textron)

Chapter 4

Electrical Prints, Specifications, and Codes

Technical Terms

Detail drawing
Elementary diagram
Elevation
Fine Print Notes (FPN)
International Association of Electrical Inspectors (IAEI)
Lighting plan
Model code
National Electrical Code (NEC)
National Fire Protection Association (NFPA)
Occupational Safety and Health Administration (OSHA)
One-line diagram
Plan
Power plan
Schedule
Schematic diagram
Section
Single-line diagram
Specifications

Objectives

After completing this chapter, you will be able to:

- Describe several types of electrical drawings.
- Identify common electrical symbols.
- Explain the purpose of specifications.
- Explain the importance of building codes.
- Define the purpose, intent, arrangement, and key terminology of the *National Electrical Code.*
- Name various agencies that set standards concerning electrical practices and procedures.
- Identify various lab facilities that perform rigorous testing on electrical devices, equipment, and associated components for safety and performance certification.

Engineers, architects, and designers use drawings to communicate construction details to those performing the work. In order for an electrical system to be installed properly, several types of drawings and diagrams are used. Some drawings specify *what* is being installed and *where* it is being installed. Other types of diagrams show *how* a device is installed. Electricians must be familiar with all types of electrical drawings and commonly-used electrical symbols.

In addition to drawings, every construction project has a set of specifications. These specifications contain detailed instructions for construction procedures, requirements, and materials. All work performed on a project must conform to the specifications. Therefore, specifications are as important as drawings to the successful completion of a project.

Specifications normally list the building code that the project must satisfy. Inspectors periodically check the site to ensure that the codes are obeyed. Many local building codes mandate that electrical work conforms to the *National Electrical Code*. This text is based on the requirements of that code.

Electrical Print Reading

Electrical drawings show device locations and identify materials needed for an electrical installation. Drawings can also be used to show individual connections or provide a simplified view of a circuit. Standard symbols are used on drawings to represent devices and equipment.

Electrical drawings are useful not only to the electrician installing the system but to others as well. Estimators use drawings to prepare bids. Maintenance personnel use drawings to troubleshoot and repair problems. Purchasers use drawings to determine the materials to order. Manufacturers use drawings to fabricate equipment. Everyone involved in the construction process must be able to read drawings.

Construction Drawings

A typical set of drawings used in constructing an electrical system have plans, elevations, sections, and details.

These four types of drawings are used throughout construction in all building trades.

Plans are drawings showing the project viewed from above. These drawings normally contain more information than the other types of drawings.

Elevations show the side view of an area or object. An elevation is provided when the plan view cannot provide the information clearly. By studying both the plan and elevation for an area, you can visualize the construction in three dimensions.

In situations where a detail cannot be shown clearly in the plan or elevation, a sectional drawing is used. Sectional drawings (***sections***) are elevations that "cut through" an object or building. The location of the "slice" is marked on the plan with a section line. All items to one side of the section line are removed for the sectional drawing. This allows internal or hidden items to be shown.

Detail drawings show a particular device, item, or part of the system that cannot be clearly shown in a plan, elevation, or section. Detail drawings often have a larger scale than the other drawings. This allows the designer to "detail" or enlarge small items. The location of the detail is shown on a plan, elevation, or section.

Electrical Symbols and Abbreviations

Symbols and abbreviations are used on drawings to simply represent the type, size, and style of equipment to be installed. Anyone using electrical drawings must have a knowledge of the symbols and abbreviations in order to understand and interpret the drawing.

NOTE

Although symbols are somewhat standardized, standard symbols are not used universally. Different designers may use different symbols to represent the same item. Therefore, always refer to the symbol legend on the drawing.

The symbols presented in this chapter are based on ANSI Y32.9-1972, *Graphic Symbols for Electrical Wiring and Layout Diagrams Used in Architecture and Building Construction,* which is prepared by the American National Standards Institute (ANSI).

Wiring symbols

Electrical plans do not show the exact location of conduit runs. However, conductors that connect devices are shown as irregular arcs. These arcs are dashed when the conductors are concealed in the floor and solid when conductors are located in the ceiling or walls.

The number of conductors in a conduit run is shown by small marks across the conduit run line. Hot wires are shown as full marks. Often, neutral wires are shown as half marks, which may be centered on the line or entirely to one side. See **Figure 4-1.** Circuits without conductor marks are assumed to have two hot conductors.

If all the circuit lines were drawn back to the panel, the drawing would be very crowded. Therefore, an arrow at the end of a conduit line signifies a conduit run back to the panel. Normally, the panel and circuit is identified next to the arrow. The number of arrows at the end of a conduit line indicates the number of circuits whose conductors are contained in the conduit. See **Figure 4-2.**

Lighting symbols

Switches, lighting outlets, and fixtures are normally shown on the electrical plan. Switches are identified by an S symbol perpendicular to the wall where the switch is located. Sometimes, a line is drawn through the S and

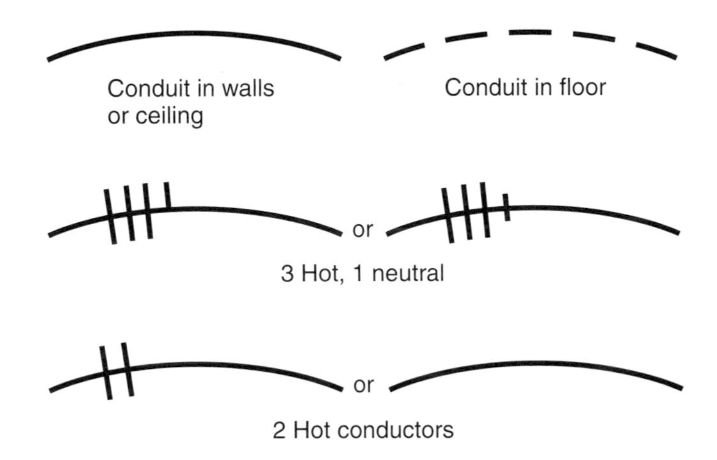

Figure 4-1. Circuits are identified by lines on the drawing. The lines do not show the exact location of the conductors; they only show which devices are connected.

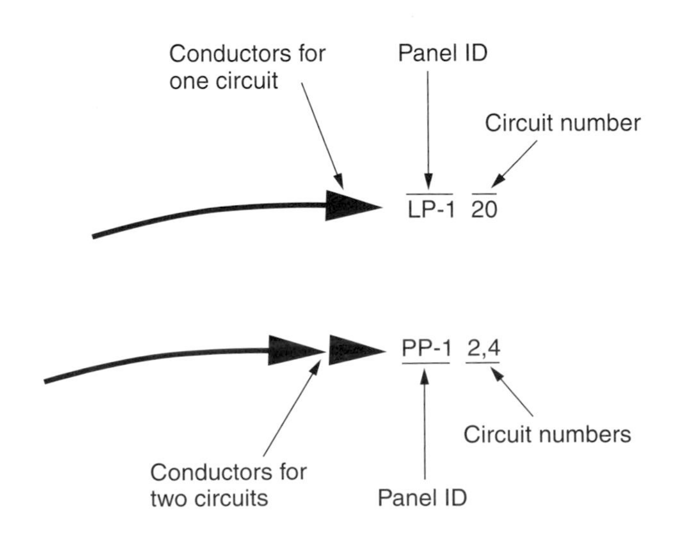

Figure 4-2. An arrow at the end of a conductor run indicates a run to the panel. The top conductors are routed to circuit 20 in panel LP-1. The bottom conduit contains conductors for circuits 2 and 4 of panel PP-1.

touching the wall. Subscript letters and numbers are added to identify various types of switches. **Figure 4-3** illustrates several switch symbols.

Symbols used for lighting outlets and fixtures vary. Some standard symbols are shown in **Figure 4-4.**

Symbol	Description
S	Single-pole switch
S_2	Double-pole switch
S_3	Three-way switch
S_4	Four-way switch
S_K	Key-operated switch
S_P	Switch and pilot lamp
S_L	Switch for low-voltage switching system
S_{LM}	Master switch for low-voltage switching system
S_D	Door switch
S_T	Time switch
S_{CB}	Circuit breaker switch
S_{MC}	Momentary contact switch
Ⓢ	Ceiling pull switch

Figure 4-3. Standard switch symbols used on electrical drawings.

Ceiling	Wall	
		Surface or pendant fixture
		Recessed fixture
		Surface or pendant exit light
		Recessed exit light
		Blanked outlet
		Junction box
		Outlet controlled by low-voltage switch
		Surface or pendant fluorescent fixture
		Recessed fluorescent fixture
		Surface or pendant continuous-row fluorescent
		Recessed continuous-row fluorescent
		Bare-lamp fluorescent strip

Figure 4-4. Standard lighting outlet and fixture symbols.

Normally, each fixture symbol is accompanied by a letter corresponding to the mark in the lighting fixture schedule. The schedule provides the exact type of fixture and lamps.

Receptacle symbols

Symbols on the electrical plan identify the locations and types of receptacles. Special or unusual receptacles are clearly marked. Circuit lines may be used to show the connections between receptacles. See **Figure 4-5.**

Additional symbols

Other standard electrical symbols are shown in **Figure 4-6.** These symbols can be used in any type of electrical drawing. Becoming familiar with these common symbols will improve your print reading skills.

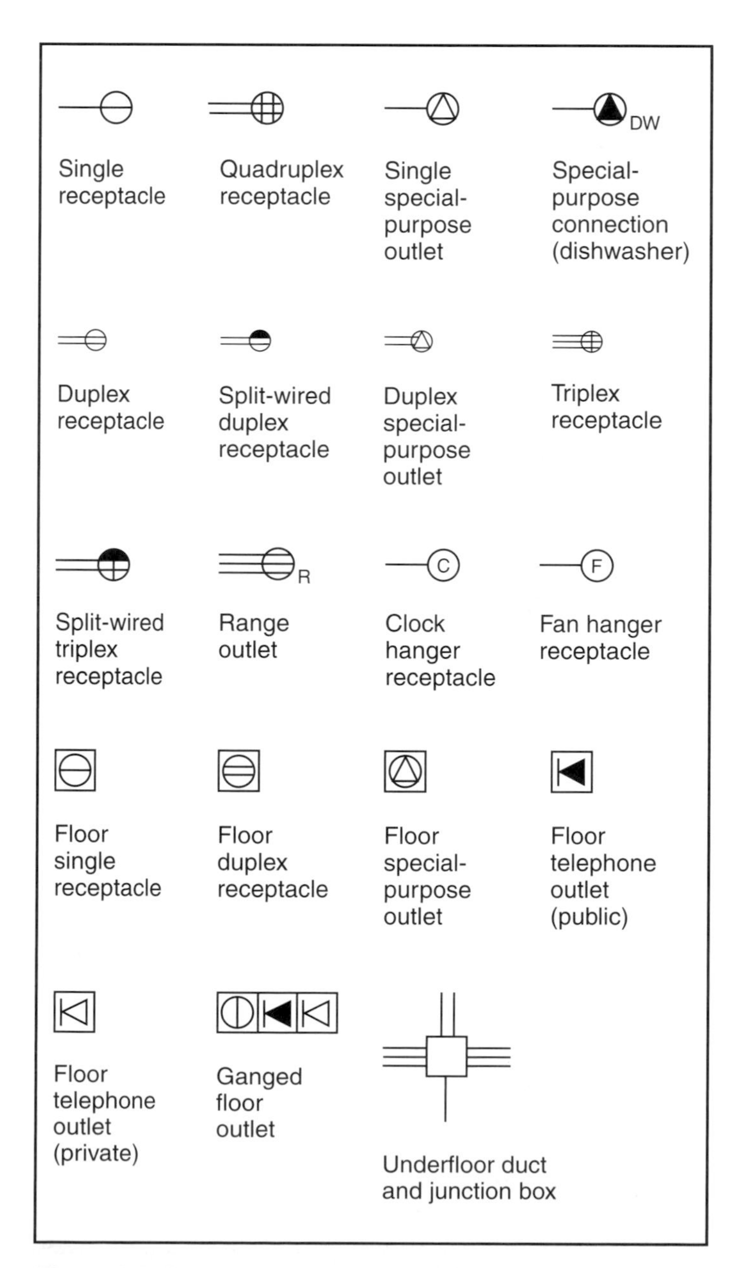

Figure 4-5. Standard receptacle symbols.

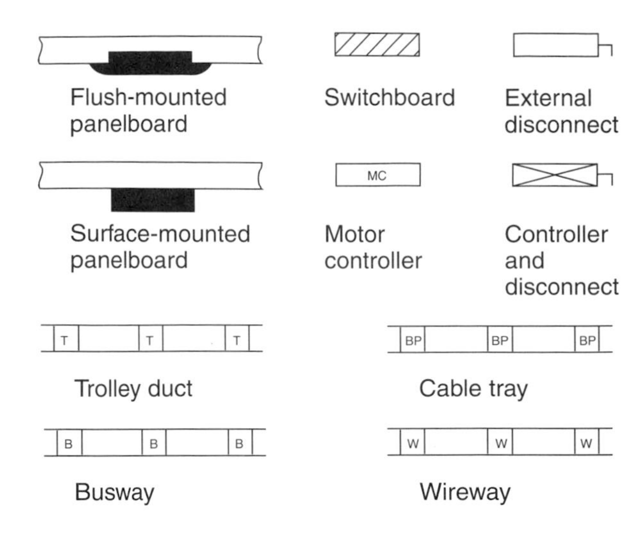

Figure 4-6. Miscellaneous electrical symbols.

Categories of Electrical Drawings

There are several types of electrical diagrams and drawings. The type of drawing used is determined by what the drawing intends to show. The following are some of the most common types of electrical drawings:

- Electrical plans
- Schematic diagrams (elementary diagrams)
- Single-line drawings (one-line diagrams)
- Schedules

Electrical plans

Electrical plans show the basic floor plan (walls, doors, windows, and stairs) along with the installed electrical devices. For small projects, a single electrical plan may be sufficient to show all the electrical equipment and connections. However, for most commercial projects, two electrical plans are used. See **Figure 4-7.**

An electrical ***lighting plan*** shows switches, lighting outlets and fixtures, and the conductors connecting them. An electrical ***power plan*** shows receptacle outlets and equipment with electrical feeds. Panels are often shown on both types of plans.

Electrical plans include marks that reference schedules, elevations, sections, and detail drawings. Other types of electrical diagrams are included to help clarify circuits and connections.

Electrical schedules

Electrical ***schedules*** are tables that list items such as fixtures, power requirements, panelboard information, receptacles, transformers, and motors. Schedules save print reading time by placing information in one easy-to-find location. This helps with project estimating,

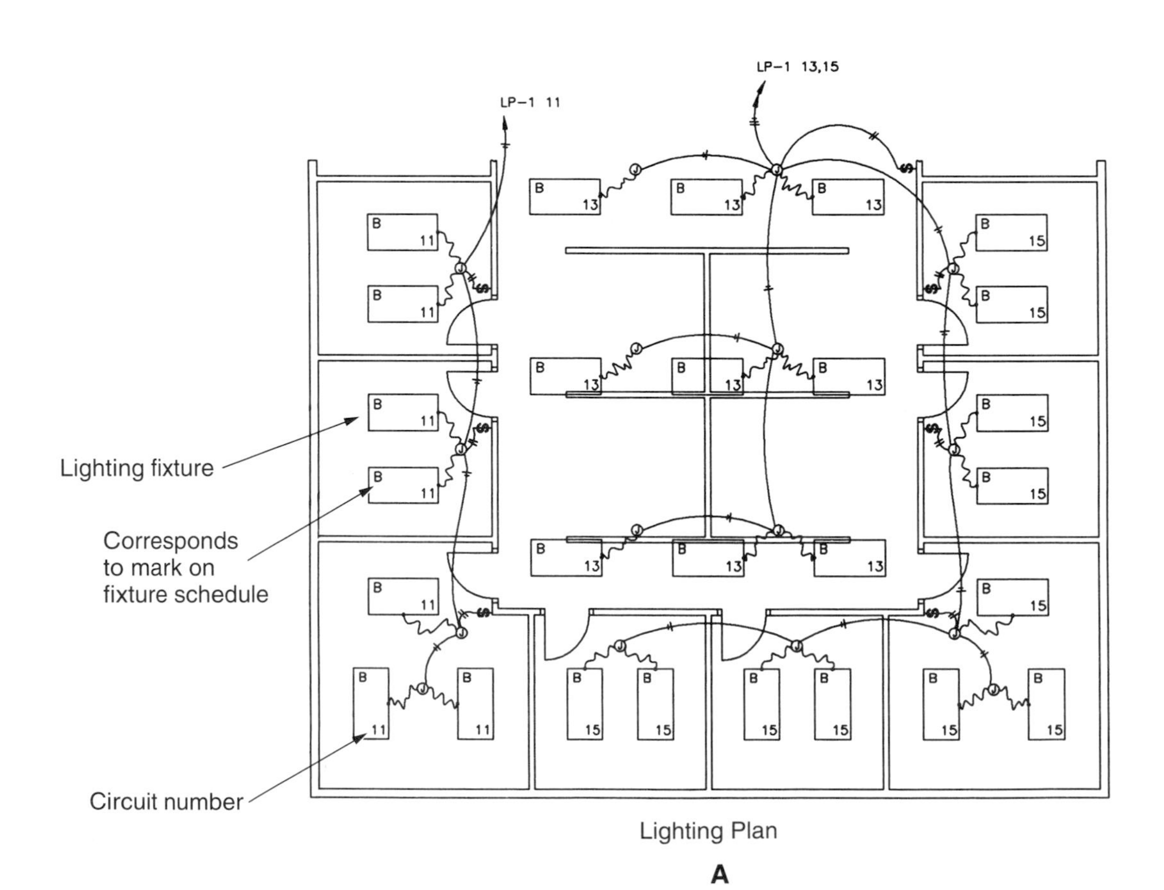

Figure 4-7. For most commercial electrical projects, two electrical plans are used: an electrical lighting plan and an electrical power plan.

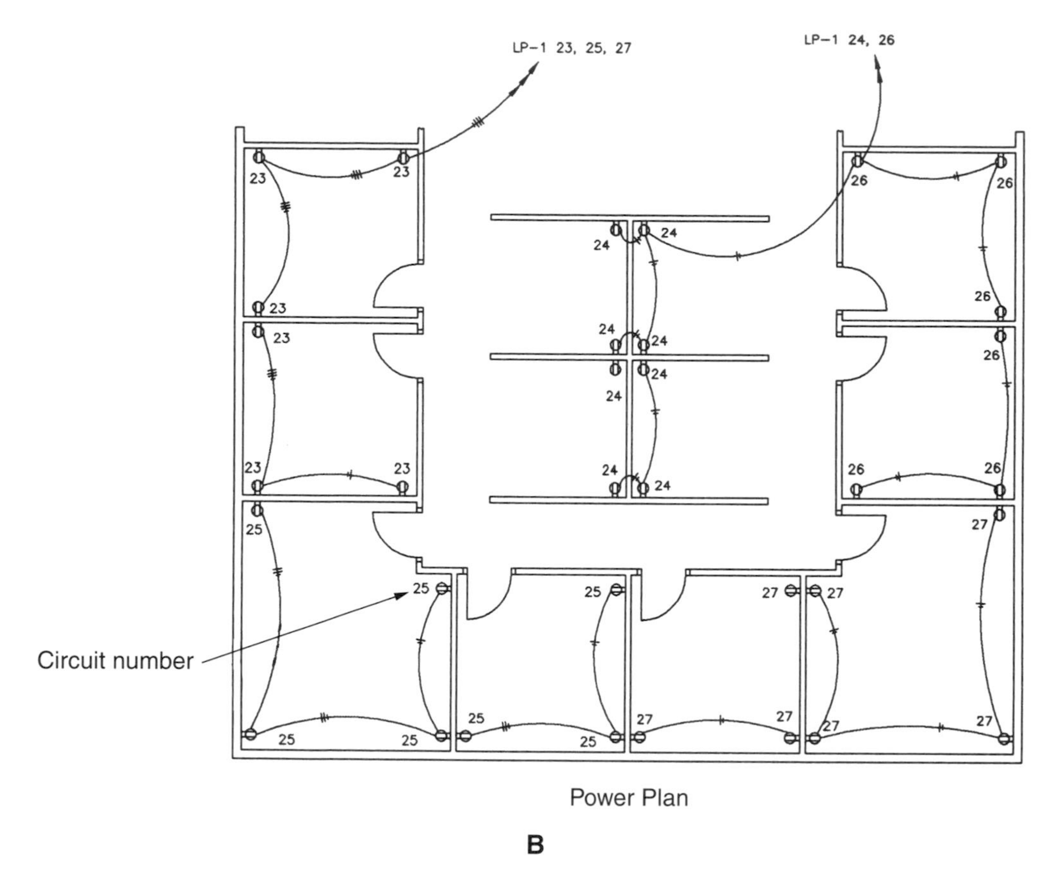

Figure 4-7. *(Continued)*

purchasing, and construction. The following are some examples of schedules:

- **Lighting fixture schedule**—This type of schedule lists the lighting fixtures used in the project. See **Figure 4-8.** Each row in the schedule lists basic information for a single type of fixture. The information includes the number of fixtures required, the manufacturer, model number, and number and type of lamps. Each type of lighting fixture also has a unique mark (normally identified as a letter). The marks, which are included on the lighting plan, identify the locations of the fixtures.
- **Equipment schedule**—An equipment schedule lists the large electrical equipment being installed. The information provided in this schedule varies depending on the specific type of equipment listed. Equipment schedules may

LIGHT FIXTURE SCHEDULE				
MARK	QUAN.	DESCRIPTION	MANUFACTURER	LAMP
A	16	WALLMOUNT SURFACE	PRESCOLITE #4382	2 - 60W
B	6	CEILING/WALL MOUNT	LIGHTOLIER - CHR - KEYLESS	6 - 40W
C	18	CEILING MOUNT CAN - WHITE	PRESCOLITE #1102	75W R - 30
D	5	CEILING/WALL MOUNT	CERAMIC KEYLESS	75W
E	9	W.P. FLOODLITE HOLDERS		100W FLOOD
F	1	PENDANT	PRESCOLITE #94401	150W
G	2	STEP LIGHT WALL RECESSED	PRESCOLITE #3761	2 - 40W
H	1	SPHEROID PENDANT	PRESCOLITE #576	150W
J	1	RECESSED IN FLOOR	PRESCOLITE #79-832	75W R - 30
K	6	2 EXPOSED FLUOR. TUBE	5 - 36" & 1 - 24"	30W & 25W

Figure 4-8. This lighting fixture schedule lists all of the lighting fixtures needed for the project. The schedule helps estimators and purchasers determine the quantity and types of products needed.

include the voltage provided to the equipment, size and type of overcurrent protection, and the number and sizes of conductors feeding the equipment. Motors, heating units, and cooling units are often listed in equipment schedules.

- **Panel schedule**—A panel schedule, **Figure 4-9,** lists the basic characteristics of a lighting or power panel. The information normally includes the panel voltage, the panel amperage rating, and a chart listing the circuits, circuit breakers, and equipment served by each circuit.
- **Miscellaneous schedules**—The size of the project determines the number and size of schedules. For large projects, schedules may be included to aid estimating and purchasing. A receptacle schedule and a switch schedule may be included. These schedules are similar to a lighting fixture schedule.

Schematic diagrams

Schematic diagrams, also called ***elementary diagrams,*** show the arrangement and connections within a circuit. These diagrams show the function and order of items in a circuit without specifying device size or location.

Single-line diagrams

Single-line diagrams use single lines and symbols to show the course of a circuit and its components. These diagrams are also called ***one-line diagrams.*** An example is shown in **Figure 4-10.**

Electrical Specifications

Specifications are the written detailed description of the project requirements. The specifications are essential, along with the set of construction drawings, to complete the contractual obligations of an electrical construction project.

Many specifications are arranged using the Construction Specification Institute's MasterFormat™. Specification writers use this numbering system to organize the various tasks involved in a construction project. The MasterFormat consists of many divisions, each addressing a general construction topic. Some of these divisions are listed in **Figure 4-11.**

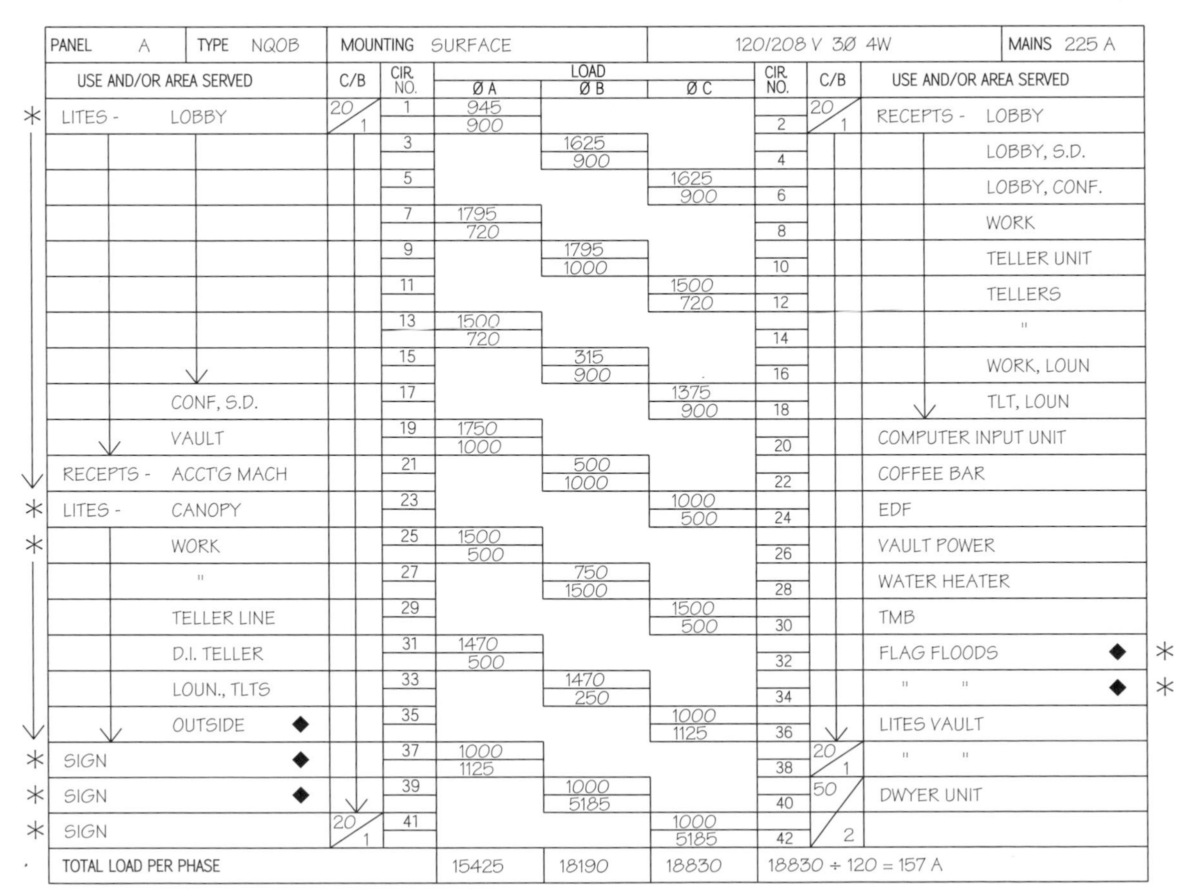

	PANEL A	TYPE NQOB	MOUNTING SURFACE				120/208 V 3Ø 4W			MAINS 225 A	
	USE AND/OR AREA SERVED		C/B	CIR. NO.	LOAD ØA	LOAD ØB	LOAD ØC	CIR. NO.	C/B	USE AND/OR AREA SERVED	
*	LITES -	LOBBY	20/1	1	945 / 900			2	20/1	RECEPTS - LOBBY	
				3		1625 / 900		4		LOBBY, S.D.	
				5			1625 / 900	6		LOBBY, CONF.	
				7	1795 / 720			8		WORK	
				9		1795 / 1000		10		TELLER UNIT	
				11			1500 / 720	12		TELLERS	
				13	1500 / 720			14		"	
				15		315 / 900		16		WORK, LOUN	
		CONF, S.D.		17			1375 / 900	18		TLT, LOUN	
		VAULT		19	1750 / 1000			20		COMPUTER INPUT UNIT	
	RECEPTS -	ACCT'G MACH		21		500 / 1000		22		COFFEE BAR	
*	LITES -	CANOPY		23			1000 / 500	24		EDF	
*		WORK		25	1500 / 500			26		VAULT POWER	
		"		27		750 / 1500		28		WATER HEATER	
		TELLER LINE		29			1500 / 500	30		TMB	
		D.I. TELLER		31	1470 / 500			32		FLAG FLOODS ◆	*
		LOUN., TLTS		33		1470 / 250		34		" " ◆	*
		OUTSIDE ◆		35			1000 / 1125	36		LITES VAULT	
*	SIGN ◆			37	1000 / 1125			38	20/1	" "	
*	SIGN ◆			39		1000 / 5185		40	50/2	DWYER UNIT	
*	SIGN		20/1	41			1000 / 5185	42			
	TOTAL LOAD PER PHASE				15425	18190	18830	18830 ÷ 120 = 157 A			

* INDICATES LOAD SHOWN IS ACTUAL x 1.25 PER CODE

◆ INDICATES CIRCUIT CONTROLLED BY TIMECLOCK

Figure 4-9. The panel schedule for a 225-amp, 120/208-volt, three-phase, four-wire, surface-mounted panel. The area or equipment supplied by each circuit is listed.

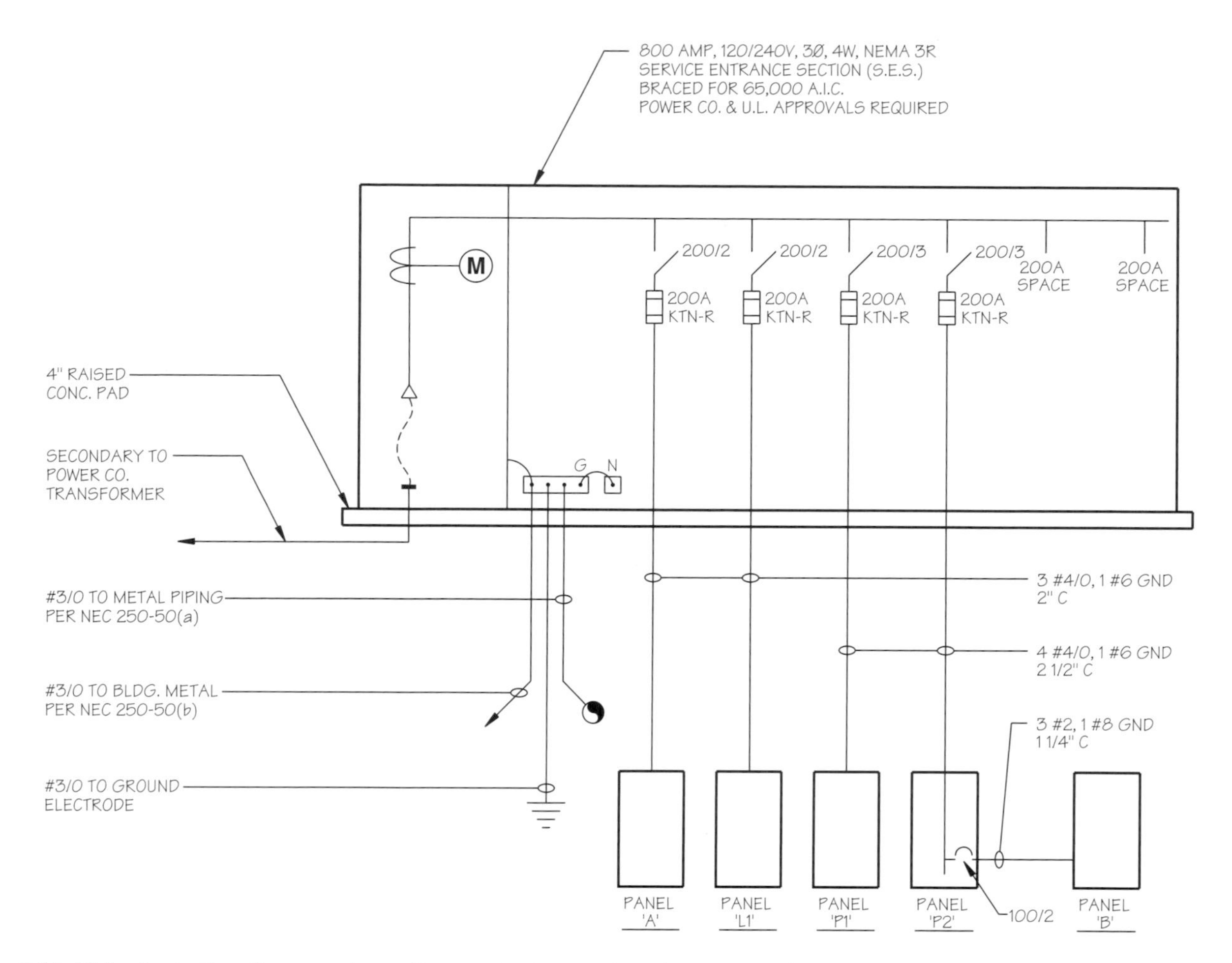

Figure 4-10. This single-line diagram shows the conductors and equipment used to transmit power from a transformer to panels. The panels are used to feed the building loads.

CSI MasterFormat Divisions

Division 20	Reserved
Division 21	Fire Suppression
Division 22	Plumbing
Division 23	Heating, Ventilating, and Air Conditioning
Division 24	Reserved
Division 25	Integrated Automation
Division 26	Electrical
Division 27	Communications
Division 28	Electronic Safety and Security
Division 29	Reserved

Figure 4-11. The CSI MasterFormat separates construction topics into broad divisions.
(Construction Specification Institute)

Within the MasterFormat, electrical work is found in Division 26. See **Figure 4-12.** Specific electrical topics are addressed within sections of the division. A complete list of the divisions and sections is in the MasterFormat Reference on page 231.

Building Codes

Localities adopt building codes to ensure that a building meets a specified quality and level of safety. Code requirements make a building safer in the event of natural disasters, such as fire, flood, or earthquake. The intention of building codes is to prevent injury and loss of life and reduce building damage when the building is subjected to these forces of nature. Building code requirements tend to be more strict for buildings designed for public use.

Building codes address the types of construction methods and materials that can be used. Unspecified materials and methods are acceptable if they are approved by the proper authority. Codes will also specify specific materials and methods that are unacceptable for a given type of construction.

MasterFormat Division 26—Electrical

26 00 00	**ELECTRICAL**
26 01 00	Operation and Maintenance of Electrical Systems
26 05 00	Common Work Results for Electrical
26 06 00	Schedules for Electrical
26 08 00	Commissioning of Electrical Systems
26 09 00	Instrumentation and Control for Electrical Systems
26 10 00	**MEDIUM-VOLTAGE ELECTRICAL DISTRIBUTION**
26 11 00	Substations
26 12 00	Medium-Voltage Transformers
26 13 00	Medium-Voltage Switchgear
26 18 00	Medium-Voltage Circuit Protection Devices
26 20 00	**LOW-VOLTAGE ELECTRICAL TRANSMISSION**
26 21 00	Low-Voltage Overhead Electrical Power Systems
26 22 00	Low-Voltage Transformers
26 23 00	Low-Voltage Switchgear
26 24 00	Switchboards and Panelboards
26 25 00	Enclosed Bus Assemblies
26 26 00	Power Distribution Units
26 27 00	Low-Voltage Distribution Equipment
26 28 00	Low-Voltage Circuit Protective Devices
26 29 00	Low-Voltage Controllers
26 30 00	**FACILITY ELECTRICAL POWER GENERATING AND STORING EQUIPMENT**
26 31 00	Photovoltaic Collectors
26 32 00	Packaged Generator Assemblies
26 33 00	Battery Equipment
26 35 00	Power Filters and Conditioners
26 36 00	Transfer Switches
26 40 00	**ELECTRICAL AND CATHODIC PROTECTION**
26 41 00	Facility Lightning Protection
26 42 00	Cathodic Protection
26 43 00	Transient Voltage Suppression
26 50 00	**LIGHTING**
26 51 00	Interior Lighting
26 52 00	Emergency Lighting
26 53 00	Exit Signs
26 54 00	Classified Location Lighting
26 55 00	Special Purpose Lighting
26 56 00	Exterior Lighting
26 60 00	Unassigned
26 70 00	Unassigned
26 80 00	Unassigned
26 90 00	Unassigned

Figure 4-12. Division 26 of the CSI MasterFormat addresses electrical work. The division is further divided into specific sections. See the MasterFormat Reference on page 231.

Many building codes specify model codes to be followed. ***Model codes*** are extensive codes established by national organizations. The organization that establishes the model code continually researches and updates the code to incorporate new technology, materials, and methods.

The model electrical code used throughout the United States is the *National Electrical Code.* Most local building codes require work to be performed according to this model code. The local codes may include additional requirements in specific areas.

National Electrical Code

The ***National Electrical Code*** (***Code*** or ***NEC***) governs nearly all electrical construction performed in the United States. There is no national law that requires electrical construction to conform to the *Code*, but most building codes mandate *Code* compliance.

The first *Code* was prepared in 1897. In 1911, the *Code* became more formally organized under the administrative sponsorship of the ***National Fire Protection Association (NFPA).*** The present format of the *Code* was set up in 1959, using article numbers preceding section numbers. A revised edition is issued every three years.

Code content and arrangement

The *Code* consists of an introduction and nine chapters. These chapters are divided into four sections:

- **General installations**—Chapters 1–4. Chapter 1 defines terms you will need to know to understand the *Code.* Chapter 2 discusses wiring design and protection. Chapter 3 provides rules governing wiring methods. Chapter 4 covers cords, cables, fixtures, fixture wires, motors, compressors, and battery-operated electric vehicles. The table in **Figure 4-13** lists the articles found in these chapters.

Section One—General Installations

Article	Title
90	Introduction
100	Definitions
110	Requirements for Electrical Installations
200	Use and Identification of Grounded Conductors
210	Branch Circuits
215	Feeders
220	Branch-Circuit, Feeder, and Service Calculations
225	Outside Branch Circuits and Feeders
230	Services
240	Overcurrent Protection
250	Grounding and Bonding
280	Surge Arresters
285	Transient Voltage Surge Suppressors: TVSSs
300	Wiring Methods
310	Conductors for General Wiring
312	Cabinets, Cutout Boxes, and Meter Socket Enclosures
314	Outlet, Device, Pull and Junction Boxes; Conduit Bodies; Fittings; and Handhole Enclosures
320	Armored Cable: Type AC
322	Flat Cable Assemblies: Type FC
324	Flat Conductor Cable: Type FCC
326	Integrated Gas Spacer Cable: Type IGS
328	Medium Voltage Cable: Type MV
330	Metal-Clad Cable: Type MC
332	Mineral-Insulated, Metal-Sheathed Cable: Type MI
334	Nonmetallic-Sheathed Cable: Types NM, NMC, NMS
336	Power and Control Tray Cable: Type TC
338	Service-Entrance Cable: Types SE and USE
340	Underground Feeder and Branch-Circuit Cable: Type UF
342	Intermediate Metal Conduit: Type IMC
344	Rigid Metal Conduit: Type RMC
348	Flexible Metal Conduit: Type FMC
350	Liquidtight Flexible Metal Conduit: Type LFMC
352	Rigid Nonmetallic Conduit: Type RNC
353	High Density Polyethylene Conduit: Type HDPE Conduit
354	Nonmetallic Underground Conduit with Connectors: Type NUCC
356	Liquidtight Flexible Nonmetallic Conduit: Type LFNC
358	Electrical Metallic Tubing: Type EMT
360	Flexible Metallic Tubing: Type FMT
362	Electrical Nonmetallic Tubing: Type ENT
366	Auxiliary Gutters
368	Busways
370	Cablebus
372	Cellular Concrete Floor Raceways
374	Cellular Metal Floor Raceways
376	Metal Wireways
378	Nonmetallic Wireways
380	Multioutlet Assembly
382	Nonmetallic Extensions
384	Strut-Type Channel Raceway
386	Surface Metal Raceways
388	Surface Nonmetallic Raceways
390	Underfloor Raceways
392	Cable Trays
394	Concealed Knob-and-Tube Wiring
396	Messenger Supported Wiring
398	Open Wiring on Insulators
400	Flexible Cords and Cables
402	Fixture Wires
404	Switches
406	Receptacles, Cord Connectors, and Attachment Plugs (Caps)
408	Switchboards and Panelboards
409	Industrial Control Panels
410	Luminaires (Lighting Fixtures), Lampholders, and Lamps
411	Lighting Systems Operating at 30 Volts or Less
422	Appliances
424	Fixed Electric Space-Heating Equipment
426	Fixed Outdoor Electric De-icing and Snow-Melting Equipment
427	Fixed Electric Heating Equipment for Pipelines and Vessels
430	Motors, Motor Circuits, and Controllers
440	Air-Conditioning and Refrigerating Equipment
445	Generators
450	Transformers and Transformer Vaults (Including Secondary Ties)
455	Phase Converters
460	Capacitors
470	Resistors and Reactors
480	Storage Batteries
490	Equipment, Over 600 Volts, Nominal

Figure 4-13. This table lists the articles contained in the General Installation section of the *Code.*

- **Special installations**—Chapters 5–7. Chapter 5 is concerned with areas where spark generation by electrical equipment may cause fire or explosion due to volatile atmosphere or material. Chapter 6 covers electric sign wiring, outline lighting, cranes, hoists, elevators, escalators, and moving sidewalks. It also covers welding machines, sound recording equipment, X-ray equipment, and machine tool wiring. Chapter 7 deals with emergency lighting circuits. See **Figure 4-14.**
- **Communications**—Chapter 8. This section contains the requirements for communication circuits, radio and television equipment, and antenna distribution systems. See **Figure 4-15.**
- **Tables and annexes**—Chapter 9 contains ten tables. The appendices contain additional

Section Two—Special Installations

Article	Title
500	Hazardous (Classified) Locations, Classes I, II, and III, Divisions 1 and 2
501	Class I Locations
502	Class II Locations
503	Class III Locations
504	Intrinsically Safe Systems
505	Class I, Zone 0, 1, and 2 Locations
506	Zone 20, 21, and 22 Locations for Combustible Dusts, Fibers, and Flyings
510	Hazardous (Classified) Locations—Specific
511	Commercial Garages, Repair and Storage
513	Aircraft Hangars
514	Motor Fuel Dispensing Facilities
515	Bulk Storage Plants
516	Spray Application, Dipping, and Coating Processes
517	Health Care Facilities
518	Assembly Occupancies
520	Theaters, Audience Areas of Motion Picture and Television Studios, Performance Areas, and Similar Locations
525	Carnivals, Circuses, Fairs, and Similar Events
530	Motion Picture and Television Studios and Similar Locations
540	Motion Picture Projection Rooms
545	Manufactured Buildings
547	Agricultural Buildings
550	Mobile Homes, Manufactured Homes, and Mobile Home Parks
551	Recreational Vehicles and Recreational Vehicle Parks
552	Park Trailers
553	Floating Buildings
555	Marinas and Boatyards
590	Temporary Installations
600	Electric Signs and Outline Lighting
604	Manufactured Wiring Systems
605	Office Furnishings (Consisting of Lighting Accessories and Wired Partitions)
610	Cranes and Hoists
620	Elevators, Dumbwaiters, Escalators, Moving Walks, Wheelchair Lifts, and Stairway Chair Lifts
625	Electric Vehicle Charging System
630	Electric Welders
640	Audio Signal Processing, Amplification, and Reproduction Equipment
645	Information Technology Equipment
647	Sensitive Electronic Equipment
650	Pipe Organs
660	X-Ray Equipment
665	Induction and Dielectric Heating Equipment
668	Electrolytic Cells
669	Electroplating
670	Industrial Machinery
675	Electrically Driven or Controlled Irrigation Machines
680	Swimming Pools, Fountains, and Similar Installations
682	Natural and Artificially Made Bodies of Water
685	Integrated Electrical Systems
690	Solar Photovoltaic Systems
692	Fuel Cell Systems
695	Fire Pumps
700	Emergency Systems
701	Legally Required Standby Systems
702	Optional Standby Systems
705	Interconnected Electric Power Production Sources
720	Circuits and Equipment Operating at Less than 50 Volts
725	Class 1, Class 2, and Class 3 Remote-Control, Signaling, and Power-Limited Circuits
727	Instrumentation Tray Cable: ITC
760	Fire Alarm Systems
770	Optical Fiber Cables and Raceways
780	Closed-Loop and Programmed Power Distribution

Figure 4-14. The Special Installation section of the *Code*.

Section Three—Communications Systems

Article	Title	Article	Title
800	Communications Circuits	830	Network-Powered Broadband Communications Systems
810	Radio and Television Equipment		
820	Community Antenna Television and Radio Distribution Systems		

Figure 4-15. Chapter 8 of the *Code* contains the requirements for communication systems.

information related to the *Code.* These items are listed in **Figure 4-16.**

Within the *Code,* chapters are divided into articles. Each article is identified with a three-digit number. The first digit of the article number identifies the chapter in which the article is located. For example, *Article 300—Wiring Methods* is the first article in Chapter 3. Each article addresses a specific topic.

Most articles consist of several parts, which are identified by letters. The parts help to organize the article and make it easier for the user to locate a specific item. The parts are then divided into sections, with the sections further divided into subsections.

Almost all references in the *Code* and in this text are section references. For example, a reference to *Section 600.5(B)(1)* of the *Code* directs the user to subpart *(1)* of subsection *(B)* of *Section 600.5. Section 600.5* is the fifth section found in *Article 600.* See **Figure 4-17.**

Finding *Code* rules

The *Code* provides two ways to find specific information: the table of contents at the beginning of the book and the index at the end of the book.

The table of contents is an overall outline of the *Code* arranged by chapters. The chapters are listed along with their component articles. If the article has two or more parts, these are shown as well.

The index is an alphabetical listing by subject matter that indicates the article and section (and pages) where the information is located. This is the best place to begin a search on a specific topic.

Section Four—Tables and Annexes

Table	Title
1	Percent of Cross Section of Conduit and Tubing for Conductors
2	Radius of Conduit and Tubing Bends
4	Dimensions and Percent Area of Conduit and Tubing (Areas of Conduit or Tubing for the Combinations of Wires Permitted in Table 1, Chapter 9)
5	Dimensions of Insulated Conductors and Fixture Wires
5A	Compact Aluminum Building Wire Nominal Dimensions and Areas
8	Conductor Properties
9	AC Resistance and Reactance for 600 Volt Cables, 3-Phase, 60 Hz, 75°C (167°F)—Three Single Conductors in a Conduit
11(A)	Class 2 and Class 3 Alternating-Current Power Source Limitations
11(B)	Class 2 and Class 3 Direct-Current Power Source Limitations
12(A)	PLFA Alternating-Current Power Source Limitations
12(B)	PLFA Direct-Current Power Source Limitations

Annex	Title
A	Product Safety Standards
B	Application Information for Ampacity Calculation
C	Conduit and Tubing Fill Tables for Conductors and Fixture Wires of the Same Size
D	Examples
E	Types of Construction
F	Cross-Reference Tables
G	Administration and Enforcement

Figure 4-16. Reference tables are found in Chapter 9 of the *Code.* Additional information is contained in the appendices.

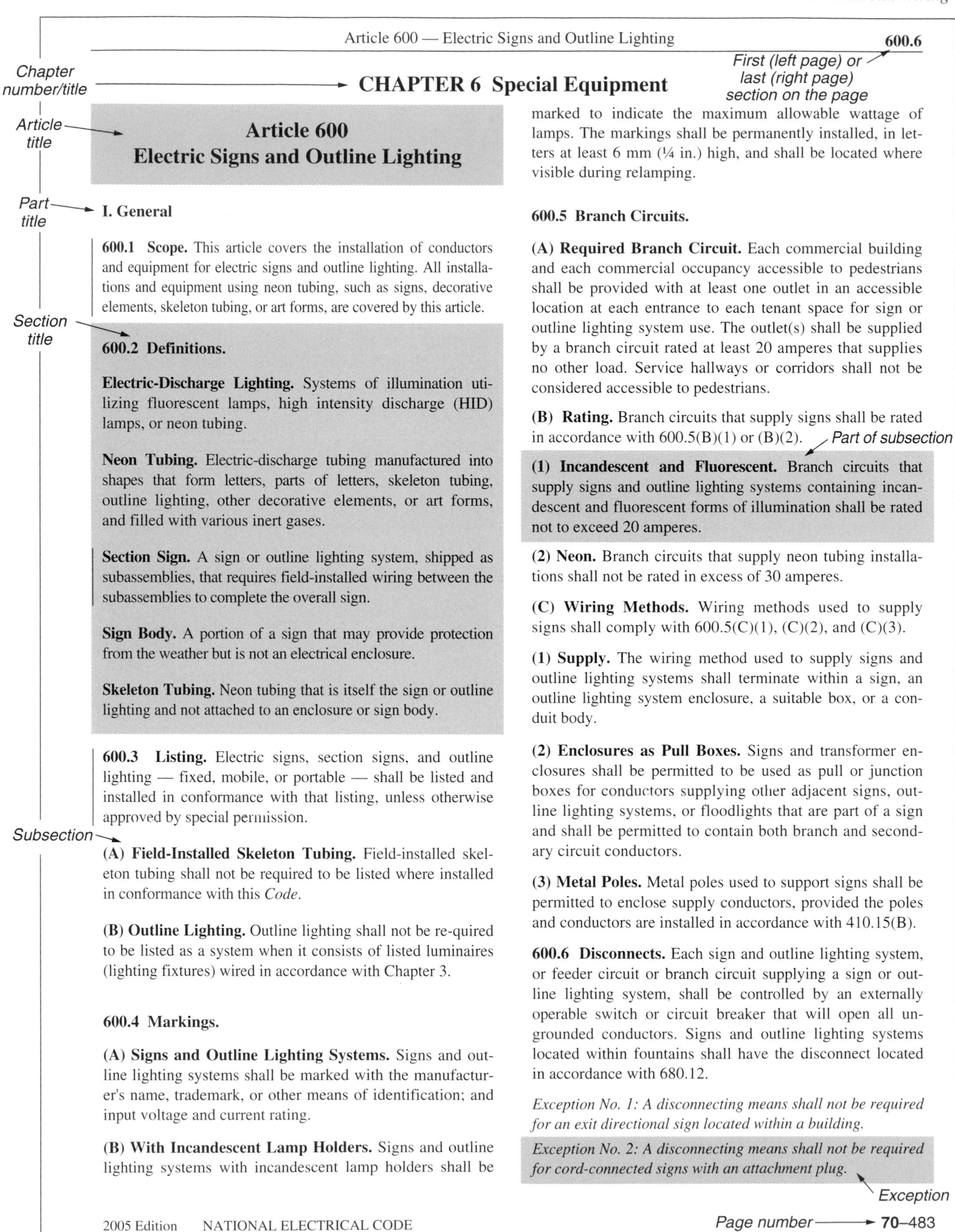

Article 600 — Electric Signs and Outline Lighting 600.6

CHAPTER 6 Special Equipment

Article 600
Electric Signs and Outline Lighting

I. General

600.1 Scope. This article covers the installation of conductors and equipment for electric signs and outline lighting. All installations and equipment using neon tubing, such as signs, decorative elements, skeleton tubing, or art forms, are covered by this article.

600.2 Definitions.

Electric-Discharge Lighting. Systems of illumination utilizing fluorescent lamps, high intensity discharge (HID) lamps, or neon tubing.

Neon Tubing. Electric-discharge tubing manufactured into shapes that form letters, parts of letters, skeleton tubing, outline lighting, other decorative elements, or art forms, and filled with various inert gases.

Section Sign. A sign or outline lighting system, shipped as subassemblies, that requires field-installed wiring between the subassemblies to complete the overall sign.

Sign Body. A portion of a sign that may provide protection from the weather but is not an electrical enclosure.

Skeleton Tubing. Neon tubing that is itself the sign or outline lighting and not attached to an enclosure or sign body.

600.3 Listing. Electric signs, section signs, and outline lighting — fixed, mobile, or portable — shall be listed and installed in conformance with that listing, unless otherwise approved by special permission.

(A) Field-Installed Skeleton Tubing. Field-installed skeleton tubing shall not be required to be listed where installed in conformance with this *Code*.

(B) Outline Lighting. Outline lighting shall not be re-quired to be listed as a system when it consists of listed luminaires (lighting fixtures) wired in accordance with Chapter 3.

600.4 Markings.

(A) Signs and Outline Lighting Systems. Signs and outline lighting systems shall be marked with the manufacturer's name, trademark, or other means of identification; and input voltage and current rating.

(B) With Incandescent Lamp Holders. Signs and outline lighting systems with incandescent lamp holders shall be marked to indicate the maximum allowable wattage of lamps. The markings shall be permanently installed, in letters at least 6 mm (¼ in.) high, and shall be located where visible during relamping.

600.5 Branch Circuits.

(A) Required Branch Circuit. Each commercial building and each commercial occupancy accessible to pedestrians shall be provided with at least one outlet in an accessible location at each entrance to each tenant space for sign or outline lighting system use. The outlet(s) shall be supplied by a branch circuit rated at least 20 amperes that supplies no other load. Service hallways or corridors shall not be considered accessible to pedestrians.

(B) Rating. Branch circuits that supply signs shall be rated in accordance with 600.5(B)(1) or (B)(2).

(1) Incandescent and Fluorescent. Branch circuits that supply signs and outline lighting systems containing incandescent and fluorescent forms of illumination shall be rated not to exceed 20 amperes.

(2) Neon. Branch circuits that supply neon tubing installations shall not be rated in excess of 30 amperes.

(C) Wiring Methods. Wiring methods used to supply signs shall comply with 600.5(C)(1), (C)(2), and (C)(3).

(1) Supply. The wiring method used to supply signs and outline lighting systems shall terminate within a sign, an outline lighting system enclosure, a suitable box, or a conduit body.

(2) Enclosures as Pull Boxes. Signs and transformer enclosures shall be permitted to be used as pull or junction boxes for conductors supplying other adjacent signs, outline lighting systems, or floodlights that are part of a sign and shall be permitted to contain both branch and secondary circuit conductors.

(3) Metal Poles. Metal poles used to support signs shall be permitted to enclose supply conductors, provided the poles and conductors are installed in accordance with 410.15(B).

600.6 Disconnects. Each sign and outline lighting system, or feeder circuit or branch circuit supplying a sign or outline lighting system, shall be controlled by an externally operable switch or circuit breaker that will open all ungrounded conductors. Signs and outline lighting systems located within fountains shall have the disconnect located in accordance with 680.12.

Exception No. 1: A disconnecting means shall not be required for an exit directional sign located within a building.

Exception No. 2: A disconnecting means shall not be required for cord-connected signs with an attachment plug.

2005 Edition NATIONAL ELECTRICAL CODE 70–483

Figure 4-17. This figure shows different parts of the *Code* organization.

Code intent

Within *Article 90—Introduction*, the purpose and scope of the *Code* is defined. The clear intent is to provide information for the safeguarding of persons and property against electrical hazards. Compliance with the *Code* prevents fire hazards, but will not necessarily result in an efficient system adequate for the intended use.

NEC NOTE 90.1(C)

This *Code* is not intended as a design specification nor an instruction manual for untrained persons.

The rules and regulations found within the *Code* are not mandatory or legally required unless the local authority accepts the *Code* for such purposes. The *Code* is recognized throughout the United States (as well as many other countries) as a legal basis of safe electrical design and practice. It is used extensively in courts and by insurance agencies in making legal judgments. The *Code* must be understood by all those involved in the industry.

In scope, the *Code* is applicable to almost all electrical work, both indoors and outdoors. This includes, but is not limited to, residences, apartments, office buildings, marinas, theaters, floating buildings, mobile homes, and industrial facilities. The *Code* addresses not only design basics and requirements, but installation rules, equipment requirements, and product acceptability.

Code enforcement

The *Code* is mandatory in most localities. This makes it the legally required document for those involved in design and installation of electrical systems.

Code enforcement is usually performed by local inspection agencies. Inspectors are authorized to interpret, modify, or waive specific *Code* requirements. Many inspectors are members of the ***International Association of Electrical Inspectors (IAEI)***. They have tremendous responsibility and must exercise their task with great care, consideration, and expertise.

Numerous *Code* references are made within this text. If you do not have a current copy of the *Code*, get one. Copies can be obtained at your local bookstore or directly from the National Fire Protection Association:

National Fire Protection Association
One Batterymarch Park
Quincy, MA 02169-7471

In addition, the NFPA publishes other useful and relevant standards:

- NFPA 30—Flammable and Combustible Liquids Code
- NFPA 70B—Electrical Equipment Maintenance
- NFPA 70E—Electrical Safety Requirements for Employee Workplaces
- NFPA 72—National Fire Alarm Code
- NFPA 77—Static Electricity
- NFPA 99—Health Care Facilities
- NFPA 101—Life Safety Code
- NFPA 780—Installation of Lightning Protection Systems

Exceptions and Fine Print Notes (FPN)

As with any set of rules and regulations, the *Code* has numerous exceptions and notes that further explain or modify rules under certain conditions. Sections of the *Code* may include exceptions and Fine Print Notes (FPN). See **Figure 4-18.**

In general, exceptions reverse, modify, or eliminate a requirement under specific conditions. The ***Fine Print Notes (FPN)*** serve to elaborate, enhance, and explain a *Code* section.

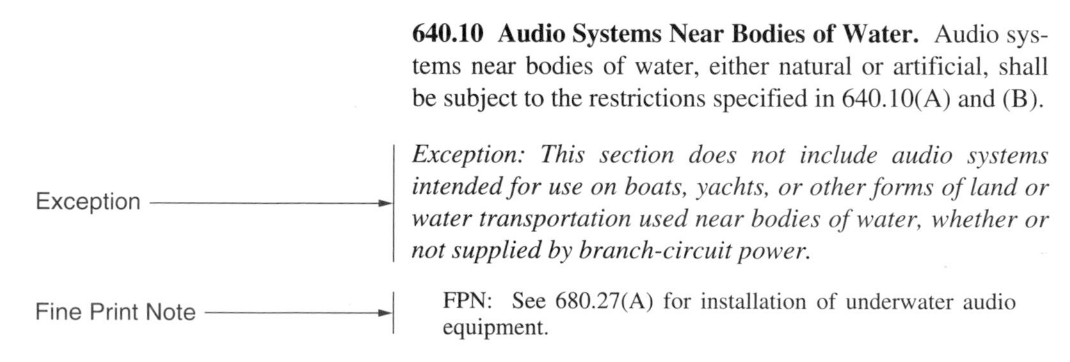
640.10 Audio Systems Near Bodies of Water. Audio systems near bodies of water, either natural or artificial, shall be subject to the restrictions specified in 640.10(A) and (B).

Exception: This section does not include audio systems intended for use on boats, yachts, or other forms of land or water transportation used near bodies of water, whether or not supplied by branch-circuit power.

FPN: See 680.27(A) for installation of underwater audio equipment.

Figure 4-18. This *Code* section specifies that audio systems near water must meet additional requirements. The exception excludes systems designed for use on boats. The Fine Print Note references information for underwater audio equipment.

Figure 4-19. Engineers test electrical appliances and equipment to ensure that the products are safe. (Underwriters Laboratories, Inc.)

NEC NOTE 90.5

Mandatory rules are indicated by the use of the terms *shall* or *shall not.* Permissive rules are characterized by the use of the terms *shall be permitted* or *shall not be permitted.* Explanatory material is indicated by an *FPN* or *fine print note.*

Significant *Code* terms

Article 100 of the *Code* defines key terms. These definitions should be carefully studied. Some important terms include the following:

- **Approved**—Acceptable to the authority having jurisdiction.
- **Labeled**—Has an attached label of an organization that is the authority having jurisdiction with product evaluation and by whose labeling the manufacturer indicates compliance with appropriate standards or performance.
- **Listed**—A product or service that either meets appropriate designated standards or has been tested and found suitable for a specified purpose and are included in a list published by an organization that is the authority having jurisdiction. See **Figure 4-19.**

OSHA Standards

The ***Occupational Safety and Health Administration (OSHA)*** under Subpart S, Part 1910 of the Code of Federal Regulations provides "Design Safety Standards for Electrical Systems." This is a legal document derived from the *National Electrical Code*.

The purpose of OSHA's electrical document is to provide a standard that secures a high degree of safety for workers by regulating the use of electrical equipment. Unlike the *Code,* OSHA's standard relates only to personnel safety. It is not a standard for design and installation.

The requirements in OSHA's regulations are applicable to all electrical installations and systems. Violations of these rules are cause for citation and fines. Everyone involved in electrical design and installation—particularly in commercial or industrial areas—must understand the OSHA standard.

As part of the electrical standard, OSHA requires that electrical equipment and materials be listed, labeled, accepted, or certified, providing that such equipment is available. A nonlisted, nonlabeled, noncertified component may be used only if it is in a class not tested by a nationally recognized testing laboratory. Should that be the case, then the object must be inspected and tested by a recognized organization.

There are many nationally recognized testing laboratories:

- Underwriters Laboratories, Inc. (UL)
- Factory Mutual Research (FM)
- United States Testing Co., Inc. (UST)
- Canadian Standards Association (CSA)

Figure 4-20 shows typical labels used by these organizations.

A product is listed only for its intended purpose and should be used only for its intended application. Also, just because one component of a piece of equipment is listed does not mean the entire item meets standards. Be sure each component and the entire unit meet the minimum standard for safety. For example, a listed appliance cord does not mean the appliance itself is listed. Check the equipment to see if it has been listed as well.

Figure 4-20. Labeled devices carry the logo of the testing agency. (Reproduced with the permission of Underwriters Laboratories, Inc. and ETL SEMKO)

Other Standards

There are other agencies that set standards used by the electrical industry. These organizations often help amend, enhance, and contribute to the NEC. Individuals representing these organizations are usually part of the committees that review and modify *Code* sections. These organizations include the following:

- National Electrical Manufacturers Association (NEMA)
- National Electrical Contractors Association (NECA)
- International Association of Electrical Inspectors (IAEI)
- International Brotherhood of Electrical Workers (IBEW)
- Institute of Electrical and Electronics Engineers (IEEE)
- American National Standards Institute (ANSI)
- Illuminating Engineering Society of North America (IESNA)

Review Questions

Answer the following questions. Do not write in this book.

1. Explain the difference between an elevation and a sectional drawing.
2. List three uses for construction drawings.
3. Which division of the Construction Specification Institute's MasterFormat addresses electrical work?
4. When writing a specification using the CSI MasterFormat as a basis, by what number would the section addressing transformers be identified?
5. What is the intention of building codes?
6. Briefly describe the intent of the *National Electrical Code.*
7. Will compliance with the *Code* always guarantee an efficient electrical installation? Why or why not?
8. Who can modify or waive *Code* requirements?
9. How often is a revised, updated edition of the *Code* issued?
10. Briefly define the terms *labeled* and *listed* as they pertain to electrical equipment, devices, and materials.
11. Compare the terms *shall* and *should,* as used in the *Code.*
12. What is the intent of OSHA's electrical document, "Design Safety Standards for Electrical Systems"?
13. List five organizations that set standards for the electrical industry.

Identify the following electrical symbols.

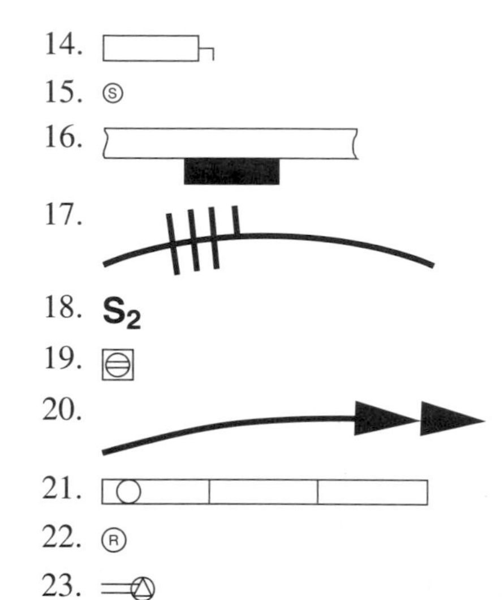

USING THE NEC

Refer to the National Electrical Code to answer the following questions. Do not write in this book.

1. Use the contents of the *Code* to determine which article in Chapter 5 contains information regarding the installation of electrical systems at state fairs.
2. How many exceptions follow *Section 725.25*?
3. Use the index of the *Code* to determine which articles contain information regarding bowling alleys.
4. Use the contents of the *Code* to determine which article in Chapter 4 contains information regarding the installation of lamp holders.
5. What does the Fine Print Note following *Section 430.61* reference?
6. Use the index of the *Code* to determine which section contains the minimum sizes of service drops.
7. Which article contains definitions of the terms used throughout the *Code*?
8. Part 2 of subheading (A) of *Section 210.8* (referred to as *Section 210.8(A)(2)*) states that ground-fault circuit-interrupters are required for receptacles in residential garages. Exception No. 1 allows some receptacles to be excluded from this requirement. What type of receptacles are excluded?
9. Use the index of the *Code* to determine which section contains the definition of *electric discharge lighting.*
10. According to the exception to *Section 550.17(A),* what type of mobile home devices do not need to be subjected to a dielectric strength test?

An electrician installs Phillips ALTO™ low-mercury florescent lamps at the John G. Shedd Aquarium. (Edward G. Lines, Jr., John G. Shedd Aquarium)

Chapter 5

Wiring Methods

Technical Terms

Armored cable (BX)
Auxiliary gutters
Busways
Cable tray
Electrical metallic tubing (EMT)
Electrical nonmetallic tubing (ENT)
Flexible metal conduit (Greenfield)
Intermediate metal conduit (IMC)
Liquidtight flexible conduit
Metal-clad cable
Multiconductor cable
Nonmetallic-sheathed cable (romex)
Rigid metal conduit
Rigid nonmetallic conduit
Service-entrance cable
Surface raceway
Wireways

Objectives

After completing this chapter, you will be able to:

- List the wiring methods available for commercial installation and the rules regarding each method of wiring.
- Size wireways to satisfy *Code* requirements.
- Identify fittings, connectors, supports, and other integral hardware unique to a particular wiring method.
- Select the correct wiring method based on *Code* requirements.
- Calculate wireway size.

This chapter is an introduction to conduits, raceways, busways, wireways, junction boxes, gutters, busbars, pull boxes, device boxes, and a host of other wiring methods and related components. These items compose the system through which electricity is routed.

The *Code* recognizes many wiring methods for use in buildings. These wiring methods fall into several main categories:

- Raceways
- Conductors
- Cable
- Cable trays
- Busways
- Gutters
- Wireways

Some of the wiring methods have very specific applications, while others can be used in a greater variety of conditions. The table in **Figure 5-1** lists wiring methods and their corresponding *Code* articles. This is not a complete list of all wiring methods approved in the *Code*. The reader is encouraged to look into all methods of wiring and techniques covered in the *Code*.

Wiring Methods in the *Code*

Wiring Methods	*Code* Article
Armored Cable	320
Auxiliary Gutters	366
Busways	368
Cable Trays	392
Electrical Metallic Tubing	358
Electrical Nonmetallic Tubing	362
Flexible Metal Conduit	348
Flexible Metallic Tubing	360
Intermediate Metal Conduit	342
Liquidtight Flexible Conduit	350, 356
Metal-Clad Cable	330
Multioutlet Assembly	380
Nonmetallic-Sheathed Cable	334
Rigid Metal Conduit	344
Rigid Nonmetallic Conduit	352
Service-Entrance Cable	338
Surface Raceway	386, 388
Underground Feeder Cable	340
Wireways	376, 378

Figure 5-1. *Code* sections for common wiring methods.

General Rules and Requirements

The *Code*, under *Article 300—Wiring Methods,* addresses the acceptable methods of installing conductors. Choosing the right method for a wiring job depends on the environment where the wiring is to be installed. Certain wiring methods are only permitted in a limited range of conditions. Others are acceptable in a broad variety of situations.

Regardless of the specific wiring method, there are some general rules common to many of the methods. These provisions should be understood before beginning any wiring installation. A brief look at these general provisions follows:

- Whether in cable or as single individual units, conductors should be used within the voltage and temperature range for which they are designed.

WARNING

Conductors placed in service where the voltage or temperature may exceed their specific rating will represent a serious hazard to the electrical system, equipment, structure, and personnel.

- Normally, single conductors may only be installed as part of a wiring method recognized by the *Code*. Thus, routing individual conductors without the protection or support of an approved wiring method is strictly prohibited.
- For a circuit, the current-carrying conductors, neutral conductor (where used), and equipment grounding conductor must all be run within the same conduit, cable, duct, tray, or enclosure. Failure to comply with this requirement could lead to inductance problems.

NEC NOTE 300.3(B)

All conductors of the same circuit and, where used, the grounded conductor and all equipment grounding conductors shall be contained within the same raceway, cable tray, trench, cable, or cord. See exceptions in *Code*.

- Circuits of different voltage may be run in the same raceways or occupy the same spaces (enclosures, equipment) providing that the voltage is less than 600 volts and the conductors each have insulation ratings greater than or equal to that of the circuit with the highest voltage rating. For example, conductors rated for 300 volts can be run in the same raceway as conductors rated for 600 volts as long as the maximum circuit voltage is 300 volts.
- Measures must be taken to protect conductors against physical damage. When running cable through bored holes in framing members, the edge of the hole should be 1 1/4″ away from the nearest edge of the member. If a cable or raceway is placed along notches made in wood framing members, the notches must be protected from nails and screws by using a metal plate (minimum 1/16″ thick) to cover the notches. If cables or nonmetallic raceway is run perpendicular to exposed studs, rafters, or joists and subject to damage, strips of 1 × 1 or 1 × 2 wood stock should be placed along either side to protect the wiring from damage. **Figure 5-2** illustrates methods of protecting conductors from physical damage.

Rules for Buried Conductors

Conduit, cables, and other raceways that are buried must meet specific criteria. The conductors must be protected so that damage does not occur.

Table 300.5 of the *Code* lists the minimum burial depth for conductors 600 volts nominal or less. The table in **Figure 5-3** is a condensed version of *Table 300.5.* Deeper burial is not uncommon due to soil conditions, interferences with other utilities, and structural footings.

Cables under buildings must be installed in raceway and the raceway must extend past the exterior walls of the building. Further, where buried conductors and cables emerge from the ground, protection must be provided by raceways that extend from the minimum cover distance to at least 8′ above grade or to the point of entry into a building. See **Figure 5-4.** At the point where underground conductors emerge from a raceway, the raceway should be fitted with a bushing or sealed.

NEC NOTE 300.5(D)

Where the enclosure or raceway is subject to physical damage, the conductors shall be installed in rigid metal conduit, intermediate metal conduit, or Schedule 80 rigid nonmetallic conduit, or equivalent.

Regardless of the method of wiring used, all underground installations must be grounded and bonded as required by the *Code*. Refer to Chapter 10 of this text and *Article 250—Grounding* of the *Code.*

Splices and taps are permitted in buried conductors and cables. Be sure to use materials that are suitable for underground use. All splicing materials must be suitable for the conditions and environment in which they are installed.

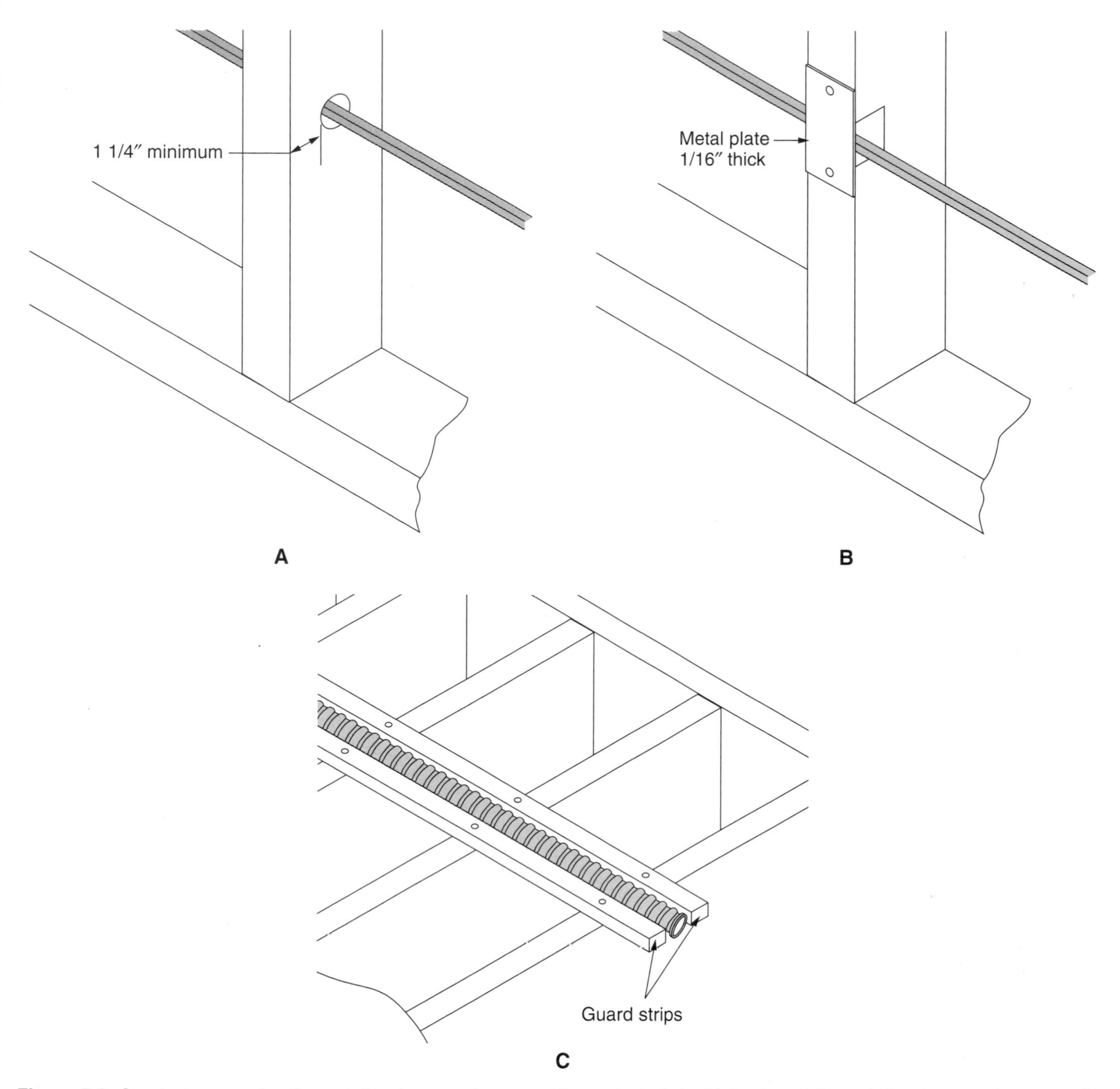

Figure 5-2. Conductors running through framing members must be protected. A—Holes bored through framing members must be at least 1 1/4″ from the edge of the member. B—Notches must be covered by a metal plate to protect conductors from nails and screws. C—Guard strips must be at least as high as the cable they protect.

Minimum Burial Depths (600V or less)

	Cables or Conductors	IMC or Rigid Metal Conduit	Rigid Nonmetallic Conduit
Below streets, alleys, and parking lots	24″	24″	24″
Below building slab or foundation (in raceway)	0″	0″	0″
In trench below 2″ thick concrete or equivalent	18″	6″	12″
Under one- or two-family dwelling driveway	18″	18″	18″

Figure 5-3. Condensed version of *Table 300.5* from the *Code*.

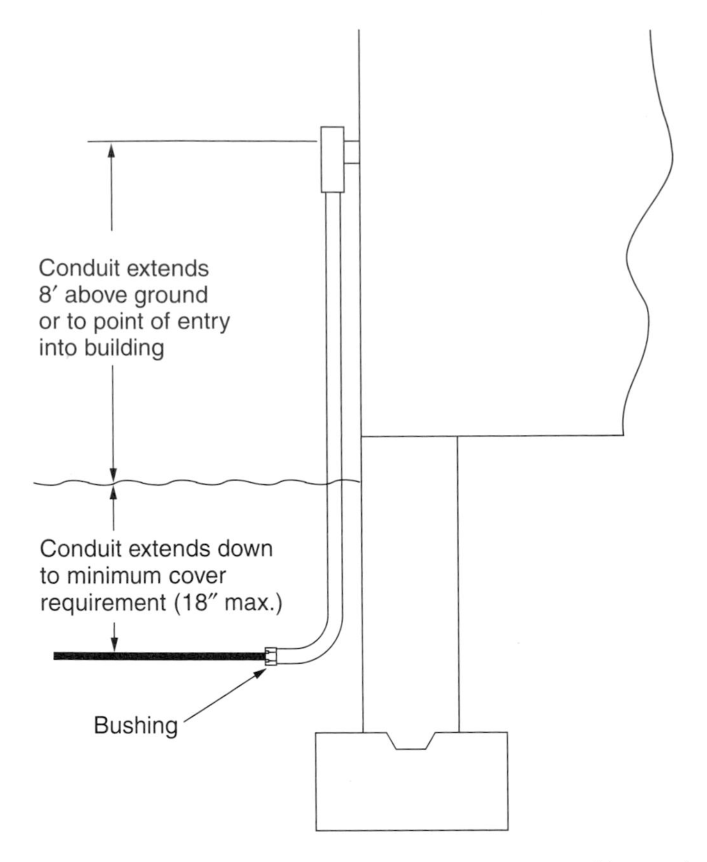

Figure 5-4. Conduit must be installed to protect cables and conductors emerging from underground.

Securing and Supporting Wiring Systems

All wiring must be secured to and supported by structural members. Normally, wiring systems are not used to support other wiring systems or nonelectric equipment. However, *Section 300.11(B)* of the *Code* specifies some conditions under which one raceway can be supported by another raceway.

Mechanical and electrical continuity is required. Metal conduit, cable sheaths, cable armor, and nonmetallic raceway must be continuous between boxes, enclosures, fittings, and cabinets. All connections between the components must be secure prior to pulling conductors. The grounding conductor should be independent of the device connection, so if the device is removed the continuity is not interrupted.

Conductor support in vertical raceway

In buildings where raceways are run vertically, the conductors must be supported at assigned intervals. One cable support at the top of the raceway run and at intervals indicated in the table in **Figure 5-5** are required. This table is a condensed version of *Table 300.19(A)* from the *Code.*

Support of vertical conductors can be accomplished in several ways, including the following:

- Placing insulated wedges between the cable and the inside wall of the raceway or using insulated clamps within an enclosure.
- Installing junction boxes with supports installed and attached to the cable, **Figure 5-6.**
- Installing junction boxes and offsetting the cable at least 90° and using tie wires within the box.

NEC NOTE 300.19(B)(3)

When the cable is supported by bending it at least 90° in a junction box and securing with tie wires, support intervals cannot exceed 20% of the normal maximum support distance. The bend in the conductor must extend horizontally at least twice the diameter of the cable.

Wiring within Air-Handling Spaces

No wiring is permitted within air-handling ducts used for the purpose of transporting dust, flammable vapors, or cooking equipment ventilation. In plenums used for environmental air only, wiring methods employing type MI or MC cable are permitted. Flexible

Vertical Spacing for Conductor Supports

Conductor Size	Maximum Distance between Supports (in feet)	
	Copper Conductors	Aluminum Conductors
18 AWG through 8 AWG	100	100
6 AWG through 1/0 AWG	100	200
2/0 AWG through 4/0 AWG	80	180
Over 4/0 AWG through 350 kcmil	60	135
Over 350 kcmil through 500 kcmil	50	120

Figure 5-5. Condensed version of *Table 300.19(A)* of the *Code.*

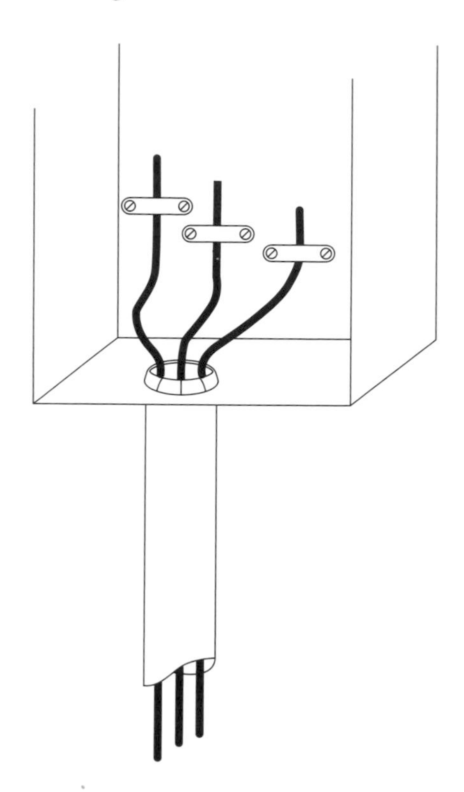

Figure 5-6. Cable must be supported vertically within the maximum support distance. Clamping the conductors to the junction box is one method of vertical support.

metal conduit and liquidtight flexible metal conduit are also permitted in short lengths (not more than 4′) to connect sensors, louvers, and other devices permitted in these plenums.

For other types of air-handling ducts, refer to *Code Section 300.22(C)* and the local inspection authority, which may have special requirements applicable to the specific conditions.

Temporary Installations

The provisions of *Article 590* of the *Code* are specifically tailored to apply to temporary electrical wiring methods, which may be less exacting than a permanent wiring system. Temporary wiring installations are allowed for the purpose of providing power and lighting to facilities during construction. Temporary wiring can also be used for testing, experimental, and developmental purposes. Upon completion of the activity, the temporary lighting must be removed.

Bear in mind that *Article 590* simply modifies *Code* requirements, and that except for those specifically modified under that article, all other requirements of the *Code* apply. Some of the modifications and specific requirements for temporary wiring are as follows:

- All lamps for general lighting will be protected from breakage by a guard over or around the lampholder.
- Splices in conductors do not require junction or splice boxes if the conductors are part of a multiconductor cable or open conductors.
- Ground-fault circuit-interrupters are required for all 125-volt, single-phase, 15- and 20-amp receptacles that are not fed from permanent wiring circuits.
- Regularly scheduled maintenance checks will be performed on equipment grounding conductors. The checks will be performed at not more than three month intervals and will verify continuity, any damaged condition, and proper polarity relative to the grounding conductor.

NEC NOTE 590.3(B)

Temporary electrical power and lighting for holiday decoration and similar purposes is allowed for a period not to exceed 90 days.

Wiring Methods

There are three broad classes of wiring methods: cable, raceways, and cable trays. All of the methods are used to connect the power supply, devices, and switches in an electrical circuit.

Cable consists of several conductors wrapped by a flexible outer covering. Raceways are enclosures installed between equipment or devices that are to be connected by conductors. The conductors or cables are placed inside the raceway, which protects and supports the wiring. Unlike raceways, cable trays are not enclosed; they are simply trays on which cables are laid.

Multiconductor Cable

Multiconductor cables are flexible assemblies of conductors having an overall protective covering. There are essentially four major types of multiconductor cable assemblies:

- Service-entrance cable
- Armored cable
- Nonmetallic-sheathed cable
- Metal-clad cable

Service-entrance cable (SE, USE, and ASE)

Service-entrance cable has conductors that can be used not only for service-entrance wiring, but also for indoor applications. Type SE, as shown in **Figure 5-7,** is unarmored, moisture resistant, and flame retardant. It can be composed of two or three insulated conductors and an additional bare conductor. It is readily available in AWG sizes 12 AWG through 4/0 AWG.

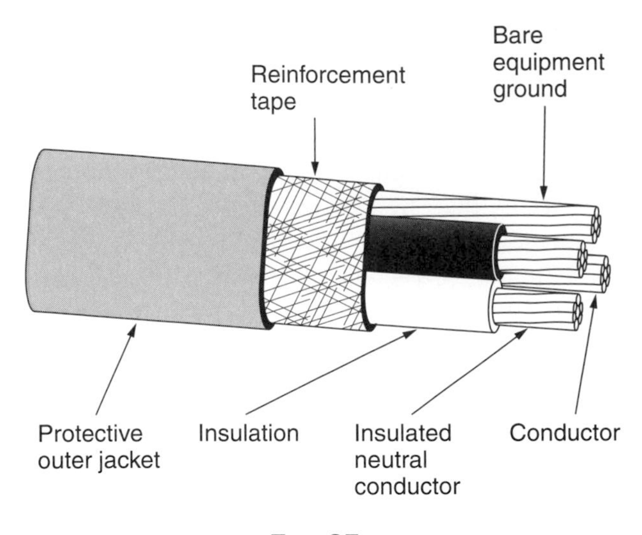

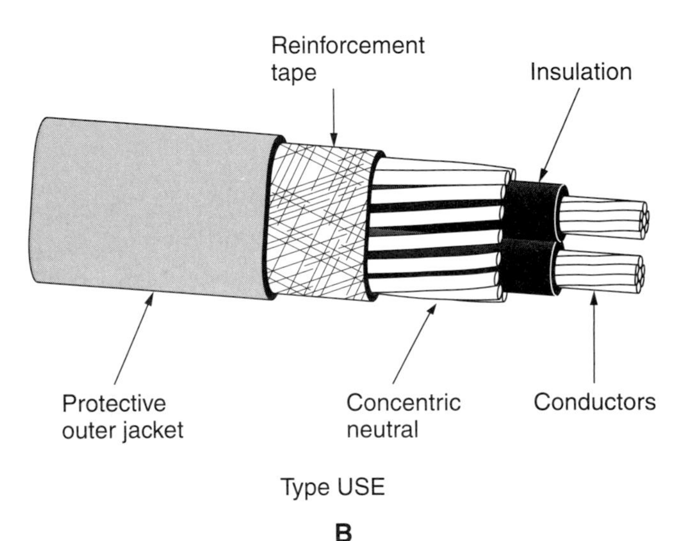

Figure 5-7. Service-entrance cable can be used for general indoor wiring as well as at the service entrance. A—Type SE cable. B—Type USE cable is used for underground applications.

Type USE is basically the same as type SE except the outer jacketing is better suited for direct burial. It is highly moisture and corrosion resistant. Type ASE has an armored jacket, which provides additional protection.

As already noted, service-entrance cable can be used for interior wiring or for service-entrance cable between the utility supply and main service disconnect. When it is used for interior wiring, the grounded neutral conductor must be insulated, with the following exceptions:

- When used as a feeder between two buildings.
- When used as the circuit conductor or branch wiring to clothes dryers, counter-mounted cooktops, ranges, and ovens.

Service-entrance cable may not be used for interior wiring of buildings of any type exceeding three stories high.

Armored cable (AC and ACL)

Type AC or ACL (often called "BX" cable) is used in both dry and wet locations, but is not permitted to be buried, see **Figure 5-8.** This type of cable, once commonly used in many applications, has some limitations for commercial use. It may not be used in places of assembly, studios, movie theaters, hazardous locations, commercial garages, areas with vapors and corrosive agents, lifts, cranes, hoists, elevators, or battery rooms.

Armored cable can be used in the following situations:

- As flexible connections to motors or vibrating equipment (up to 24″ length).
- As fixture whips (up to 6′ length).
- In dry locations.
- Concealed behind walls (may be fished behind walls in old work).
- Exposed along wall surfaces.
- Where exposed to weather or moisture (type ACL only).

Armored cable must be properly supported at intervals of 4′ - 6″ and within 12″ of terminal boxes or fittings. When routing BX cable through wall, floor, or ceiling members, the member must be drilled through the center or notched and covered with a metal plate to protect the cable from nails. The cable can be run along the sides of studs, joists, or rafters without further protection. Guard strips made of 1 × 1 or 1 × 2 stock are used to protect BX or AC cable when it is run along attic floor joists.

CAUTION

Antishort insulating bushing must be placed between conductors and the outer armor wherever the cable is cut. The bushings protect the conductor insulation from any sharp edges on the cut armor.

Nonmetallic-sheathed cable (NM, NMC, and UF)

Often referred to as ***romex,*** nonmetallic-sheathed cable is frequently used as the preferred wiring method in small commercial establishments, as well as in residential structures. This cable is made up of two to four insulated conductors plus a green insulated or bare grounding conductor with an overall nonmetallic sheath. See **Figure 5-9.**

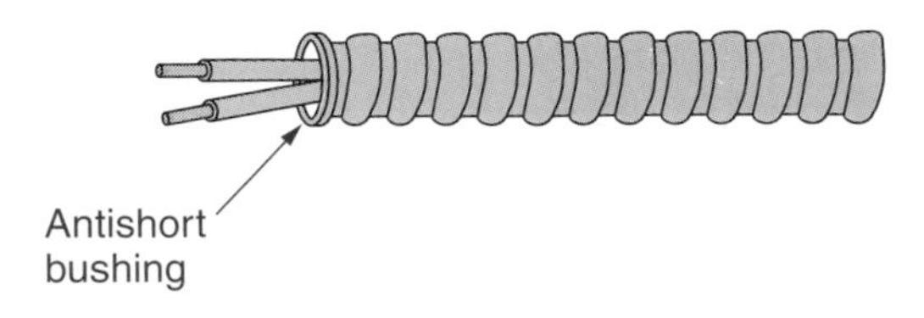

Figure 5-8. Types AC and ACL cable are also commonly referred to as BX cable.

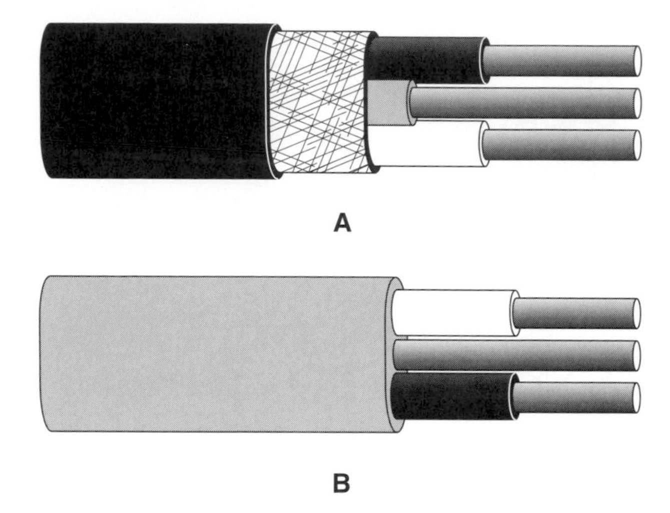

Figure 5-9. Nonmetallic-sheathed cable. A—Type NM is often called Romex. B—Type UF cable is used both indoors and outdoors.

Type NM is used strictly indoors while types NMC and UF are suitable for either indoor or outdoor applications. Further, type UF can be directly buried in the ground.

Nonmetallic cable must be supported every 4′ - 6″ and within 12″ of a terminal box, junction box, outlet, or fitting. It may be installed alongside or through studs, joists, rafters, and other building members with the same protection requirements and restrictions as with armored cable. This type of cable may also be installed in unfinished basements, attics, and crawl spaces, and may be terminated in metal or nonmetallic boxes.

Metal-clad cable (MC)

This type of cable is a heavy-duty commercial and industrial assembly consisting of one or more conductors individually insulated and enclosed in a metallic sheath that consists of corrugated tubing or interlocking tape. This type of cable looks similar to armored cable (BX). MC cable is available in sizes 14 AWG up to 1000 kcmil. Its use is permitted in a wide range of applications:

- Indoors or outdoors.
- Exposed or concealed.
- For direct burial.
- As open run cable, or in a raceway or a conduit tray.
- As aerial cable on a messenger wire.
- In hazardous locations.
- In wet and dry locations.
- For feeders, branch circuits, and service conductors.
- For power, control, lighting, and other circuit applications.

Type MC cable cannot be used in destructive corrosive environments, such as buried in concrete or earth that exposes the cable to harmful chemicals and contaminants.

Raceways

A wiring method using raceways is more secure and safe than a method using only cables. This is a result of the added protection provided by the raceway. Unfortunately, a method using raceways is also more expensive.

There are many different types of raceway. The selection of a particular type depends on the specific application.

NEC NOTE 100

Raceway: An enclosed channel of metal or nonmetallic materials designed expressly for holding wires, cables, or busbars, with additional functions as permitted in this *Code.*

Rigid metal conduit (RMC)

This extremely strong and versatile wiring method may be installed inside, outside, in damp, wet, dry locations and in almost any kind of building. It can be installed exposed or concealed in nearly all locations. When connected and joined properly, the metal-to-metal continuity is excellent and provides good grounding for equipment fed by the conduit system.

Rigid metal conduit is manufactured in 10′ lengths, but other sizes can be ordered. It can be field cut and threaded. Threading must have a taper equivalent to 3/4″ to every foot. Bushings should be used at ends where wires enter or exit the conduit to prevent damage to the insulation.

There are numerous connectors and fittings used with rigid metal conduit. See **Figure 5-10.** Most of these fittings are also used with intermediate metal conduit.

Generally, the conduit should be supported every 10′ and within 3′ of a junction, outlet, or fitting. The total number of bends in a run of rigid metal conduit must not exceed 360°, **Figure 5-11.**

NEC NOTE 344.30(B)(3)

For vertical rigid metal conduit risers from machinery with threaded coupling, the distance between supports can be increased to 20′, provided that both ends of the riser are securely fastened and no means of intermediate support is readily available.

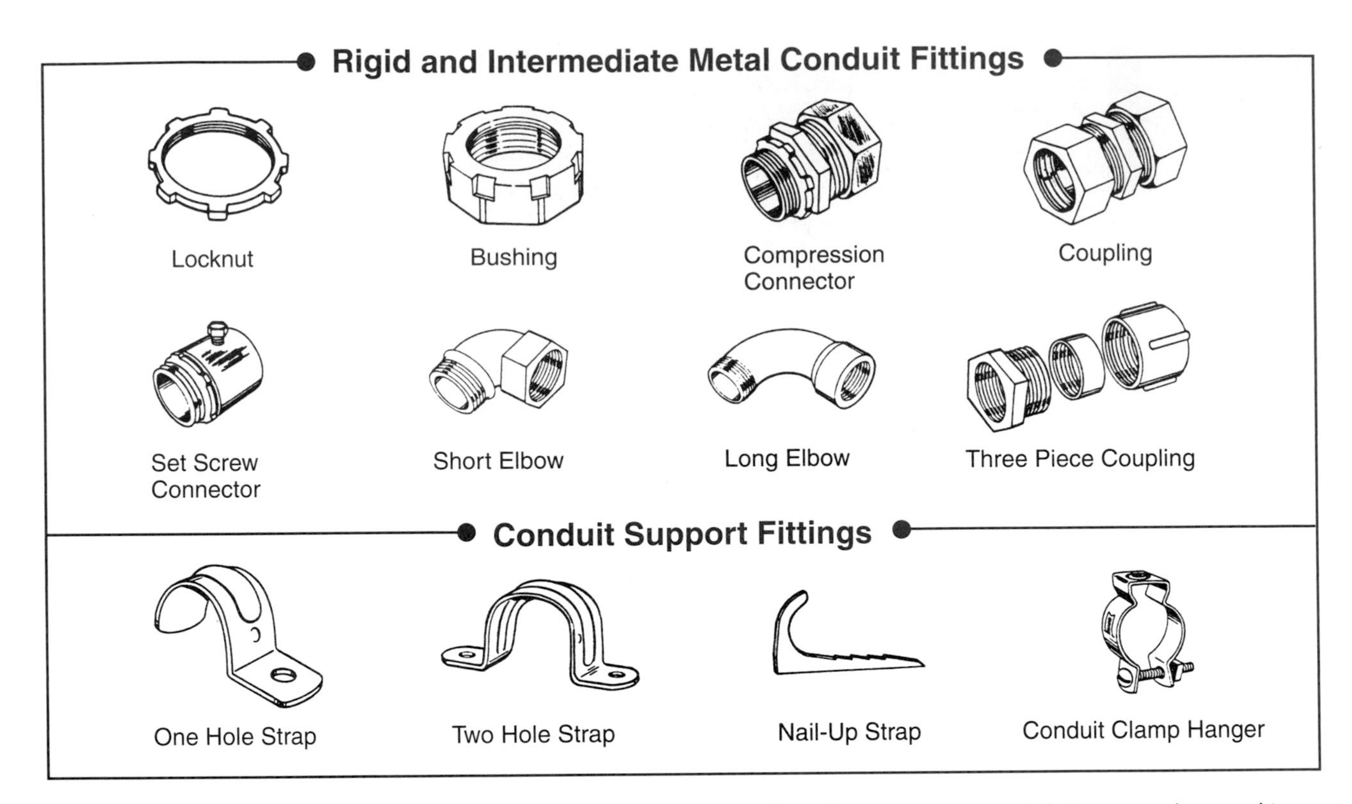

Figure 5-10. Many types of fittings are used with rigid metal conduit and IMC. The conduit support fittings are also used to support cables. (RACO, Inc.)

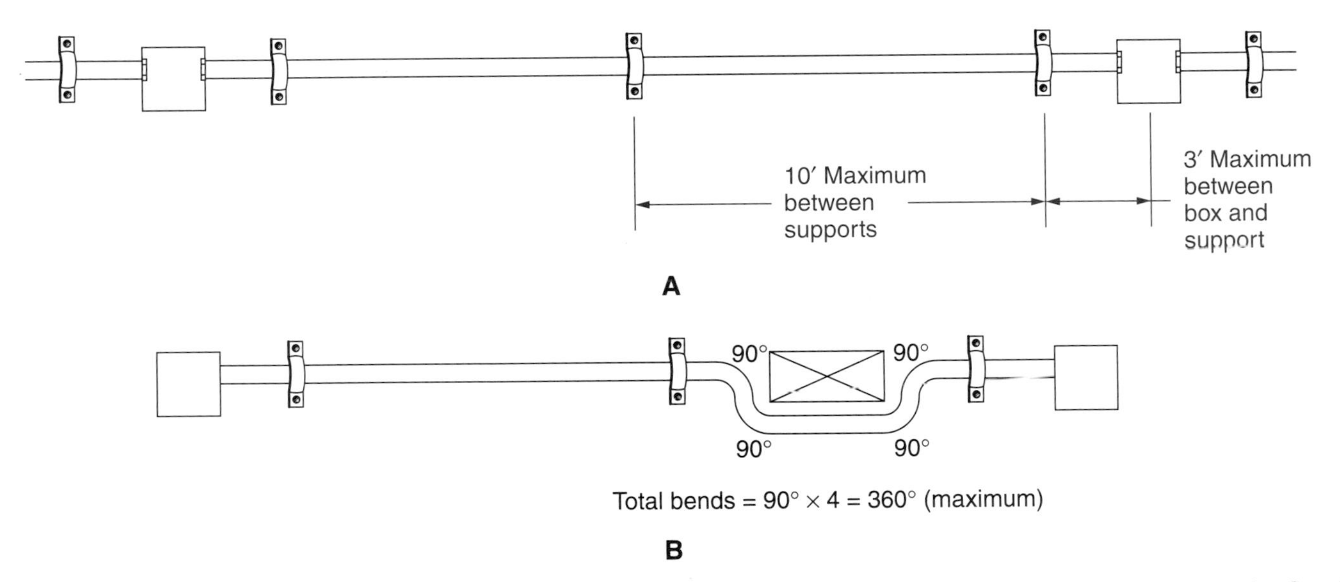

Figure 5-11. Conduit support and bending limitations for rigid metal conduit, IMC, EMT, and rigid nonmetallic conduit. A—Conduit must be supported within 3′ of a box. The maximum distance between supports is 10′. B—The total number of bends in a conduit must not exceed 360°. Here, the maximum number of bends have been made.

Intermediate metal conduit (IMC)

IMC has a thinner wall than rigid metal conduit, but is nearly as strong. It is approximately 25% lighter than rigid conduit and is less costly. IMC can be threaded and reamed in the field.

This conduit has the same permitted uses as rigid metal conduit. It is manufactured in sizes of 1/2″ to 4″. IMC also serves as an excellent equipment ground.

Just like rigid metal conduit, IMC must be supported every 10′ and within 3′ of every outlet and fitting. The total number of bends in each run must be less than 360°.

Electrical metallic tubing (EMT)

EMT, or thinwall conduit, is about half as heavy as rigid metal conduit. This is due to the fact that EMT has a much thinner wall (about 60% less) than rigid metal

Conduit Properties

Trade Size	Rigid Steel Conduit				EMT			
	OD	ID	Wall Thickness	Weight (lb/1000′)	OD	ID	Wall Thickness	Weight (lb/1000′)
1/2	0.840	0.622	0.109	820	0.706	0.622	0.042	295
3/4	1.050	0.824	0.113	1120	0.922	0.824	0.049	445
1	1.315	1.049	0.133	1600	1.163	1.049	0.057	650
2	2.375	2.067	0.154	3500	2.197	2.067	0.065	1410
3	3.500	3.068	0.216	7120	3.500	3.356	0.072	2700
4	4.500	4.026	0.237	10,300	4.500	4.334	0.083	4000

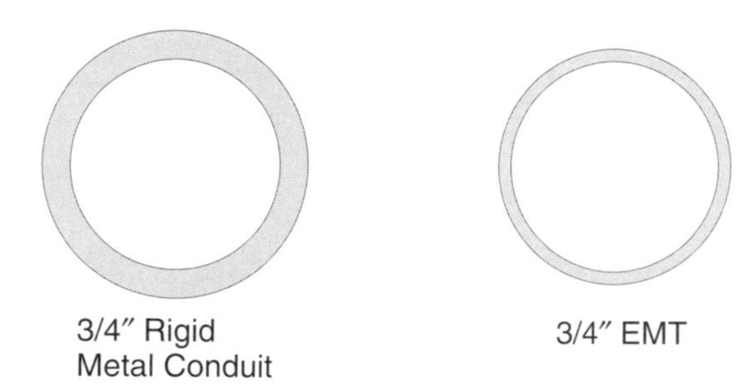

Figure 5-12. This table compares several sizes of rigid metal conduit and EMT. Cross sections of nominal 3/4″ raceway are shown. The EMT is lighter and more affordable, while the rigid metal conduit is more sturdy and durable.

conduit, although their outer diameters are about the same. See **Figure 5-12.** IMC is also thicker than EMT.

This advantage of lighter weight is offset by the loss in ability to withstand physical damage. EMT does not enjoy as wide a variety of permitted usage as does rigid metal or intermediate metal conduit. Still, EMT may be used in most locations.

EMT is not threaded. Its couplings and connectors can be set screw, compression, or indenter types. See **Figure 5-13.** It is supported at intervals of no more than

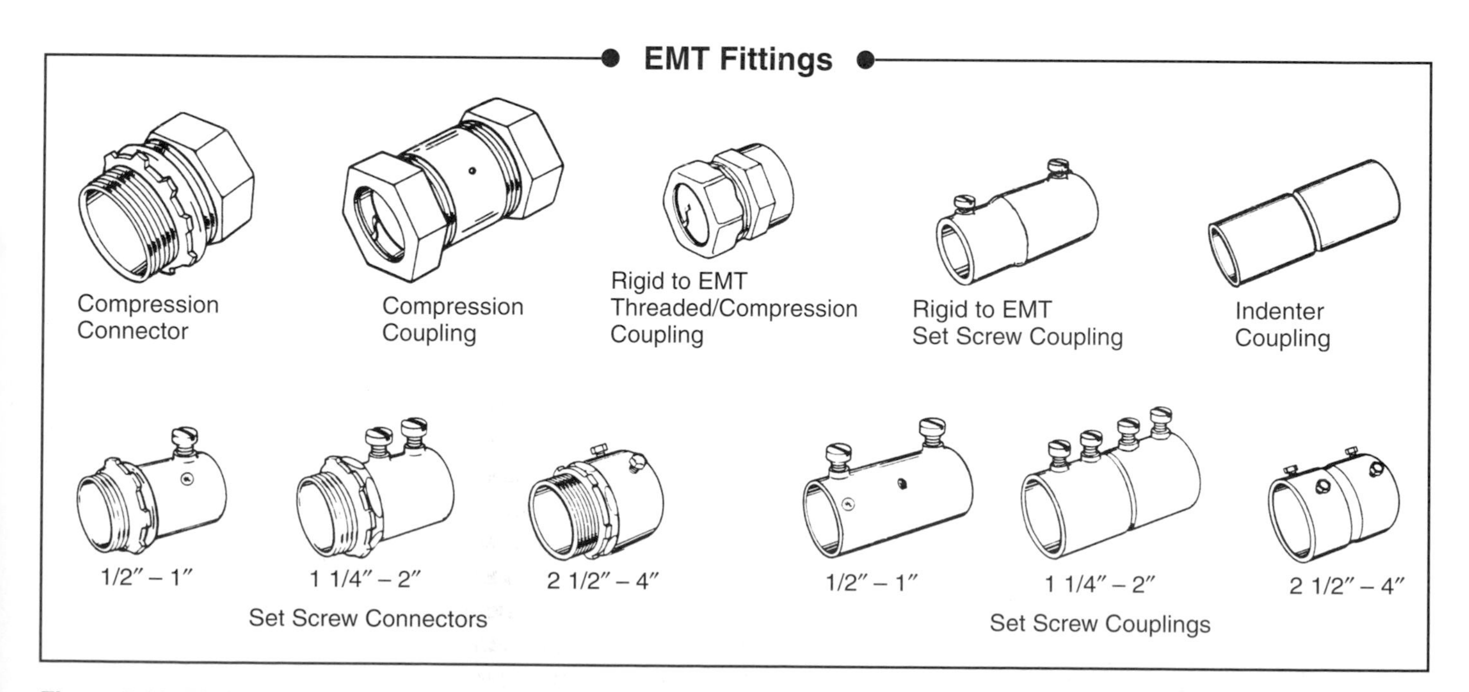

Figure 5-13. EMT couplings and connectors. (RACO, Inc.)

10′ and within 3′ of outlets and fittings. The number of conductors permitted in EMT is the same as for rigid or IMC, and is determined from the tables in *Chapter 9* of the *Code*. As with rigid metal conduit and IMC, the total bends between raceway ends must not exceed 360°.

Flexible metal conduit (FMC)

Flexible metal conduit is very similar in appearance to armored cable. The primary differences are that FMC does not come with conductors and the armor is more closely interlocked.

This type of conduit may be installed in dry or wet locations (providing the conductors are "W" rated and the flex is liquidtight), hoistways, hazardous areas (Class I, Division 2), and oil and gasoline areas if the conductor insulation is suitable for the purpose.

FMC is manufactured in sizes 3/8″ to 4″ diameter, the 3/8″ size being permitted for use in connections not over 6′ in length for fixture whips, motor connections, under plaster extensions, and manufactured wiring systems. In such lengths or less, the FMC may serve as the grounding conductor. Flexible metal conduit fittings are shown in **Figure 5-14.**

FMC must be supported at intervals less than 4′ - 6″ and within 12″ of each end. It should also be firmly supported at every bend. Lengths of FMC less than 3′ do not need to be supported, **Figure 5-15.** As with other forms of conduit, the total bends between termination points must not exceed 360°.

The maximum number of conductors permitted in 1/2″ to 4″ FMC is determined by using the tables in *Chapter 9* of the *Code*. For 3/8″ FMC, the number of conductors is indicated in *Table 348.22* of the *Code*.

NOTE

Conduit sizing is discussed in Chapter 6 of this text.

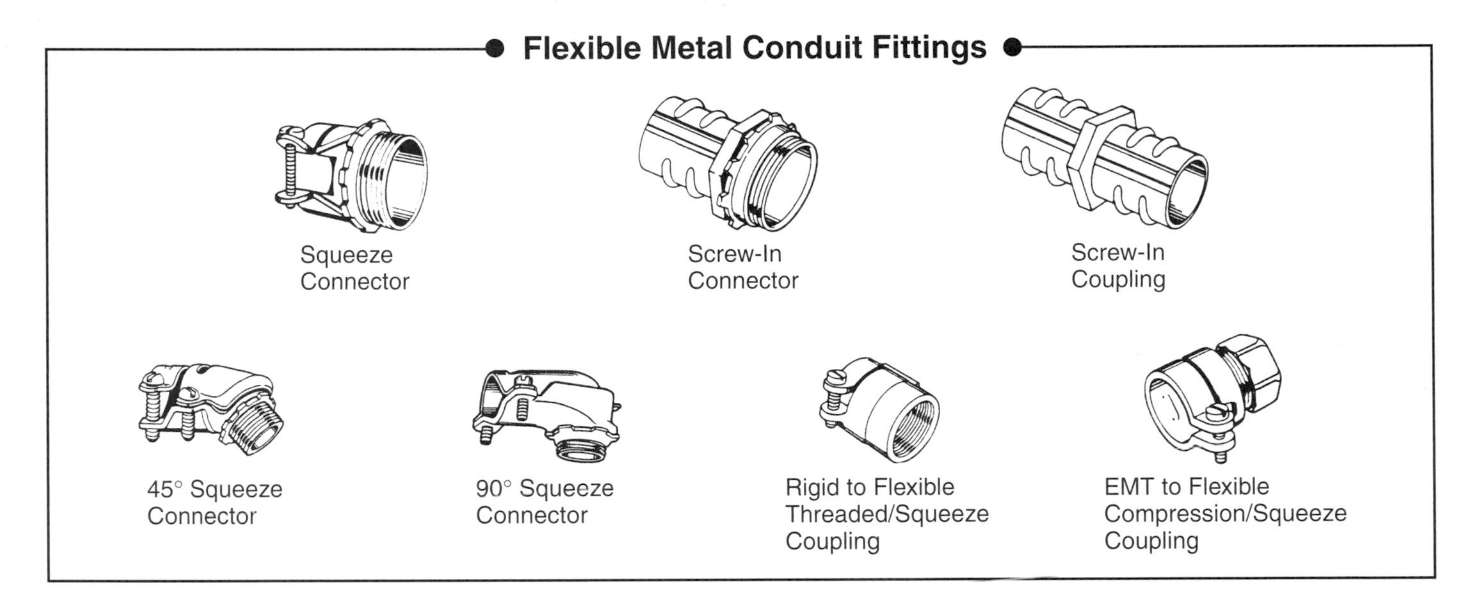

Figure 5-14. FMC connectors are tightened around the conduit to maintain a solid fit. (RACO, Inc.)

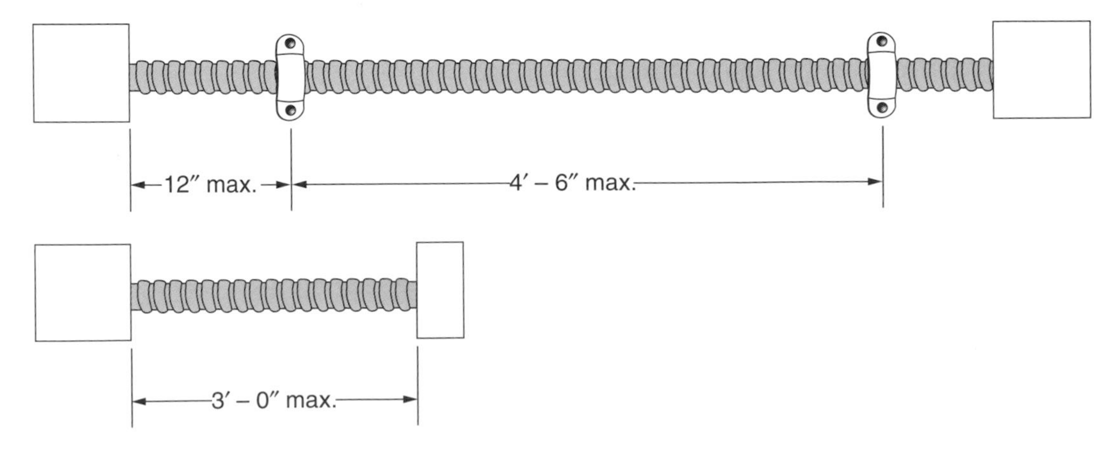

Figure 5-15. Support requirements for flexible metal conduit. No support is needed for lengths less than 3′. For longer lengths, conduit is supported within 12″ of boxes, with a maximum space of 4′ – 6″ between supports.

Liquidtight flexible metal conduit (LFMC)

Liquidtight flexible metal conduit is identical to flexible metal conduit except it has an outer plastic jacket that makes it impervious to liquids. It is manufactured in sizes from 3/8″ to 4″ in diameter. The connectors are watertight as well.

Requirements for number of conductors, bends permitted, and supporting liquidtight metal flex are the same as for standard flexible metal conduit. As with standard metal flex, lengths of 3′ and less need not be supported.

Uses of liquidtight flexible metal conduit include direct burial, concrete embedment, exposed surfaces, and through walls (concealed). Sizes of 3/8″ to 1 1/4″ liquidtight flexible metal conduit are suitable for equipment grounding in lengths not exceeding 6′; longer lengths and sizes 2″ or larger require a separate grounding conductor to be run inside the conduit.

Liquidtight flexible nonmetallic conduit (LFNC) shares all the same rules as its metallic counterpart except for the following:

- The maximum permitted size is 2″.
- Lengths of more than 6′ are prohibited.
- The grounding conductor, where required, is run inside or outside the flex.

Rigid nonmetallic conduit (RNC)

There are several types of nonmetallic conduit on the market today, such as polyethylene, styrene, PVC (polyvinyl chloride), fiber, and soapstone. Applications and restrictions are specific to each type.

Article 352 of the *Code* addresses the rules regarding the installation of rigid nonmetallic conduit. Careful attention to this article is necessary to properly choose and install this type of conduit. There are many considerations before choosing the best type for a particular job. Several key rules are contained in this article:

- All rigid nonmetallic conduits can be installed underground, but some types must be encased in concrete.
- Only Schedule 40 and Schedule 80 rigid polyvinyl chloride (PVC) are approved for above-ground use.
- Fiber conduit may be installed underground only.
- Only PVC is acceptable for nonmetallic conduit installation within buildings.
- Rigid nonmetallic conduit is not permitted as an equipment support means.

Electrical nonmetallic tubing (ENT)

ENT is a very versatile wiring method. The tubing is composed of a corrugated nonmetallic material (typically plastic). The tubing can be bent by hand, making it very easy to install, **Figure 5-16.** ENT is available in 1/2″ to 2″ sizes.

Support for ENT must be provided within 3′ of each box and every 3′ along the length of tubing. Compared to other wiring methods, ENT requires more support due to its flexibility.

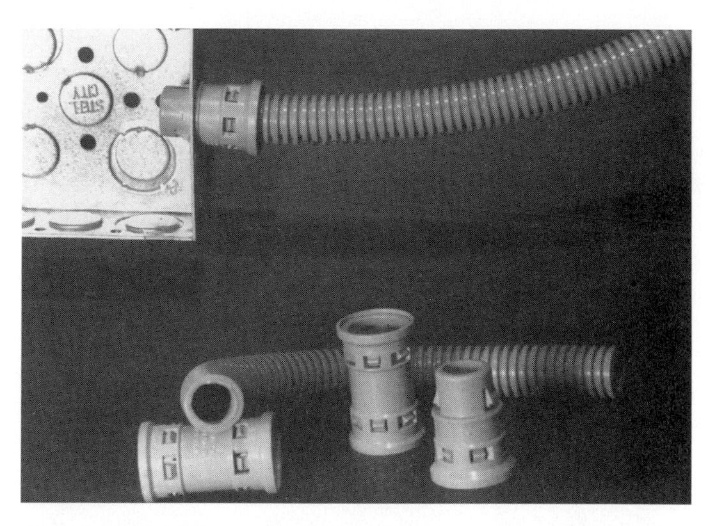

Figure 5-16. ENT is easy to install and can be used in a wide variety of applications.

Article 362—Electrical Nonmetallic Tubing specifies the uses for ENT. The following is a partial list of these uses:

- For exposed work in areas not subject to physical damage in buildings three stories and less.
- Concealed in walls, floors, and ceilings.
- Above suspended ceilings with a 15-minute finish rating.
- Embedded in concrete.

CAUTION

Some types of ENT may become brittle in cold temperatures. If the tubing becomes brittle, it could be easily damaged.

Surface raceway

In many situations there arises a need to add new circuits in an area where the existing wiring is not accessible. The wiring could be embedded in concrete or in conduit behind walls.

In such conditions, the needed changes can be made using surface raceway, **Figure 5-17.** Surface raceway consists of one- or two-piece channels and are easily installed. These raceways are used not only for power outlets and lighting, but are also permitted, when provided with internal barriers, to route communication wiring and fire protection cable.

The number of conductors permitted in surface raceway is not defined by the *Code* but is provided by the manufacturer. As with wireway and auxiliary gutters, splices and taps are permitted (in those raceways having removable covers) so long as they do not occupy more than 75% of the cross-sectional area. The one-piece unit must be installed and secured prior to pulling conductors. There are numerous types of fittings, switches, receptacles, elbows, and adapters available for surface raceway.

Figure 5-17. Surface raceway can be used to supply power outlets, lighting, and communication outlets. (Wiremold)

Wireways

Wireways are rectangular sheet metal enclosures with removable covers. Fittings, such as elbows, clips, and end pieces, enable the sections to be joined to form an overall wiring system. Concentric knockouts are provided in each side and at the ends.

When installed vertically, wireways are supported every 15′. Each wireway system must be complete prior to installing the conductors. Entry into a wireway can be made using other wiring methods, such as mineral-insulated metal-sheathed cable (MI), metal-clad cable (MC), rigid conduit, intermediate conduit, electrical metallic tubing, electrical nonmetallic tubing, and rigid nonmetallic conduit.

The total cross-sectional area of all conductors in a wireway must be less than 20% of the cross-sectional area of the wireway. Splices and taps are permitted in the wireway, but must take up less than 75% of the cross-sectional area. See **Figure 5-18.**

Sample Problem 5-1

Problem: What is the minimum cross-sectional area for a wireway that will accommodate four 2/0 AWG THW, ten 6 AWG THWN, and ten 10 AWG THWN insulated copper conductors?

Solution: First, determine the cross-sectional area of all of the conductors. Cross-sectional areas of conductors are found in *Chapter 9, Table 5* in the *Code*.

2/0 AWG THW: $4 \times 0.2624 \text{ in}^2 = 1.05 \text{ in}^2$
6 AWG THWN: $10 \times 0.0507 \text{ in}^2 = 0.51 \text{ in}^2$
10 AWG THWN: $10 \times 0.0211 \text{ in}^2 = 0.21 \text{ in}^2$
Total area = 1.77 in^2

The wireway size must be large enough that 1.77 in^2 is less than 20% of the total area. If 1.77 in^2 is 20%, dividing 1.77 in^2 by 0.20 will equal the area equivalent to 100%. This is the minimum wireway cross section:

$$1.77 \text{ in}^2 \div 0.20 = 8.85 \text{ in}^2$$

Therefore, the wireway must have a cross section of 8.85 in^2 or larger.

Type	Size (AWG or kcmil)	Approximate Diameter mm	Approximate Diameter in.	Approximate Area mm²	Approximate Area in.²
Type: RHH*, RHW*, RHW-2*, THHN, THHW, THW, THW-2, TFN, TFFN, THWN, THWN-2, XF, XFF					
THHW, THW, AF, XF, XFF	10	5.232	0.206	21.48	0.0333
RHH*, RHW*, RHW-2*	8	6.756	0.266	35.87	0.0556
TW, THW, THHW, THW-2, RHH*, RHW*, RHW-2*	6	7.722	0.304	46.84	0.0726
	4	8.941	0.352	62.77	0.0973
	3	9.652	0.380	73.16	0.1134
	2	10.46	0.412	86.00	0.1333
	1	12.50	0.492	122.6	0.1901
	1/0	13.51	0.532	143.4	0.2223
	2/0	14.68	0.578	169.3	0.2624
	3/0	16.00	0.630	201.1	0.3117
	4/0	17.48	0.688	239.9	0.3718
TFN, TFFN	18	2.134	0.084	3.548	0.0055
	16	2.438	0.096	4.645	0.0072
THHN, THWN, THWN-2	14	2.819	0.111	6.258	0.0097
	12	3.302	0.130	8.581	0.0133
	10	4.166	0.164	13.61	0.0211
	8	5.486	0.216	23.61	0.0366
	6	6.452	0.254	32.71	0.0507
	4	8.230	0.324	53.16	0.0824
	3	8.941	0.352	62.77	0.0973
	2	9.754	0.384	74.71	0.1158
	1	11.33	0.446	100.8	0.1562
	1/0	12.34	0.486	119.7	0.1855
	2/0	13.51	0.532	143.4	0.2223
	3/0	14.83	0.584	172.8	0.2679
	4/0	16.31	0.642	208.8	0.3237
	250	18.06	0.711	256.1	0.3970
	300	19.46	0.766	297.3	0.4608

Additional information and requirements concerning the design and use of wireways can be found under *Article 376* and *Article 378* of the *Code*. Some key points within this article include the following:

- Wireways are not to be installed where subject to damage.
- Wireways are not to be placed in corrosive environments.
- Wireways are not to be concealed.
- Wireways are normally restricted to no more than thirty current-carrying conductors. More than thirty current-carrying conductors can be contained in a wireway if their ampacity rating is derated per *Section 310.15(B)(2)(a)* of the *Code*.

NEC NOTE 376.22, 378.22

Conductors for signaling circuits or controller conductors between a motor and its starter and used only for starting duty shall not be considered as current-carrying conductors.

Auxiliary gutters

Essentially identical in construction and appearance to wireways, ***auxiliary gutters*** are primarily used to extend or supplement wiring spaces at load centers, transformers, and metering cabinets. Auxiliary gutters cannot extend beyond 30′ from the equipment they supplement. The same rules regarding fill (20%) and splices or taps (75%) apply as with wireways.

Busways

Busways are sheet metal enclosures into which conductors are installed at the factory. These conductors, which are actually copper or aluminum busbars, are supported by insulating material. Busways commonly come in up to 10′ sections that are bolted together. When installed horizontally, busways require supports every 5′ unless otherwise designated. Busway systems are often used in commercial buildings as the primary wiring method.

There are three types of busways available:

- Feeder busways
- Plug-in busways
- Trolley busways

A plug-in busway is illustrated in **Figure 5-19.** Although the initial cost of busway material is higher than other wiring methods, the installation labor costs are much lower. This often makes it a more cost-effective system.

There are numerous wiring methods that are permitted for use with busways as taps or branch circuits:

- Rigid nonmetallic conduit (PVC)
- Electrical nonmetallic tubing (ENT)
- Intermediate metal conduit (IMC)
- Electrical metallic tubing (EMT)
- Flexible metal conduit
- Rigid metal conduit
- Armored cable (AC)
- Surface metal raceway

Wireway Cross-Sectional Area

Wireway Size (in.)	Cross-Sectional Area (in²)		
	100%	75%	20%
2 1/2 × 2 1/2	6.25	4.68	1.25
4 × 4	16	12	3.20
6 × 6	36	27	7.20
8 × 8	64	48	12.80

Figure 5-18. This table lists allowable fill for common wireway sizes.

Figure 5-19. Plug-in busways provide many locations for an electrical hookup. (Star Products Division, US Trolley Corp.)

Overcurrent protection of busways is required at the supply end. If a smaller busway is tapped from a larger busway, overcurrent protection at the tap point should protect the smaller busway. An exception allows a short tap (50′ or less) with an ampere rating of at least 1/3 that of the overcurrent device protecting the larger busway to be exempt from the required overcurrent protection at the tap point.

Further information and requirements concerning busways can be found within *Article 368* of the *Code.*

CAUTION

Busways must not be used where subject to mechanical or physical damage, where there are corrosive surroundings (such as in battery rooms), where it is damp or wet, where there are explosive gases or vapors, where there is ignitable dust or fibers, where embedded in concrete, underground, in any outdoor or hazardous area—except where expressly approved for the purpose.

Figure 5-20. Cable trays are used to support cables in industrial settings. (PW Industries, Inc.)

Cable Trays

Cable trays are open cable-supporting assemblies used in a variety of commercial and industrial buildings. Cable trays are not enclosed, so they do not fit the description of raceways, but they have the same function. Tray systems are fully recognized as an approved method for wiring.

Cable trays resemble troughs, open at the top (although covers are often used), with ventilated bottom sections, **Figure 5-20.** There are two main types of cable trays: trough and ladder.

There are many specific rules regarding the types of wire and cable that can be used in a tray system, and how those wires and cables must be arranged within the tray. *Article 392* of the *Code* permits trays to be used as a support system for wiring methods that can be used without a tray. Uses permitted and not permitted are as follows:

- Where single conductor building wire is used in a tray, only size 1/0 AWG or larger is permitted and must be marked as suitable for tray installation. Further, this only applies to industrial installations; only multiconductor cable is permitted in a tray within commercial premises.
- A metallic cable tray is acceptable as the equipment grounding conductor for the circuits within the tray.
- Nonmetallic cable trays are permitted in areas where there are corrosive conditions.
- Multiconductor cables rated 600 volts or less may be placed in the same cable tray. Tray cable rated over 600 volts can be placed in the same tray with cables rated under 600 volts if a noncombustable barrier is installed to separate the high and low voltage cables.
- Cables within a cable tray can be spliced.
- Single-conductor cable can be used only if multiconductor cable is not available.

Multioutlet Assemblies

The *Code* addresses multioutlet assemblies in *Article 380.* Essentially, these are two-piece assemblies. The top piece (cover) is prepunched to acccpt receptacles at close (6″ and up) intervals. The receptacles can be factory installed.

This method of wiring is particularly useful and commonly found in laboratories, workshops, stores, schools, and offices. It is installed exposed on the surface like other surface raceway. It may pass through walls provided there is no receptacle within the partition and the covers on either side can be easily removed.

Review Questions

Answer the following questions. Do not write in this book.

1. What types of rigid nonmetallic conduit can be used for above-ground applications?
2. What is the maximum interval for vertical support of a 4 AWG copper conductor?

3. Regarding cable trays, what is allowed in industrial applications that is not allowed in commercial applications?
4. Describe surface raceway.
5. Which *Code* article contains the requirements for underground feeder cable?
6. What is normally the maximum number of conductors that can be placed in a single wireway?
7. Can the conductors for a single circuit be located in different conduit runs?
8. Which conduit has thicker walls: IMC or EMT?
9. Compare the support requirements of ENT with other wiring methods. What is the reason for the difference?
10. Under what conditions can conductors with different voltage ratings be located within the same conduit?
11. What are the four main types of cable?
12. What are the requirements for protecting conduit running through wall studs?
13. What percentage of wireway area can be filled by conductor splices?
14. What type of cable is called *romex*?
15. What is the minimum burial depth of rigid metal conduit below a residential driveway?
16. In what situations is service-entrance cable used?
17. Is direct burial of type USE cable allowed?
18. When underground wiring is used, how far above ground level must the raceway extend at the point where the conductors exit the ground?
19. What methods can be used to support vertical conductors?
20. Which type of cable does FMC resemble?
21. What is the maximum total cross-sectional area of conductors that can be placed in a 4″ × 4″ rectangular wireway?
22. A cable splice that will occupy 8.8 in^2 is proposed to be located in a 3″ × 4″ wireway. Will this splice satisfy wireway space limitation requirements?

USING THE NEC

Refer to the National Electrical Code to answer the following questions. Do not write in this book.

1. *Section 392.3* defines two areas in which nonmetallic cable trays can be used. What are these two areas?
2. *Table 300.5* lists minimum burial depths. In general, which needs to be buried *deeper*, IMC or PVC conduit?
3. *Section 300.4* defines requirements for protecting wiring methods. What must be installed to protect electrical nonmetallic tubing running through metal framing members?
4. An inspector cited a violation for supporting one conduit with a hanger attached to another conduit. Which *Code* section does this practice violate?
5. Can an IMC wiring system be installed in an air duct containing flammable vapors?
6. Is a 2″ × 4″ rectangular wireway large enough to safely contain six 3/0 AWG XHHW, six 12 AWG XHHW, and four 14 AWG XHHW conductors?
7. Metal-clad cable is permitted for use in wet locations if any one of three conditions are met. What are these conditions?
8. A wireway containing six 2/0 AWG THHN and two 14 AWG THHN conductors is needed for a project. The wireway must be 2″ wide. What depth of wireway should be ordered (assuming the depth must be ordered in 1″ increments)?
9. What is the maximum allowable interval between supports for strut-type channel raceway?
10. The length of rigid nonmetallic conduit changes due to thermal contraction and expansion. Expansion fittings are used to compensate for these changes. What is the maximum allowable length change in a straight run of conduit without expansion fittings?
11. When lighting busways and trolley busways are not provided with covers, what is the minimum height above the floor at which they can be installed?
12. Of metal wireways and nonmetallic wireways, which can be used where subject to corrosive vapors?

Although residential and commercial installations are similar in theory, commercial electrical work involves larger equipment, conduit, and conductors. This fixture, located in the Von Braun Civic Center in Huntsville, Alabama, is much larger than any fixtures encountered in residential work. (Armstrong World Industries, Inc.)

Chapter 6
Conductors

Technical Terms

American Wire Gage (AWG)
Bare conductors
Branch-circuit conductors
Cable
Circular mil
Conductor
Feeders
Insulated conductor
Main feeders
Service-entrance conductors
Shielding
Subfeeders
Voltage drop

Objectives

After completing this chapter, you will be able to:

- Recall the *Code* rules regarding conductors for general wiring.
- Identify the different functions of conductors in an electrical system.
- List the factors that affect conductor ampacity rating.
- Explain the cause of voltage drop and compute the voltage drop of a conductor.
- Select different types of conductors.
- Size conductors based on circuit load using the *Code*.
- Use the *Code* to adjust conductor ampacity based on ambient temperature and number of conductors.
- Designate conduit sizes based on permissible percentage fill of conduit.

In this chapter, you will explore the part of the electrical system that provides the pathway and connecting means between the electrical source and the utilization equipment.

Conductors

Conductor refers to copper, aluminum, or copper-clad aluminum wire that carries current. If this wire is encased in a noncurrent-carrying material, it is referred to as an ***insulated conductor,*** or ***cable.*** A cable is actually a complete wire assembly having a conductor, insulation, shielding (if needed), and an outer protective covering. Cables can contain one conductor or many separately insulated conductors. See **Figure 6-1.**

> **NEC NOTE** **100**
>
> **Conductor:** *Insulated.* A conductor encased within material of composition and thickness that is recognized by this *Code* as electrical insulation.

There are three classifications of conductors discussed in this text. Their uses are illustrated in **Figure 6-2.**

- **Service-entrance conductors**—These conductors deliver electrical energy from the utility company distribution supply conductors to the service-entrance equipment. These are the conductors between the point of supply and the main service disconnect.
- **Feeder conductors**—These conductors, called ***feeders,*** conduct electrical current from the main service disconnect to the final circuit overcurrent device, usually at the branch circuit distribution panel. Feeders are further categorized as ***main feeders*** (those originating at the service disconnect) and ***subfeeders*** (those originating at distribution panels other than service equipment and connecting to other panels in the system).

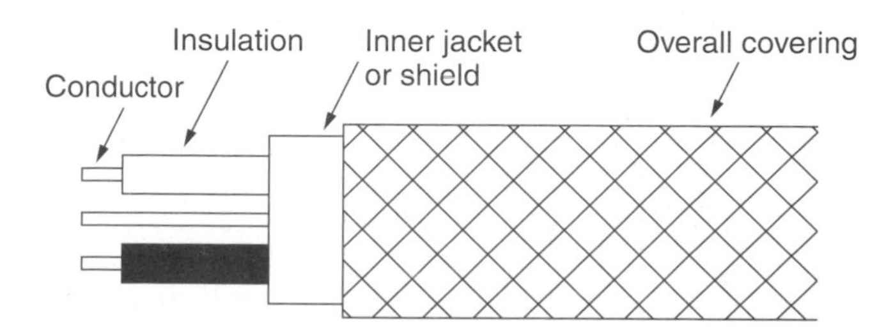

Figure 6-1. This cable has insulated conductors surrounded by a jacket, all protected by an outer covering.

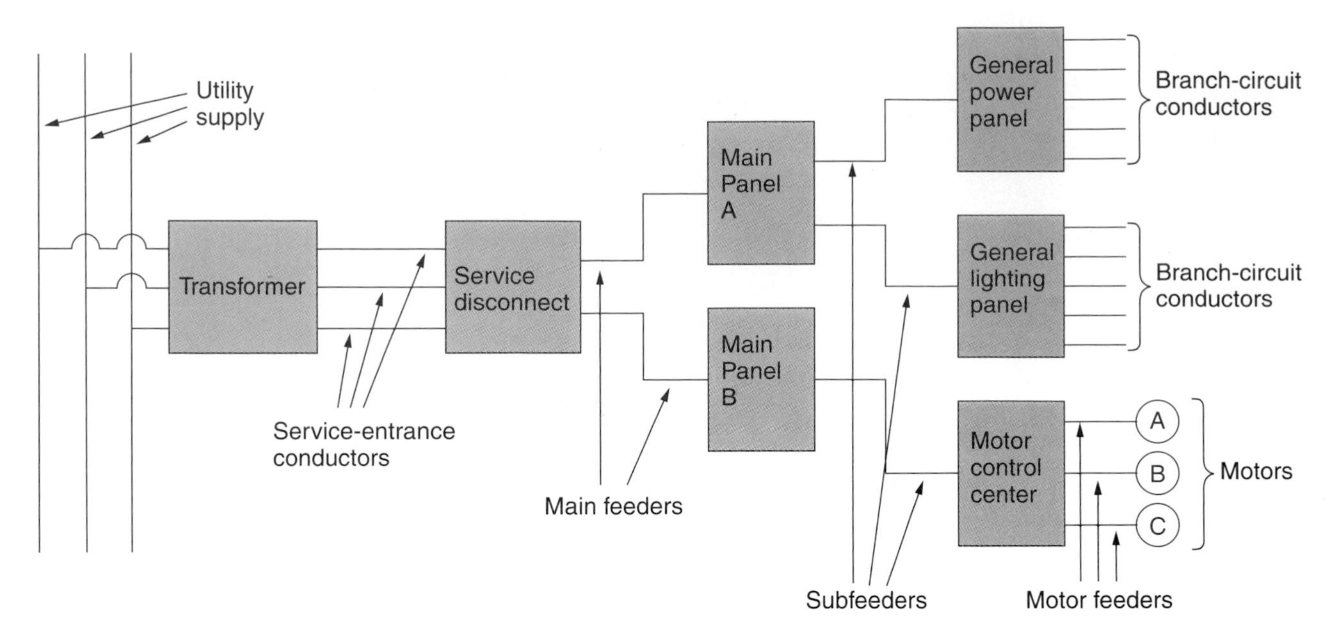

Figure 6-2. This diagram shows different conductor classifications.

- **Branch-circuit conductors**—These conductors originate at the final overcurrent device and terminate at the utilization equipment.

Conductor and Cable Classifications

Conductors are identified by letter codes (THW, THHN, and XHHW, for example). These letter designations identify the type of insulation and the appropriate environments for the conductor. Some of the letter designations indicate insulation composition:

A = Asbestos
MI = Mineral insulated
R = Rubber
T = Thermoplastic
V = Varnished cambric
X = Cross-linked synthetic polymers

Letters also identify the physical characteristics of the conductor:

H = Heat resistant to 167°F (75°C)
HH = Heat resistant to 194°F (90°C)
UF = Appropriate for direct burial
W = Appropriate for wet conditions
C = Corrosion resistant

Letter designations also indicate the type of outer jacket, sheath outer covering, or overall outer layer:

AC = Armored cable (having an interlocked metal sheath)
L = Lead covered
MC = Metal clad
N = Nylon jacketed
NM = Nonmetallic sheath

Numerous other letter designations are included in *Table 310.13, Conductor Application and Insulation,* of the *Code*.

Letter designations are marked on the product for easy identification. For example, THHN is a conductor having a thermoplastic, high heat-resistant (194°F) insulation and nylon (or equivalent) jacketing. THW indicates a thermoplastic heat-resistant insulation that can be used in wet locations. UF is suitable as an underground (directly buried) feeder. USE is underground service-entrance cable.

Conductor size designation

The size of a conductor is determined by its cross-sectional area. There are two methods of expressing conductor size: American Wire Gage and circular mils.

Within the ***American Wire Gage*** (***AWG***) system, the smaller the number, the larger the wire. For power wiring, 14 AWG is the smallest size used in commercial wiring. Control wiring may be smaller sizes, such as 16 AWG or 18 AWG. The AWG system designates wires as large as 4/0 AWG (four-aught).

Larger wires are designated by their actual cross-sectional area in terms of kcmils (thousands of circular mils). A mil is 0.001 inch (a thousandth of an inch) and one inch is 1000 mils. One ***circular mil*** (1 cmil) is equivalent to the area of a circle one mil in diameter. This is a very small area.

Table 8 of *Chapter 9* of the *Code* is shown in **Figure 6-3.** The first column of the table contains wire

Table 8 Conductor Properties

Size (AWG or kcmil)	Area mm²	Area Circular mils	Stranding Quantity	Stranding Diameter mm	Stranding Diameter in.	Overall Diameter mm	Overall Diameter in.	Overall Area mm²	Overall Area in.²	Copper Uncoated ohm/km	Copper Uncoated ohm/kFT	Copper Coated ohm/km	Copper Coated ohm/kFT	Aluminum ohm/km	Aluminum ohm/kFT
			Conductors							**Direct-Current Resistance at 75°C (167°F)**					
18	0.823	1620	1	—	—	1.02	0.040	0.823	0.001	25.5	7.77	26.5	8.08	42.0	12.8
18	0.823	1620	7	0.39	0.015	1.16	0.046	1.06	0.002	26.1	7.95	27.7	8.45	42.8	13.1
16	1.31	2580	1	—	—	1.29	0.051	1.31	0.002	16.0	4.89	16.7	5.08	26.4	8.05
16	1.31	2580	7	0.49	0.019	1.46	0.058	1.68	0.003	16.4	4.99	17.3	5.29	26.9	8.21
14	2.08	4110	1	—	—	1.63	0.064	2.08	0.003	10.1	3.07	10.4	3.19	16.6	5.06
14	2.08	4110	7	0.62	0.024	1.85	0.073	2.68	0.004	10.3	3.14	10.7	3.26	16.9	5.17
12	3.31	6530	1	—	—	2.05	0.081	3.31	0.005	6.34	1.93	6.57	2.01	10.45	3.18
12	3.31	6530	7	0.78	0.030	2.32	0.092	4.25	0.006	6.50	1.98	6.73	2.05	10.69	3.25
10	5.261	10380	1	—	—	2.588	0.102	5.26	0.008	3.984	1.21	4.148	1.26	6.561	2.00
10	5.261	10380	7	0.98	0.038	2.95	0.116	6.76	0.011	4.070	1.24	4.226	1.29	6.679	2.04
8	8.367	16510	1	—	—	3.264	0.128	8.37	0.013	2.506	0.764	2.579	0.786	4.125	1.26
8	8.367	16510	7	1.23	0.049	3.71	0.146	10.76	0.017	2.551	0.778	2.653	0.809	4.204	1.28
6	13.30	26240	7	1.56	0.061	4.67	0.184	17.09	0.027	1.608	0.491	1.671	0.510	2.652	0.808
4	21.15	41740	7	1.96	0.077	5.89	0.232	27.19	0.042	1.010	0.308	1.053	0.321	1.666	0.508
3	26.67	52620	7	2.20	0.087	6.60	0.260	34.28	0.053	0.802	0.245	0.833	0.254	1.320	0.403
2	33.62	66360	7	2.47	0.097	7.42	0.292	43.23	0.067	0.634	0.194	0.661	0.201	1.045	0.319
1	42.41	83690	19	1.69	0.066	8.43	0.332	55.80	0.087	0.505	0.154	0.524	0.160	0.829	0.253
1/0	53.49	105600	19	1.89	0.074	9.45	0.372	70.41	0.109	0.399	0.122	0.415	0.127	0.660	0.201
2/0	67.43	133100	19	2.13	0.084	10.62	0.418	88.74	0.137	0.3170	0.0967	0.329	0.101	0.523	0.159
3/0	85.01	167800	19	2.39	0.094	11.94	0.470	111.9	0.173	0.2512	0.0766	0.2610	0.797	0.413	0.126
4/0	107.2	211600	19	2.68	0.106	13.41	0.528	141.1	0.219	0.1996	0.0608	0.2050	0.0626	0.328	0.100
250		—	37	2.09	0.082	14.61	0.575	168	0.260	0.1687	0.0515	0.1753	0.0535	0.2778	0.0847
300		—	37	2.29	0.090	16.00	0.630	201	0.312	0.1409	0.0429	0.1463	0.0446	0.2318	0.0707
350		—	37	2.47	0.097	17.30	0.681	235	0.364	0.1205	0.0367	0.1252	0.0382	0.1984	0.0605
400		—	37	2.64	0.104	18.49	0.728	268	0.416	0.1053	0.0321	0.1084	0.0331	0.1737	0.0529
500		—	37	2.95	0.116	20.65	0.813	336	0.519	0.0845	0.0258	0.0869	0.0265	0.1391	0.0424
600		—	61	2.52	0.099	22.68	0.893	404	0.626	0.0704	0.0214	0.0732	0.0223	0.1159	0.0353
700		—	61	2.72	0.107	24.49	0.964	471	0.730	0.0603	0.0184	0.0622	0.0189	0.0994	0.0303
750		—	61	2.82	0.111	25.35	0.998	505	0.782	0.0563	0.0171	0.0579	0.0176	0.0927	0.0282
800		—	61	2.91	0.114	26.16	1.030	538	0.834	0.0528	0.0161	0.0544	0.0166	0.0868	0.0265
900		—	61	3.09	0.122	27.79	1.094	606	0.940	0.0470	0.0143	0.0481	0.0147	0.0770	0.0235
1000		—	61	3.25	0.128	29.26	1.152	673	1.042	0.0423	0.0129	0.0434	0.0132	0.0695	0.0212
1250		—	91	2.98	0.117	32.74	1.289	842	1.305	0.0338	0.0103	0.0347	0.0106	0.0554	0.0169
1500		—	91	3.26	0.128	35.86	1.412	1011	1.566	0.02814	0.00858	0.02814	0.00883	0.0464	0.0141
1750		—	127	2.98	0.117	38.76	1.526	1180	1.829	0.02410	0.00735	0.02410	0.00756	0.0397	0.0121
2000		—	127	3.19	0.126	41.45	1.632	1349	2.092	0.02109	0.00643	0.02109	0.00662	0.0348	0.0106

Conductor resistance

AWG sizes

Circular mils sizes (kcmil)

Area of conductor, including voids for stranded conductors

Notes:

1. These resistance values are valid **only** for the parameters as given. Using conductors having coated strands, different stranding type, and, especially, other temperatures changes the resistance.
2. Formula for temperature change: $R_2 = R_1[1 + \alpha (T_2 - 75)]$ where: $\alpha_{cu} = 0.00323$, $\alpha_{AL} = 0.00330$ at 75° C.
3. Conductors with compact and compressed stranding have about 9 percent and 3 percent, respectively, smaller bare conductor diameters than those shown. See Table 5A for actual compact cable dimensions.
4. The IACS conductivities used: bare copper = 100%, aluminum = 61%.
5. Class B stranding is listed as well as solid for some sizes. Its overall diameter and area is that of its circumscribing circle.

FPN: The construction information is per NEMA WC8-1992 or ANSI/UL. 1581-1998. The resistance is calculated per National Bureau of Standards Handbook 100, dated 1966, and Handbook 109, dated 1972.

Figure 6-3. This table, from *Chapter 9* of the *Code,* provides conductor size and resistance.

size. From 18 AWG through 4/0 AWG, the sizes are expressed as AWG. The corresponding area in circular mils is shown in the second column. This area is the actual area of the conducting material; for stranded conductors, the voids between strands are *not* included. The sixth column contains the overall area expressed in square inches. This area includes the voids in stranded conductors, so it is slightly larger than the circular mils area. See **Figure 6-4.**

Conductors larger than 4/0 AWG are expressed in thousand circular mils (kcmil). Conductors that are 250 kcmil (250,000 cmils) are larger than 4/0 AWG conductors (211,600 cmils). The relation between inches and mils can be easily calculated:

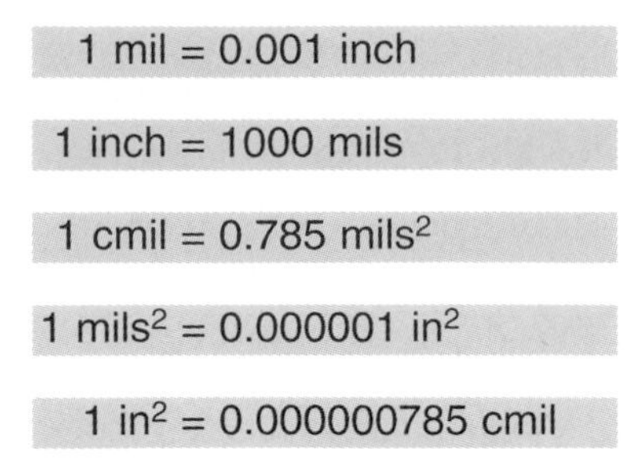

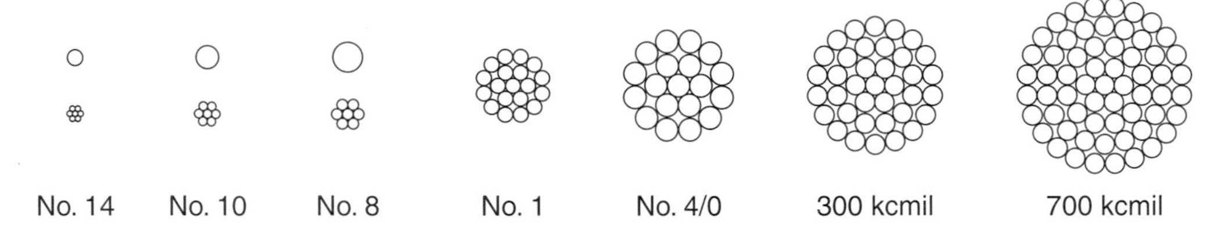

Figure 6-4. These cross sections show the relative sizes of various conductors.

Sample Problem 6-1

Problem: Determine the area of a 6 AWG copper conductor in square inches.

Solution: From *Table 8*, the area of a 6 AWG conductor is 26240 cmils.
To convert this to square inches, multiply cmils by 0.000000785.

$$26240 \text{ cmils} \times 0.000000785 = 0.0206 \text{ in}^2$$

This value differs from the 0.027 in² shown in *Table 8* because the calculated value is the actual area of the strands. The area in the table includes the voids between strands.

Table 8 Conductor Properties

Size (AWG or kcmil)	Area		Conductors: Stranding				Overall				Direct-Current Resistance at 75°C (167°F): Copper				Aluminum	
				Diameter			Diameter		Area		Uncoated		Coated			
	mm²	Circular mils	Quantity	mm	in.		mm	in.	mm²	in.²	ohm/km	ohm/kFT	ohm/km	ohm/kFT	ohm/km	ohm/kFT
18	0.823	1620	1	—	—		1.02	0.040	0.823	0.001	25.5	7.77	26.5	8.08	42.0	12.8
18	0.823	1620	7	0.39	0.015		1.16	0.046	1.06	0.002	26.1	7.95	27.7	8.45	42.8	13.1
16	1.31	2580	1	—	—		1.29	0.051	1.31	0.002	16.0	4.89	16.7	5.08	26.4	8.05
16	1.31	2580	7	0.49	0.019		1.46	0.058	1.68	0.003	16.4	4.99	17.3	5.29	26.9	8.21
14	2.08	4110	1	—	—		1.63	0.064	2.08	0.003	10.1	3.07	10.4	3.19	16.6	5.06
14	2.08	4110	7	0.62	0.024		1.85	0.073	2.68	0.004	10.3	3.14	10.7	3.26	16.9	5.17
12	3.31	6530	1	—	—		2.05	0.081	3.31	0.005	6.34	1.93	6.57	2.01	10.45	3.18
12	3.31	6530	7	0.78	0.030		2.32	0.092	4.25	0.006	6.50	1.98	6.73	2.05	10.69	3.25
10	5.261	10380	1	—	—		2.588	0.102	5.26	0.008	3.984	1.21	4.148	1.26	6.561	2.00
10	5.261	10380	7	0.98	0.038		2.95	0.116	6.76	0.011	4.070	1.24	4.226	1.29	6.679	2.04
8	8.367	16510	1	—	—		3.264	0.128	8.37	0.013	2.506	0.764	2.579	0.786	4.125	1.26
8	8.367	16510	7	1.23	0.049		3.71	0.146	10.76	0.017	2.551	0.778	2.653	0.809	4.204	1.28
6	13.30	26240	7	1.56	0.061		4.67	0.184	17.09	0.027	1.608	0.491	1.671	0.510	2.652	0.808
4	21.15	41740	7	1.96	0.077		5.89	0.232	27.19	0.042	1.010	0.308	1.053	0.321	1.666	0.508
3	26.67	52620	7	2.20	0.087		6.60	0.260	34.28	0.053	0.802	0.245	0.833	0.254	1.320	0.403

Conductor Material

There are three types of conducting material used for general wiring: copper, aluminum, and copper-clad aluminum. Copper has excellent electrical and physical properties. It has been used as a conductor material from the earliest efforts to conduct electricity.

NEC NOTE 100

Copper-Clad Aluminum Conductors: Conductors drawn from a copper-clad aluminum rod with the copper metallurgically bonded to an aluminum core. The copper forms a minimum of 10 percent of the cross-sectional area of a solid conductor or each strand of a stranded conductor.

Aluminum is a good conductor, but must be installed with additional care and preparation. Aluminum must be cleaned just prior to terminating due to the rapid oxidation of the freshly exposed surface. Exposed aluminum at terminals should be coated with an antioxidant. In addition, aluminum connections tend to deform and loosen more than copper connections. If installed properly, aluminum conductors are dependable and economical.

Aluminum has a greater resistance than copper, requiring a larger size to equal the same ampacity as copper—generally, two sizes larger. For example, a 2/0 AWG copper conductor will carry 200 amperes, whereas a 4/0 AWG aluminum conductor would be required to handle the same current.

WARNING

Improperly connected aluminum conductors can deteriorate with time and cause problems in the electrical system. Always use devices approved for use with aluminum conductors and follow proper installation procedures.

Flexible Cords and Cables

Flexible cords fall into one of three categories:

- For extra-hard usage.
- For hard usage.
- Not for hard usage.

Table 400.4 of the *Code* lists some properties of various types of cords. The ampacities of the different cords are found in *Table 400.5(A)*. When using these tables, note that the equipment grounding conductor, neutral, and bonding conductors are not counted when determining ampacity ratings of cords.

Cords and cables are permitted to be used as pendant wiring, fixture wiring, appliance connections, portable lamp (utility light) connections, data processing equipment connections, elevator wiring and cabling, crane cabling, and hoist wiring.

Cords and cables may not be used as a substitute for permanent wiring. Generally, they may not be concealed, be permanently attached, or pass through walls, floors, or ceilings. They should be continuous and unspliced and should not be used as taps.

Cords generally do not require a separate overcurrent protective device. The branch-circuit fuse may serve as the sole protection for the cord. The grounded neutral conductor in a cord must be identified with a white or natural gray separator. The equipment grounding conductor, which serves to ground the noncurrent-carrying parts of the cord supplied equipment, must always have a green color in the insulation, covering, or braiding. Except for double-insulated equipment, a grounding conductor must be provided in flexible cords.

The table in **Figure 6-5** lists some devices that may be cord and plug connected.

Cord Requirements

Device/Equipment	NEC Reference	Remarks
Kitchen waste disposal	422.16(B)(1)	18″–36″ length
Dishwasher	422.16(B)(2)	36″–48″ length
Trash compactor	422.16(B)(2)	36″–48″ length
Wall-mounted oven	422.16(B)(3)	Cord and plug may not be the disconnect means
Countertop cooking unit	422.16(B)(3)	Cord and plug may not be the disconnect means
Cooking range	422.33(B)	Cord and plug may serve as disconnect
Portable motor (1/3 hp or less)	430.42(C)	Cord and plug may serve as disconnect
Mobile home	550.10(D)	21′–36 1/2′ maximum length
Battery-powered emergency lighting	700.12	No longer than 36″
Room air conditioner	440.64	No longer than 10′

Figure 6-5. Rules for appliances with cords are found throughout the *Code*.

Code Requirements for Conductors

In the *Code, Article 310—Conductors for General Wiring* addresses the general requirements for conductors, conductor designations, temperature considerations, environmental conditions, size limitations, ampacity, installation methods, material usage, insulation, and other factors concerning conductors. Refer to *Article 310* when planning any residential, commercial, or industrial wiring installation.

Bare Conductors

A ***bare conductor*** is a conductor with no covering or electrical insulation. Per *Section 230.41* of the *Code*, conductors for general wiring must be insulated, except in the following situations:

- The grounded neutral conductor may be bare when used as a service-entrance conductor to ground or on noncurrent-carrying service-entrance components such as the service head, raceway, meter enclosure, or main service disconnect enclosure. See **Figure 6-6.**
- A bare conductor is permitted as the grounding electrode conductor for service-entrance equipment, **Figure 6-6.**
- A bare conductor is allowed for purposes of grounding the noncurrent-carrying parts of equipment such as dryers, ovens, and stoves. See *Section 250.118* of the *Code*.
- A bare conductor may be used to bond the noncurrent-carrying components (boxes, enclosures, equipment frames, and motor skids) of the electrical system.
- A bare conductor is permitted as the grounding means for load-side equipment at a subfed panel.

NEC NOTE 100

Conductor: *Bare.* A conductor having no covering or electrical insulation whatsoever.

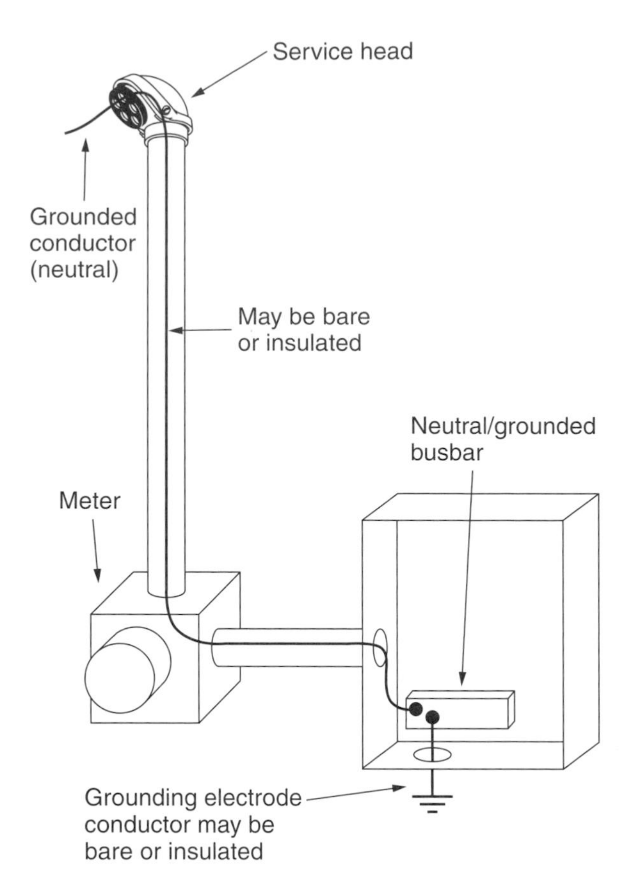

Figure 6-6. The grounded neutral service conductor and the grounding electrode conductor may be bare or insulated.

Stranded and Solid Conductors

Conductors 8 AWG and larger that will be pulled through raceways must be stranded. The conductors need flexibility when being pulled around offsets and bends. Solid conductors are too rigid in larger sizes to afford much flexibility.

NEC NOTE 310.3

Where installed in raceways, conductors of size 8 AWG and larger shall be stranded. Exception: As permitted or required elsewhere in this *Code*.

Paralleled Conductors

It is often more feasible and economical to run conductors in parallel because of large feeder load requirements. Conductors run in parallel must meet the following criteria:

- Conductors must be 1/0 AWG and larger.
- Conductors must be the same size.
- Conductors must have the same insulation type.
- Conductors must be the same material.
- Conductors must be the same length.
- Conductors must be terminated in the same manner.

Paralleled conductors can be installed in separate conduits or in a single conduit. If the conductors are installed in the same conduit, the electrician should observe *Section 310.15(B)(2)(a)* of the *Code* regarding allowable ampacities of conductors.

Conductor Size Limits

With some exceptions, *Section 310.5* of the *Code* lists the smallest conductor permitted for general wiring as 14 AWG copper or 12 AWG aluminum or copper-clad aluminum. The exceptions allow smaller sizes for flexible cords, fixture wire, fractional horsepower motors, cranes and hoists, elevator control and signal circuits, remote control circuits, protective signaling, and motor control circuits. The table in **Figure 6-7** contains minimum conductor sizes.

Shielding

The *Code* states that conductors operated above 2000 volts in permanent installations must have ozone-resistant insulation and be shielded. Further, those with metal shields must be grounded.

Direct Burial

Section 310.7 of the *Code* states that conductors used in direct burial applications must be specifically approved for the purpose. In addition, these conductors, if rated above 2000 volts, must be shielded and the shield properly grounded. However, direct burial conductors rated between 2000 volts and 5000 volts do not require shielding if the cable has an overall metallic shield or armor.

Wet Locations

Conductors used in wet conditions must be either lead-covered or marked with a "W" (RW, TW, and THW for example), indicating they are suitable for wet locations.

Minimum Conductor Size

Application	Minimum Conductor Size
Conductors for general wiring	14 AWG (copper) 12 AWG (aluminum or copper-clad)
Crane and hoist control circuits	16 AWG
Flexible cords Fixture wires	18 AWG
Elevator control and signaling circuits	20 AWG
Fire protection signaling circuits Remote control and signal circuits	14 AWG

Figure 6-7. Minimum conductor sizes for various applications.

NEC NOTE 100

Wet Location: Installations underground or in concrete slabs or masonry in direct contact with the earth; and locations subject to saturation with water or other liquids, such as vehicle washing areas; and unprotected locations exposed to weather.

Corrosive Conditions

Conductors that will be used in severe environments, in the presence of corrosive agents, and in situations where insulation deterioration is likely must be covered with insulation specifically listed and suitable for such conditions. These conductors are marked with a "C."

Temperature Limitation

Section 310.10 of the *Code* prohibits the use of any conductor, in any manner, where its operating temperature will be greater than that designated for its insulation type. Further, different conductors are not to be associated together in a way that would cause the limiting temperature of any conductor to be exceeded.

The operating temperature of a conductor is determined by the heat produced by the resistance to current flow and the rate at which this heat is dissipated to the surroundings. Greater current flow and greater resistance in the conductor produce more heat. An aluminum conductor will produce more heat than a copper conductor of the same size with the same current because the aluminum conductor has greater resistance.

Several factors affect the rate of heat dissipation. The warmer the temperature of the air surrounding the conductor, the less heat dissipated. If there are many conductors close together, they will all release heat, raising the temperature of the surrounding air, reducing the amount of heat they can release. Insulation greatly reduces the heat flow between the conductor and the surrounding air. (Caution: See *Section 310.10, FPN2.*)

Conductor Marking

All conductors and cables must be marked to show the following information:

- Maximum rated voltage.
- The proper letter(s) designation for the type of wire or cable.
- The manufacturer's name, trademark, or other distinctive marking by which the organization responsible for the product can be readily identified.
- The size of the conductor in AWG or circular-mil area.

Cables are marked with surface markings, tape, or tags. The type of marking is determined by the type of cable or conductor. Refer to *Section 310.11(B)* of the *Code* for the type of marking required for a specific cable or conductor.

Conductor Identification (Color)

Grounded conductors (neutral) 6 AWG or smaller must be insulated white, gray, or three continuous white stripes on other than green insulation along its entire length. Larger sizes may be white, gray, or three continuous white stripes or may be painted white at all junction and termination locations.

Equipment grounding conductors that are insulated must be insulated solid green or green with yellow stripes if 6 AWG or smaller. Larger sizes must either be bare at all boxes and outlets or have the insulation painted green at every box and outlet. See *Section 250.119* of the *Code*.

Ungrounded (hot) conductors may have insulation of any color other than green, white, gray, or green with yellow stripes.

NOTE

Per 110.15 of the Code, the high leg of a four-wire delta distribution system must be identified with orange markings.

Conductor Application

Application and insulation types for conductors are listed in *Table 310.13* of the *Code*. Consult this extensive table *prior* to installing any wiring. The table gives several attributes for insulated conductors:

- Trade name of the insulation.
- Letter designation assigned to that type of insulation.
- The maximum operating temperature allowable for the type of insulation or outer covering.
- The conditions suitable for application.
- The insulation composition.
- Size availability.
- Insulation thickness.
- Outer covering.

Allowable Ampacities in Conductors

Tables 310.16 through *310.19* are used to compute the size and ampacity of conductors rated under 2000 volts. As we proceed, refer to the tables, which are included in the Reference Section of this text. Each table is used for a different type of conductor:

- ***Table 310.16***—Used for up to three insulated conductors installed in raceway, as multiconductor cable within raceway, in air, or directly buried. This table is based on a surrounding temperature of 86°F (30°C).
- ***Table 310.17***—Ampacities for single insulated conductors installed in free air. The ambient temperature is also 86°F (30°C). Outside, overhead wiring is sized using this table.
- ***Table 310.18***—This table contains ampacities for three single insulated conductors in raceway or in cable whose insulation range temperatures are 302°F to 482°F (150°C to 250°C) and based on an ambient temperature of 104°F (40°C). Conductors used for general wiring have an insulation range of 140°F to 194°F (60°C to 90°C).
- ***Table 310.19***—Ampacities for single insulated conductors in free air having conductor insulation temperature ranging from 302°F to 482°F (150°C to 250°C). This table is based on an ambient temperature of 104°F (40°C).

Section 310.15 further explains and modifies the values found in *Tables 310.16* through *310.19.*

To find the ampacity of a conductor, simply read down the vertical column indicating the type of conductor, its temperature rating, and the insulation type to the horizontal row indicating its size. Normally, conductors are selected for a known load. The ampacity of the selected conductor must be greater than or equal to the specified load.

Sample Problem 6-2

Problem: Determine the rated ampacity of each of three 12 AWG type THW copper conductors installed in conduit.

Solution: In *Table 310.16*, a value of 25 amps is found in the 12 AWG row of the copper THW column. The allowable ampacity of each of the 12 AWG copper conductors is 25 amps.

(Continued on the following page.)

Sample Problem 6-2 *Continued*

Table 310.16 Allowable Ampacities of Insulated Conductors Rated 0 Through 2000 Volts, 60°C Through 90°C (140°F Through 194°F), Not More than Three Current-Carrying Conductors in Raceway, Cable, or Earth (Directly Buried), Based on Ambient Temperature of 30°C (86°F)

Size AWG or kcmil	Temperature Rating of Conductor (See Table 310.13.)						Size AWG or kcmil
	60°C (140°F)	75°C (167°F)	90°C (194°F)	60°C (140°F)	75°C (167°F)	90°C (194°F)	
	Types TW, UF	Types RHW, THHW, THW, THWN, XHHW, USE, ZW	Types TBS, SA, SIS, FEP, FEPB, MI, RHH, RHW-2, THHN, THHW, THW-2, THWN-2, USE-2, XHH, XHHW, XHHW-2, ZW-2	Types TW, UF	Types RHW, THHW, THW, THWN, XHHW, USE	Types TBS, SA, SIS, THHN, THHW, THW-2, THWN-2, RHH, RHW-2, USE-2, XHH, XHHW, XHHW-2, ZW-2	
	COPPER			ALUMINUM OR COPPER-CLAD ALUMINUM			
18	—	—	14	—	—	—	—
16	—	—	18	—	—	—	—
14*	20	20	25	—	—	—	—
12*	25	25	30	20	20	25	12*
10*	30	35	40	25	30	35	10*
8	40	50	55	30	40	45	8
6	55	65	75	40	50	60	6

Sample Problem 6-3

Problem: Three aluminum type THWN conductors have been specified to carry 65 amps through rigid nonmetallic conduit. What size conductors are needed?

Solution: Looking in the aluminum THWN column of *Table 310.16,* 65 amps is found in the 4 AWG row.

Therefore, the 4 AWG aluminum THWN conductors are the smallest conductors allowed.

Table 310.16 Allowable Ampacities of Insulated Conductors Rated 0 Through 2000 Volts, 60°C Through 90°C (140°F Through 194°F), Not More than Three Current-Carrying Conductors in Raceway, Cable, or Earth (Directly Buried), Based on Ambient Temperature of 30°C (86°F)

Size AWG or kcmil	Temperature Rating of Conductor (See Table 310.13.)						Size AWG or kcmil
	60°C (140°F)	75°C (167°F)	90°C (194°F)	60°C (140°F)	75°C (167°F)	90°C (194°F)	
	Types TW, UF	Types RHW, THHW, THW, THWN, XHHW, USE, ZW	Types TBS, SA, SIS, FEP, FEPB, MI, RHH, RHW-2, THHN, THHW, THW-2, THWN-2, USE-2, XHH, XHHW, XHHW-2, ZW-2	Types TW, UF	Types RHW, THHW, THW, THWN, XHHW, USE	Types TBS, SA, SIS, THHN, THHW, THW-2, THWN-2, RHH, RHW-2, USE-2, XHH, XHHW, XHHW-2, ZW-2	
	COPPER			ALUMINUM OR COPPER-CLAD ALUMINUM			
18	—	—	14	—	—	—	—
16	—	—	18	—	—	—	—
14*	20	20	25	—	—	—	—
12*	25	25	30	20	20	25	12*
10*	30	35	40	25	30	35	10*
8	40	50	55	30	40	45	8
6	55	65	75	40	50	60	6
4	70	85	95	55	65	75	4
3	85	100	110	65	75	85	3
2	90	115	130	75	90	100	2
1	110	130	150	85	100	115	1

Ampacity Correction Factors

The allowable ampacities listed in *Tables 310.16* through *310.19* are only applicable for a specific number of conductors (one conductor in *Tables 310.17* and *310.19*, three conductors or less in the other two tables) and a specific ambient temperature (86°F (30°C) in *Tables 310.16* and *310.17* and 104°F (40°C) for the other two tables). When there are more than three conductors in the raceway or the ambient temperature differs from the table values, corrections must be made.

The lower section of all four tables contains temperature correction factors, **Figure 6-8.** Ambient temperatures are listed on both sides of the factors: Celsius on the left and Fahrenheit on the right. The correction factor is multiplied by the allowable ampacity from the table to determine the allowable ampacity for a different ambient temperature.

The ampacities of different conductor types vary differently with temperature change. Therefore, each type of conductor has unique temperature corrections. As the ambient temperature increases, the ampacity of the conductor is reduced.

Select column matching conductor material and insulation

Table 310.16 Allowable Ampacities of Insulated Conductors Rated 0 Through 2000 Volts, 60°C Through 90°C (140°F Through 194°F), Not More than Three Current-Carrying Conductors in Raceway, Cable, or Earth (Directly Buried), Based on Ambient Temperature of 30°C (86°F)

Size AWG or kcmil	Temperature Rating of Conductor (See Table 310.13.)						Size AWG or kcmil
	60°C (140°F)	75°C (167°F)	90°C (194°F)	60°C (140°F)	75°C (167°F)	90°C (194°F)	
	Types TW, UF	Types RHW, THHW, THW, THWN, XHHW, USE, ZW	Types TBS, SA, SIS, FEP, FEPB, MI, RHH, RHW-2, THHN, THHW, THW-2, THWN-2, USE-2, XHH, XHHW, XHHW-2, ZW-2	Types TW, UF	Types RHW, THHW, THW, THWN, XHHW, USE	Types TBS, SA, SIS, THHN, THHW, THW-2, THWN-2, RHH, RHW-2, USE-2, XHH, XHHW, XHHW-2, ZW-2	
	COPPER			ALUMINUM OR COPPER-CLAD ALUMINUM			
18	—	—	14	—	—	—	—
16	—	—	18	—	—	—	—
1750	545	650	735	455	545	615	1750
2000	560	665	750	470	560	630	2000

CORRECTION FACTORS

Ambient Temp. (°C)	For ambient temperatures other than 30°C (86°F), multiply the allowable ampacities shown above by the appropriate factor shown below.						Ambient Temp. (°F)
21–25	1.08	1.05	1.04	1.08	1.05	1.04	70–77
26–30	1.00	1.00	1.00	1.00	1.00	1.00	78–86
31–35	0.91	0.94	0.96	0.91	0.94	0.96	87–95
36–40	0.82	0.88	0.91	0.82	0.88	0.91	96–104
41–45	0.71	0.82	0.87	0.71	0.82	0.87	105–113
46–50	0.58	0.75	0.82	0.58	0.75	0.82	114–122
51–55	0.41	0.67	0.76	0.41	0.67	0.76	123–131
56–60	—	0.58	0.71	—	0.58	0.71	132–140
61–70	—	0.33	0.58	—	0.33	0.58	141–158
71–80	—	—	0.41	—	—	0.41	159–176

* See 240.4(D)

Select ambient temperature in °C or °F

Figure 6-8. Temperature correction factors are listed at the bottom of the ampacity tables. Multiply the allowable ampacity at the standard temperature by the correction factor.

Conductors installed close together, as in a raceway, are exposed to heat generated by the surrounding cables. This heat is not readily dissipated because the air in a conduit is fairly still. *Section 310.15(B)(2)(a)* includes correction factors for more than three conductors in a raceway, **Figure 6-9.** Multiplying this correction factor by the allowable ampacity for three or less conductors obtains the actual allowable ampacity. In the *Code,* this correction factor is expressed as a percentage, rather than a decimal value.

Sample Problem 6-4

Problem: Find the maximum allowable ampacity for six 1 AWG THW copper conductors in a single conduit installed along the ceiling of a boiler room with an ambient temperature averaging 125°F (52°C).

Solution: First, find the unmodified allowable ampacity from *Table 310.16.*

Table 310.16 Allowable Ampacities of Insulated Conductors Rated 0 Through 2000 Volts, 60°C Through 90°C (140°F Through 194°F), Not More than Three Current-Carrying Conductors in Raceway, Cable, or Earth (Directly Buried), Based on Ambient Temperature of 30°C (86°F)

Size AWG or kcmil	Temperature Rating of Conductor (See Table 310.13.)						Size AWG or kcmil
	60°C (140°F)	75°C (167°F)	90°C (194°F)	60°C (140°F)	75°C (167°F)	90°C (194°F)	
	Types TW, UF	Types RHW, THHW, THW, THWN, XHHW, USE, ZW	Types TBS, SA, SIS, FEP, FEPB, MI, RHH, RHW-2, THHN, THHW, THW-2, THWN-2, USE-2, XHH, XHHW, XHHW-2, ZW-2	Types TW, UF	Types RHW, THHW, THW, THWN, XHHW, USE	Types TBS, SA, SIS, THHN, THHW, THW-2, THWN-2, RHH, RHW-2, USE-2, XHH, XHHW, XHHW-2, ZW-2	
	COPPER			ALUMINUM OR COPPER-CLAD ALUMINUM			
18	—	—	14	—	—	—	—
16	—	—	18	—	—	—	—
14*	20	20	25	—	—	—	—
12*	25	25	30	20	20	25	12*
10*	30	35	40	25	30	35	10*
8	40	50	55	30	40	45	8
6	55	65	75	40	50	60	6
4	70	85	95	55	65	75	4
3	85	100	110	65	75	85	3
2	95	115	130	75	90	100	2
1	110	130	150	85	100	115	1
1/0	125	150	170	100	120	135	1/0
2/0	145	175	195	115	135	150	4/0
1750	545	650	735	455	545	615	1750
2000	560	665	750	470	560	630	2000

CORRECTION FACTORS

Ambient Temp. (°C)	For ambient temperatures other than 30°C (86°F), multiply the allowable ampacities shown above by the appropriate factor shown below.						Ambient Temp. (°F)
21–25	1.08	1.05	1.04	1.08	1.05	1.04	70–77
26–30	1.00	1.00	1.00	1.00	1.00	1.00	78–86
31–35	0.91	0.94	0.96	0.91	0.94	0.96	87–95
36–40	0.82	0.88	0.91	0.82	0.88	0.91	96–104
41–45	0.71	0.82	0.87	0.71	0.82	0.87	105–113
46–50	0.58	0.75	0.82	0.58	0.75	0.82	114–122
51–55	0.41	0.67	0.76	0.41	0.67	0.76	123–131
56–60	—	0.58	0.71	—	0.58	0.71	132–140
61–70	—	0.33	0.58	—	0.33	0.58	141–158
71–80	—	—	0.41	—	—	0.41	159–176

* See 240.4(D).

(Continued on the following page.)

Sample Problem 6-4 *Continued*

For 1 AWG THW copper conductors, the allowable ampacity is 130 amps. Then, determine the correction factors:

Temperature Correction Factor (125°F) = 0.67

Number of Conductors Correction Factor (6) = 0.80

By multiplying the unmodified allowable ampacity by the two correction factors, the modified allowable ampacity is obtained:

130 amps × 0.67 × 0.80 = 70 amps

Ampacity Adjustment Factors

Number of Current-Carrying Conductors	Adjustment Factor
4 to 6	0.80
7 to 9	0.70
10 to 20	0.50
21 to 30	0.45
31 to 40	0.40
41 and above	0.35

Figure 6-9. Allowable ampacity is reduced when a raceway or cable contains more than three current-carrying conductors.

When working with the correction factor for multiple conductors in a conduit, not all conductors are included in the count. Equipment grounding conductors and the neutral conductor in single-phase, three-wire circuits are not counted because they normally carry no current. Therefore, they generate no heat. Neutral conductors that regularly carry current are counted, however.

Voltage Drop

Most electrical equipment is sensitive to variations in line voltage. In order for the equipment to operate effectively, it is important to supply the correct voltage. The *Code* suggests, but does not mandate, a maximum voltage drop of 3% for any feeder or branch circuit and a maximum combined voltage drop (from service entrance to final utilization outlet) of 5%.

Many factors affect voltage drop: temperature, conductor length, conductor size, and conductor material. However, the greatest cause of voltage drop is conductor resistance, which is proportional to the current. The primary remedy for voltage drop is to reduce the resistance of the conductor by increasing its size (diameter). Voltage drop can be reduced in other ways, such as reducing ambient temperature, shortening the length of the conductor, and changing the conductor material, but these methods are less practical.

Calculating Voltage Drop

Voltage drop is calculated using Ohm's law, $E = I \times R$. Conductor resistance is listed in *Table 8* of *Chapter 9* of the *Code,* which was shown in **Figure 6-3**.

NOTE

Table 8 lists dc resistance rather than impedance, which includes resistance and reactance. The voltage drop formulas assume no reactance and a power factor of one.

There are two categories of copper in the table: coated and uncoated. Tin-coated copper is often used for conductors having rubber insulation to prevent chemical reaction between the copper and the rubber. Most copper conductors are uncoated, however.

Single-phase voltage drop

Single-phase voltage drop can be calculated using the following formula:

$$VD = 2 \times L \times I \times \frac{R}{1000}$$

where

VD = Voltage drop

L = One-way length of conductor (in feet)

I = Current in the circuit (amps)

R = Conductor resistance as shown in *Table 8, Chapter 9* of the *Code* (in Ohms/1000′)

In a single-phase system, current travels along the conductor and then returns on the neutral conductor. Therefore, the current travels from the source to the equipment and then back again. The total distance traveled is twice the length of a single conductor, so a factor of two is added to the equation.

The entire product must be divided by 1000 because the resistance is based on a length of 1000′. By dividing the actual length (in feet) by 1000 and multiplying by the resistance per 1000′, the actual resistance of the conductor is determined.

The voltage drop equation can be algebraically rearranged to create an equation to determine the maximum allowable resistance to have less than a 3% voltage drop for a given conductor run.

$$R_{max} = \frac{1000 \times 0.03 \times (line\ voltage)}{2 \times L \times I}$$

Sample Problem 6-5

Problem: Determine the voltage drop on a 20-amp circuit having a length (one-way) of 170′,10 AWG THW stranded copper conductors, and a single-phase, 120-volt power supply.

Solution: From *Table 8* of *Chapter 9,* the resistance for 10 AWG stranded copper conductor (uncoated) is 1.24 ohm/kFT.

Table 8 Conductor Properties

			Conductors							Direct-Current Resistance at 75°C (167°F)					
			Stranding			Overall				Copper				Aluminum	
	Area			Diameter		Diameter		Area		Uncoated		Coated			
Size (AWG or kcmil)	mm²	Circular mils	Quantity	mm	in.	mm	in.	mm²	in.²	ohm/km	ohm/kFT	ohm/km	ohm/kFT	ohm/km	ohm/kFT
18	0.823	1620	1	—	—	1.02	0.040	0.823	0.001	25.5	7.77	26.5	8.08	42.0	12.8
18	0.823	1620	7	0.39	0.015	1.16	0.046	1.06	0.002	26.1	7.95	27.7	8.45	42.8	13.1
16	1.31	2580	1	—	—	1.29	0.051	1.31	0.002	16.0	4.89	16.7	5.08	26.4	8.05
16	1.31	2580	7	0.49	0.019	1.46	0.058	1.68	0.003	16.4	4.99	17.3	5.29	26.9	8.21
14	2.08	4110	1	—	—	1.63	0.064	2.08	0.003	10.1	3.07	10.4	3.19	16.6	5.06
14	2.08	4110	7	0.62	0.024	1.85	0.073	2.68	0.004	10.3	3.14	10.7	3.26	16.9	5.17
12	3.31	6530	1	—	—	2.05	0.081	3.31	0.005	6.34	1.93	6.57	2.01	10.45	3.18
12	3.31	6530	7	0.78	0.030	2.32	0.092	4.25	0.006	6.50	1.98	6.73	2.05	10.69	3.25
10	5.261	10380	1	—	—	2.588	0.102	5.26	0.008	3.984	1.21	4.148	1.26	6.561	2.00
10	5.261	10380	7	0.98	0.038	2.95	0.116	6.76	0.011	4.070	1.24	4.226	1.29	6.679	2.04
8	8.367	16510	1	—	—	3.264	0.128	8.37	0.013	2.506	0.764	2.579	0.786	4.125	1.26
8	8.367	16510	7	1.23	0.049	3.71	0.146	10.76	0.017	2.551	0.778	2.653	0.809	4.204	1.28
6	13.30	26240	7	1.56	0.061	4.67	0.184	17.09	0.027	1.608	0.491	1.671	0.510	2.652	0.808

Using the equation presented in the previous discussion:

$$VD = \frac{2 \times L \times I \times R}{1000}$$

$$= \frac{2 \times 170' \times 20 \text{ amps} \times 1.24 \text{ ohm/kFT}}{1000}$$

$$= 8.4 \text{ V}$$

To determine the voltage drop as a percentage of the total voltage, divide the voltage drop by the total voltage:

$$\frac{8.4 \text{ V}}{120 \text{ V}} = 0.07$$

$$= 7\%$$

The *Code* recommends a maximum 3% voltage drop. It would be advisable to rewire this circuit using a No. 6 AWG copper conductor, which would reduce the voltage drop to an acceptable level:

$$VD = \frac{2 \times 170' \times 20 \text{ amps} \times 0.491 \text{ ohm/kFT}}{1000}$$

$$= 3.3 \text{ V}$$

Checking the voltage drop as a percentage of the total voltage:

$$\frac{3.3 \text{ V}}{120 \text{ V}} = 0.0275$$

$$= 2.75\%$$

This is acceptable.

Three-phase voltage drop

In a three-phase system, voltage drop is calculated with a slightly different equation:

$$VD = \frac{\sqrt{3} \times L \times I \times R}{1000}$$

or

$$VD = \frac{2 \times 0.866 \times L \times I \times R}{1000}$$

where

VD = Voltage drop

L = One-way length of conductor (in feet)

I = Current in the circuit (amps)

R = Conductor resistance as shown in *Table 8, Chapter 9* of the *Code* (in Ohms/1000′)

In three-phase wiring, two current-carrying conductors are connected to the equipment. These two conductors are supplying current that is not in phase, so the resulting current is $\sqrt{3}$ times the nominal current. The current is grounded at the equipment so the factor of two on the length is unnecessary. The second form of the equation shown is identical to the first because $2 \times 0.866 = \sqrt{3}$.

To determine the maximum allowable conductor resistance to maintain a 3% voltage drop in a three-phase system, use the following equation:

$$R_{max} = \frac{1000 \times 0.03 \times (line\ voltage)}{\sqrt{3} \times L \times I}$$

Sample Problem 6-6

Problem: A 200-ampere panel must be fed from a source (switchgear) that is 460′ away. The switchgear supplies 208/120 V, three-phase, four-wire power. What size THW conductor (copper) is needed for the feeder? (Assume 3% maximum voltage drop.)

Solution: First, determine the maximum allowable resistance for the conductor in this three-phase system:

$$R_{max} = \frac{1000 \times 0.03 \times (line\ voltage)}{\sqrt{3} \times L \times I}$$

$$= \frac{1000 \times 0.03 \times 208\text{ V}}{\sqrt{3} \times 460' \times 200\text{ amps}}$$

$$= \frac{6240}{159{,}349}$$

$$= 0.0392\text{ ohms/kFT}$$

Checking *Table 8* of *Chapter 9* for a resistance less than 0.0392 ohms/kFT for an uncoated copper conductor, we see that 350 kcmil conductor (0.0367 ohm/kFT) is the smallest allowable size.

Table 8 Conductor Properties

			Conductors							Direct-Current Resistance at 75°C (167°F)					
			Stranding			Overall				Copper				Aluminum	
	Area			Diameter		Diameter		Area		Uncoated		Coated			
Size (AWG or kcmil)	mm²	Circular mils	Quantity	mm	in.	mm	in.	mm²	in.²	ohm/km	ohm/kFT	ohm/km	ohm/kFT	ohm/km	ohm/kFT
18	0.823	1620	1	—	—	1.02	0.040	0.823	0.001	25.5	7.77	26.5	8.08	42.0	12.8
18	0.823	1620	7	0.39	0.015	1.16	0.046	1.06	0.002	26.1	7.95	27.7	8.45	42.8	13.1
250		—	37	2.09	0.082	14.61	0.575	168	0.260	0.1687	0.0515	0.1753	0.0535	0.2778	0.0847
300		—	37	2.29	0.090	16.00	0.630	201	0.312	0.1409	0.0429	0.1463	0.0446	0.2318	0.0707
350		—	37	2.47	0.097	17.30	0.681	235	0.364	0.1205	0.0367	0.1252	0.0382	0.1984	0.0605
400		—	37	2.64	0.104	18.49	0.728	268	0.416	0.1053	0.0321	0.1084	0.0331	0.1737	0.0529
500		—	37	2.95	0.116	20.65	0.813	336	0.519	0.0845	0.0258	0.0869	0.0265	0.1391	0.0424
600		—	61	2.52	0.099	22.68	0.893	404	0.626	0.0704	0.0214	0.0732	0.0223	0.1159	0.0353

Conduit Capacity

Of the many methods of wiring used in commercial installations, conduit is used most frequently, either as the sole method or in conjunction with other systems. The required conduit size depends on the conductors to be routed. There are three factors involved when sizing conduit:

- The allowable conduit fill (percentage).
- The total cross-sectional area of the conductors.
- The number of conductors within the conduit.

These three factors determine the conduit size. Several tables from *Chapter 9* of the *Code* are needed to size conduit:

- ***Table 1***—Maximum allowable fill in conduit.
- ***Table 4***—Total and allowable areas for different conduit types and sizes.
- ***Table 5***—Conductor diameter and areas.

Given the number and size of conductors, the conduit is sized using a two-step process. First, the total area of the conductors is calculated. Then, the conduit size is selected.

Area of Conductors

Table 5 in *Chapter 9* of the *Code* contains the cross-sectional areas for conductors. The table is organized by conductor type and conductor size. Determining the area of conductors is easy. First, find the type of conductor in the left column, then find the size and read the area in the far right column. When you are determining the total area of all conductors in a conduit, repeat this process for each conductor to be contained in the conduit. Then, add the areas of all of the conductors together to determine the total conductor area.

Sample Problem 6-7

Problem: One 12 AWG THHN, one 10 AWG THHN, and three 8 AWG THHN conductors are to be run through rigid metal conduit. Determine the total area of the conductors.

Solution: Use *Table 5* to find the conductor areas.

Table 5 ***Continued***

Type	Size (AWG or kcmil)	Approximate Diameter		Approximate Area	
		mm	in.	mm²	in.²
Type: RHH*, RHW*, RHW-2*, THHN, THHW, THW, THW-2, TFN, TFFN, THWN, THWN-2, XF, XFF					
THHW, THW, AF, XF, XFF	10	5.232	0.206	21.48	0.0333
RHH*, RHW*, RHW-2*	8	6.756	0.266	35.87	0.0556
	2000	47.80	1.882	1795	2.7818
TFN,	18	2.134	0.084	3.548	0.0055
TFFN	16	2.438	0.096	4.645	0.0072
THHN,	14	2.819	0.111	6.258	0.0097
THWN,	12	3.302	0.130	8.581	0.0133
THWN-2	10	4.166	0.164	13.61	0.0211
	8	5.486	0.216	23.61	0.0366
	6	6.452	0.254	32.71	0.0507
	4	8.230	0.324	53.16	0.0824
	3	8.941	0.352	62.77	0.0973
	2	9.754	0.384	74.71	0.1158
	1	11.33	0.446	100.8	0.1562

12 AWG THHN: $1 \times 0.0133 \text{ in}^2 = 0.0133 \text{ in}^2$

10 AWG THHN: $1 \times 0.0211 \text{ in}^2 = 0.0211 \text{ in}^2$

8 AWG THHN: $3 \times 0.0366 \text{ in}^2 = 0.1098 \text{ in}^2$

Adding the areas together…

$$0.0133 + 0.0211 + 0.1098 = 0.1442 \text{ in}^2$$

The total area of the conductors is 0.1442 in^2.

Minimum Conduit Size

Table 1 in *Chapter 9* of the *Code* shows the allowable percentage of conduit area that can be filled with conductors. The area that can be filled is based on the number of conductors. The different percentages are a result of the geometry involved with circular conductors within a circular conduit. See **Figure 6-10.**

Once the allowable percentage of fill is known, *Table 4* in *Chapter 9* of the *Code* is used to find the minimum allowable size for the type of raceway being used. Raceways larger than the minimum required size can be used, but this will normally be uneconomical.

Allowable Conduit Fill

Number of Conductors	1	2	Over 2
Allowable Fill	53%	31%	40%
Diagram			

Figure 6-10. The allowable percentage of filled conduit area varies with the number of conductors.

Sample Problem 6-8

Problem: One 12 AWG THHN, one 10 AWG THHN, and three 8 AWG THHN conductors are to be run through a rigid metal conduit. Determine the smallest conduit size that can be used.

Solution: The total conductor area is 0.1442 in² (see Sample Problem 6-7). *Table 1* shows that the conductor area must be 40% (or less) of the conduit cross section.

Under Rigid Metal Conduit in *Table 4,* when filled with more than two conductors, 3/4″ conduit can carry conductors with a total area of 0.220 in². This is the smallest size that can carry at least 0.1442 in², so 3/4″ conduit is the best size.

Table 1 Percent of Cross Section of Conduit and Tubing for Conductors

Number of Conductors	All Conductor Types
1	53
2	31
Over 2	40

Table 4 ***Continued***

Article 344 — Rigid Metal Conduit (RMC)

Metric Designator	Trade Size	Nominal Internal Diameter mm	Nominal Internal Diameter in.	Total Area 100% mm²	Total Area 100% in.²	2 Wires 31% mm²	2 Wires 31% in.²	Over 2 Wires 40% mm²	Over 2 Wires 40% in.²	1 Wire 53% mm²	1 Wire 53% in.²	60% mm²	60% in.²
12	3/8	—	—	—	—	—	—	—	—	—	—	—	—
16	1/2	16.1	0.632	204	0.314	63	0.097	81	0.125	108	0.166	122	0.188
21	3/4	21.2	0.836	353	0.549	109	0.170	141	0.220	187	0.291	212	0.329
27	1	27.0	1.063	573	0.887	177	0.275	229	0.355	303	0.470	344	0.532
35	1¼	35.4	1.394	984	1.526	305	0.473	394	0.610	522	0.809	591	0.916
41	1½	41.2	1.624	1333	2.071	413	0.642	533	0.829	707	1.098	800	1.243
53	2	52.9	2.083	2198	3.408	681	1.056	879	1.363	1165	1.806	1319	2.045
63	2½	63.2	2.489	3137	4.866	972	1.508	1255	1.946	1663	2.579	1882	2.919
78	3	78.5	3.090	4840	7.499	1500	2.325	1936	3.000	2565	3.974	2904	4.499
91	3½	90.7	3.570	6461	10.010	2003	3.103	2584	4.004	3424	5.305	3877	6.006
103	4	102.9	4.050	8316	12.882	2578	3.994	3326	5.153	4408	6.828	4990	7.729
129	5	128.9	5.073	13050	20.212	4045	6.266	5220	8.085	6916	10.713	7830	12.127
155	6	154.8	6.093	18821	29.158	5834	9.039	7528	11.663	9975	15.454	11292	17.495

Review Questions

Answer the following questions. Do not write in this book.

1. List three permitted uses of bare conductors.
2. What is the advantage of stranded conductors over solid conductors?
3. List three conditions that must be met when running conductors in parallel.
4. Name three markings that are required to appear on conductors or cables in addition to the maximum rated voltage.
5. What is the most influential factor contributing to voltage drop?
6. Describe two ways of decreasing the voltage drop in a circuit.
7. Explain what an insulated cable with "XHHW" marking indicates.
8. What does the designation AWG stand for?
9. What is the maximum motor size that can have a cord and plug serve as the only means of disconnect?
10. How does temperature affect conductors?
11. What is the minimum allowable conductor size for elevator control and signaling circuits?
12. A conduit contains two conductors. What percentage of the total conduit area can the conductors occupy?

USING THE NEC

Refer to the National Electrical Code to answer the following questions. Do not write in this book.

1. Using *Table 8, Chapter 9* of the *Code*, find the area of a solid 8 AWG copper conductor.
2. What size rigid metal conduit is required for a feeder circuit consisting of four 8 AWG and two 10 AWG THW conductors?
3. A piece of equipment is supplied with 20-amp, 120-volt, single-phase power by solid, uncoated 12 AWG copper conductors. The conductors travel a length of 110′. Calculate the voltage drop. Determine voltage drop as a percent of total voltage.
4. For what application is conductor type SIS used?
5. In which wire sizes is conductor type UF available?
6. What is the nominal insulation thickness on a 6 AWG conductor in a portable power cable (type of flexible cable)?
7. What size of conductors are used in vacuum cleaner cords?
8. Two copper type THHW conductors are being designed to supply 30 amps through EMT with an ambient temperature of 86°F. What size conductors should be used?
9. A run of rigid metal conduit in an 86°F area will contain five aluminum type XHHW conductors. If the conductors supply 22 amps, what size conductors should be used?
10. What is the allowable ampacity of a 2/0 AWG copper-clad aluminum type RH conductor in an area with an ambient temperature of 135°F?
11. Twelve 14 AWG copper type THW conductors are located within a raceway. If all the conductors carry current, what is the allowable ampacity of each conductor?
12. What is the allowable ampacity of five 4 AWG copper type FEP conductors located in a raceway surrounded by 210°F air?
13. The one-way length of a single-phase, 15-amp circuit is 220′. Determine the voltage drop if the conductors are stranded, uncoated 12 AWG aluminum. What is the percentage voltage drop if it is a 120-volt circuit?
14. What size PVC conduit (Schedule 40) is needed to contain three 1/0 AWG RHW conductors?

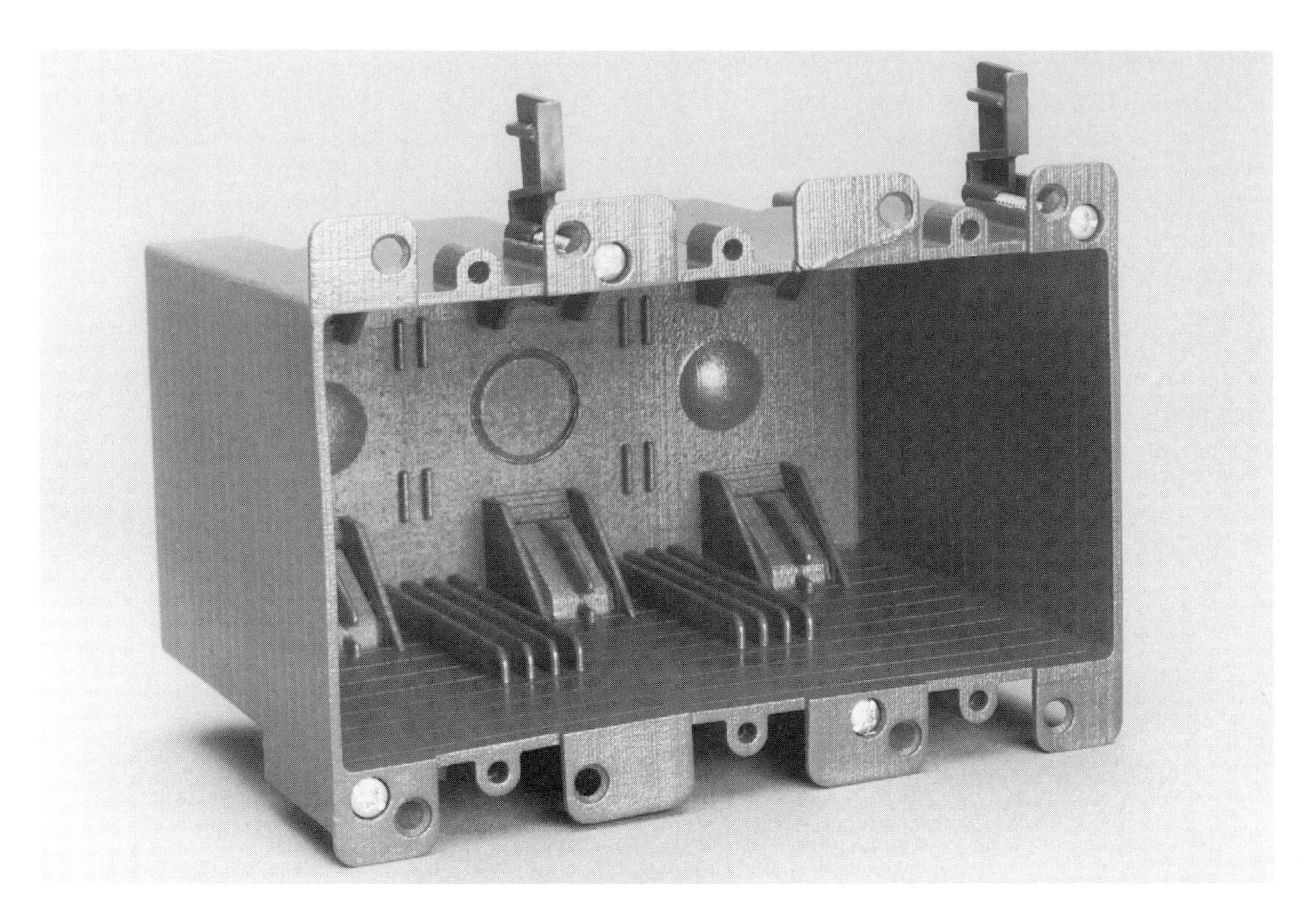

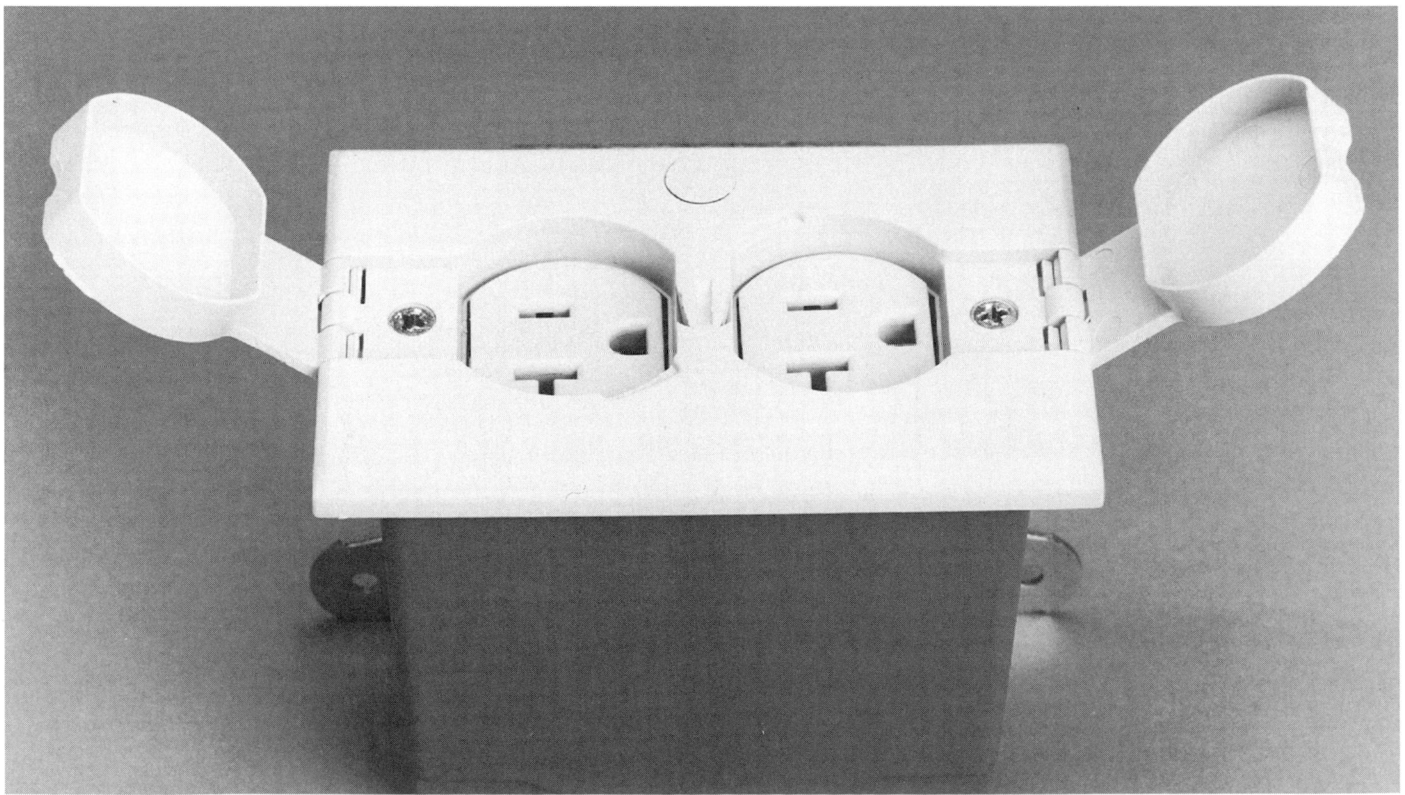

PVC boxes are used for remodeling and new installations. Top—This three-gang box is extremely useful when a new switch or receptacle is being added to an existing box location. Bottom—This box is designed for floor receptacles. Covers are provided to keep dust and debris out of the receptacle. (Lamson Home Products)

Chapter 7

Boxes and Conduit Bodies

Technical Terms

Angular pulls
Box
Conductor fill
Conductor volume
Conduit body
Enclosure
Fitting
Ganged boxes
Handy box
Straight pulls

Objectives

After completing this chapter, you will be able to:

- ❍ Identify different types of boxes.
- ❍ Select boxes for various applications.
- ❍ Explain how boxes are grounded.
- ❍ Mount and support boxes in accordance with the *Code.*
- ❍ Identify various types of conduit bodies.
- ❍ Perform box fill calculations using the *Code.*
- ❍ Compute box sizes for straight and angular pulls.

The *Code* requires all conductor joints, connections, taps, and splices to be housed inside approved enclosures. Boxes provide protection from both fire and shock. *Article 314—Outlet, Device, Pull and Junction Boxes; Conduit Bodies; Fittings; and Manholes* contains the *Code* rules in this area.

CAUTION

Electrical failures that cause insulation to overheat or arcing between conductors are more likely to occur at connections. Enclosing the connections within a box helps to prevent electrical failures from causing fires.

Enclosures

Conductor enclosures support conductors and separate them from other building materials. This prevents flexible conductors from moving and provides some protection from electrical fires. There are several types of enclosures:

- **Box**—A box is an enclosure designed to house electrical connections, junctions, and taps. Boxes are also used to mount switches, receptacles, and fixtures. See **Figure 7-1.**
- **Conduit body**—A conduit body has a removable cover that provides access to the conduit system. Conduit bodies are located at junctions and endpoints. They are used primarily for pulling wires, but are also useful for making splices, taps, and installing devices. See **Figure 7-2.**
- **Enclosure**—Electrical equipment must be enclosed to prevent personnel from contacting energized parts. An enclosure is a device that accomplishes this purpose—either a housing around equipment or a fence protecting large equipment. See **Figure 7-3.**

Figure 7-1. A box encloses and protects connections, junctions, taps, and devices.

NEC NOTE 300.14

At least 6 in. (152 mm) of free conductor shall be left at each outlet, junction, and switch point for splices or the connection of fixtures or devices.

Boxes

Although there are some exceptions, a box is required wherever conductors are spliced and connected.

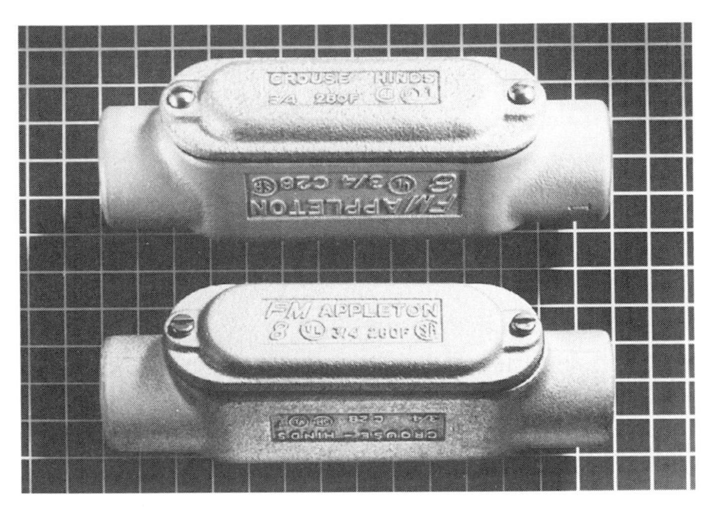

Figure 7-2. Conduit bodies provide access to the wires inside the conduit. Note removable plates on these type C fittings. (Appleton Electric Company)

Figure 7-3. The fence around this transformer is a type of enclosure. It separates personnel from the energized parts. (Jack Klasey)

Boxes serve as pull points and the location of conductor terminations, **Figure 7-4.**

There are numerous types of boxes. They are classified on the basis of shape, size, use, and material. **Figure 7-5** shows four of the most popular shapes.

Figure 7-4. Boxes are required at outlet, switch, and junction points for most wiring methods.

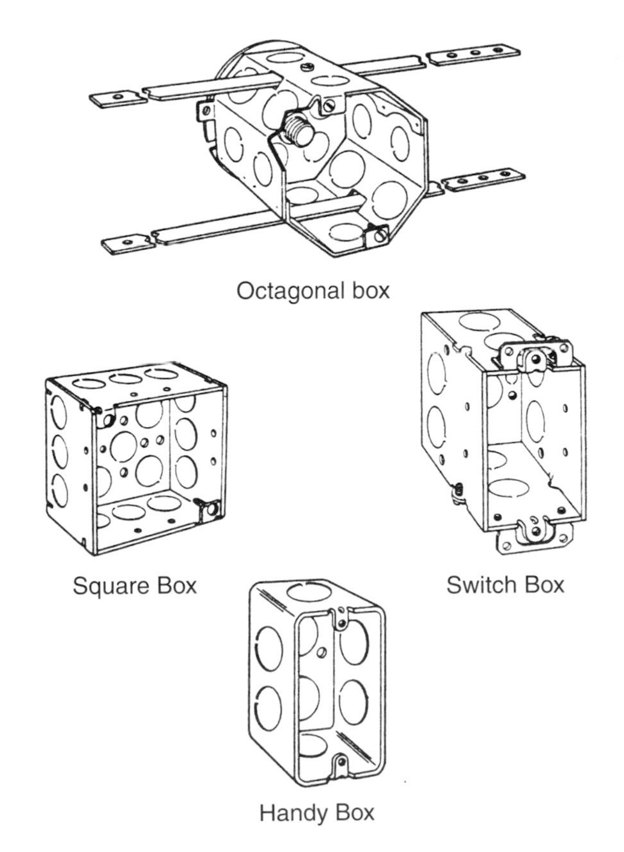

Figure 7-5. Octagon boxes are primarily used for junction boxes and fixture outlets. Square and rectangular boxes are commonly referred to as switch or device boxes. The handy box is designed for surface mounting. (RACO, Inc.)

Round boxes

Round boxes do not have a flat bearing surface on their sides. For this reason they are not permitted if the wiring method requires locknuts or bushings to attach to the box sides. Cable clamps within the box can secure BX, NM, or NMC cable. Wiring methods requiring locknuts and bushings, such as conduit, can connect to the bottom of the box, where a knockout is provided. Round boxes are often used for mounting lighting fixtures.

NEC NOTE 314.2

Round boxes shall not be used where conduits or connectors requiring the use of locknuts or bushings are to be connected to the side of the box.

Octagonal boxes

Like round boxes, octagonal boxes are primarily used for ceiling outlets. They are also well-suited for wall mounting, for hanging fixtures, and for junction and pull boxes.

Square and rectangular boxes

Generally referred to as device, receptacle, or switch boxes, square and rectangular boxes are normally mounted in walls. The rectangular boxes can be ***ganged,*** that is, joined side-to-side to create more space for additional devices and conductors. These boxes, as well as the round and octagonal boxes, can be installed in wood frames, concrete, tile walls, stucco, and brick.

Utility boxes

Utility boxes (also called ***handy boxes***) are rectangular with rounded, smooth corners. These boxes are normally surface-mounted. The rounded corners make them safer than square or rectangular boxes.

Metal boxes

All metal boxes must be grounded. The equipment grounding conductor of the branch circuit is simply connected to the metal box by a clip or screw, **Figure 7-6.** Metal boxes are approved for most wiring methods.

NEC NOTE 314.4

All metal boxes shall be grounded in accordance with the provisions of Article 250.

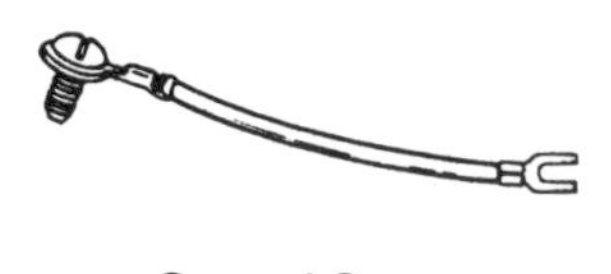

Figure 7-6. Grounding accessories for metal boxes. (RACO, Inc.)

Nonmetallic boxes

Nonmetallic boxes are only permitted with certain wiring methods:

- Open wiring on insulators.
- Concealed knob-and-tube wiring.
- Nonmetallic raceways.
- NM, NMC, and other nonmetallic jacketed cable.

Nonmetallic boxes can be used with metal conduit or metal-jacketed cable if there is an internal bonding means between all entries to the box.

Box Mounting

Boxes may be installed in walls, ceilings, floors, partitions, cabinets, and other surfaces. The mounting surface can be of almost any material or composition. Boxes should not be concealed or inaccessible.

Boxes may be mounted directly on surfaces as long as there is adequate support. If the wall surface is concrete or wood, this is not a problem. If the surface material is inadequate, then the box must be mounted within the surface, supported by a structural member, or affixed with clamps.

There are a number of ways to mount a box within a wall, ceiling, or floor. Adapters and accessories are available to aid in mounting boxes. Expansion anchors, bar hangers, and mounting bracket are often used. **Figure 7-7** illustrates some of the more common mounting devices.

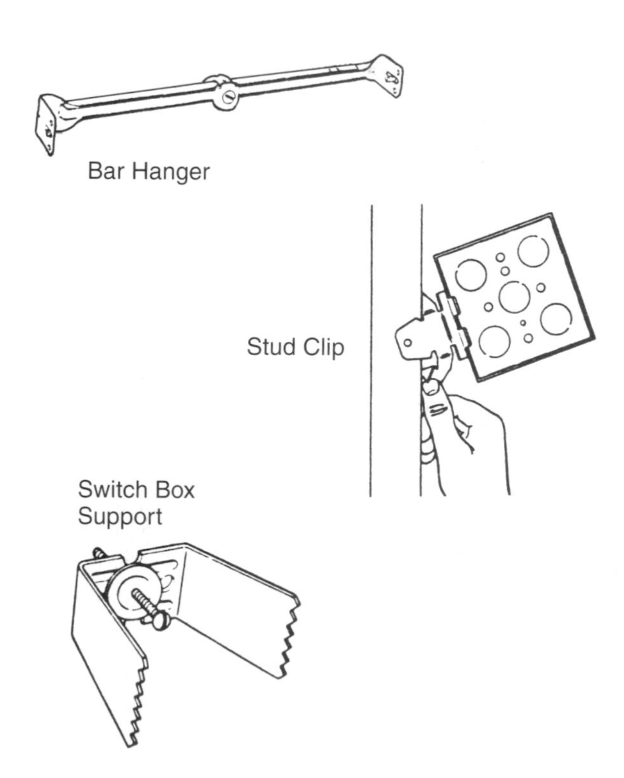

Figure 7-7. A wide variety of accessories are available to mount boxes. (RACO, Inc.)

Threaded boxes with a volume less than 100 in^3 may be supported by conduit alone. Two conduits threaded wrench-tight into the enclosure are considered adequate support. **Figure 7-8** illustrates the maximum spacing between the box and the nearest conduit support.

Boxes and enclosures can be mounted using nails or screws. If nails pass through the interior of the enclosure, they must be located within 1/4″ of the end or back of the box.

If a box is located in combustible material, it must be installed flush or project from the surface. If installed in noncombustible material, the box may be set back 1/4″ from the surface. See **Figure 7-9.**

Conduit Bodies

Conduit bodies allow access into the system at junction and terminal points. **Figure 7-10** shows many of the different types available. Conduit bodies are used in several situations:

- Making 90° turns in a conduit run.
- Pulling cables into a wiring system.
- Splicing conductors.
- Making taps from feeders or branch circuit conductors.

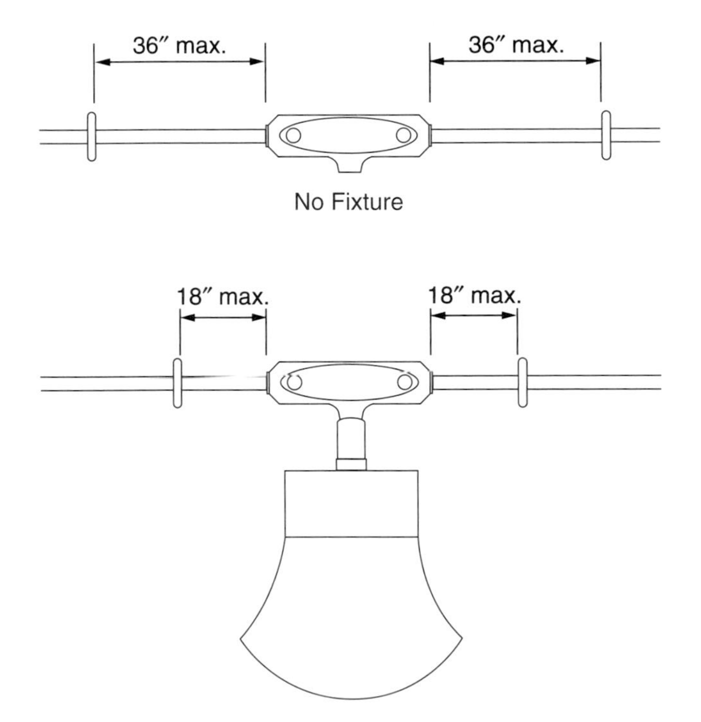

Figure 7-8. Threaded boxes can be supported by conduit alone. The maximum distance between the box and the nearest conduit support varies, depending on whether a fixture is attached to the box. When a fixture is not attached to an enclosure, the maximum spacing is 36″. When a fixture is attached to the enclosure, the maximum spacing is 18″.

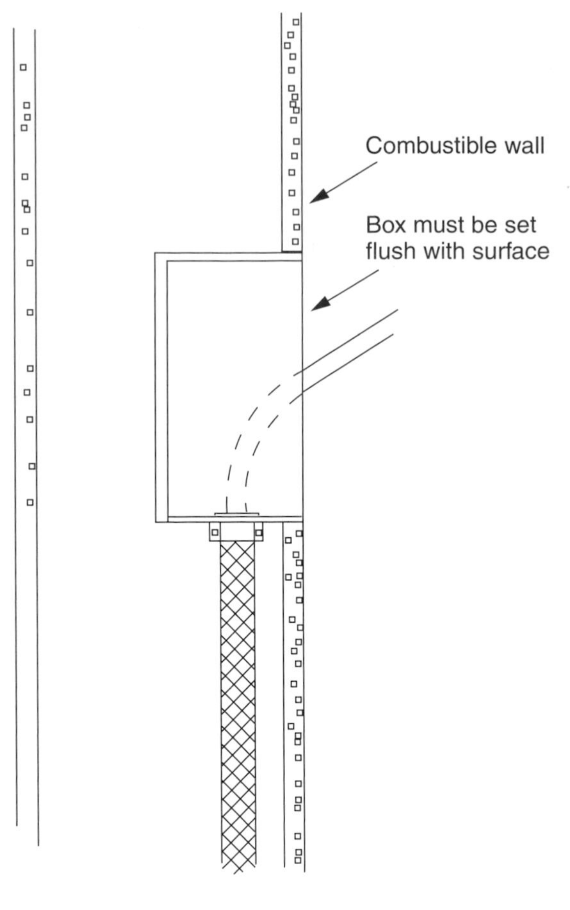

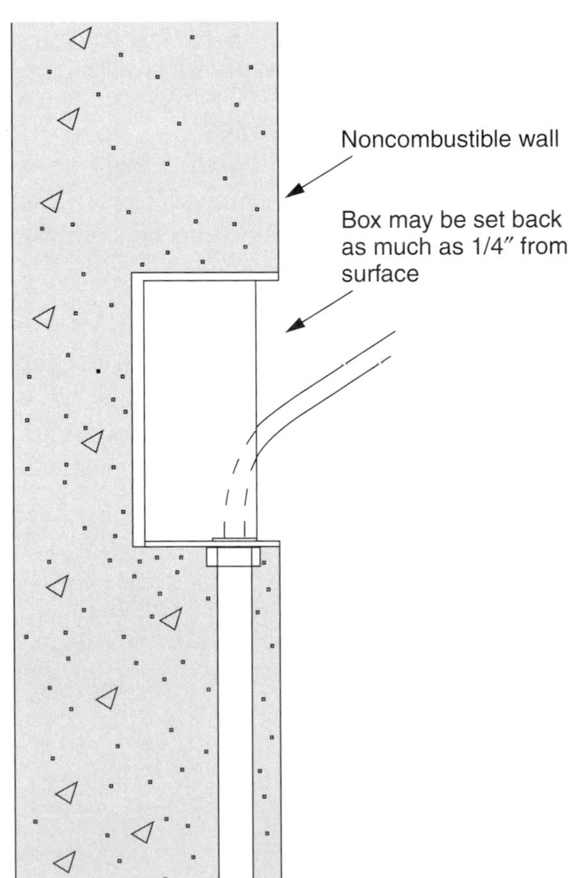

Figure 7-9. Boxes must be flush with combustible walls and can be set back 1/4″ in noncombustible walls.

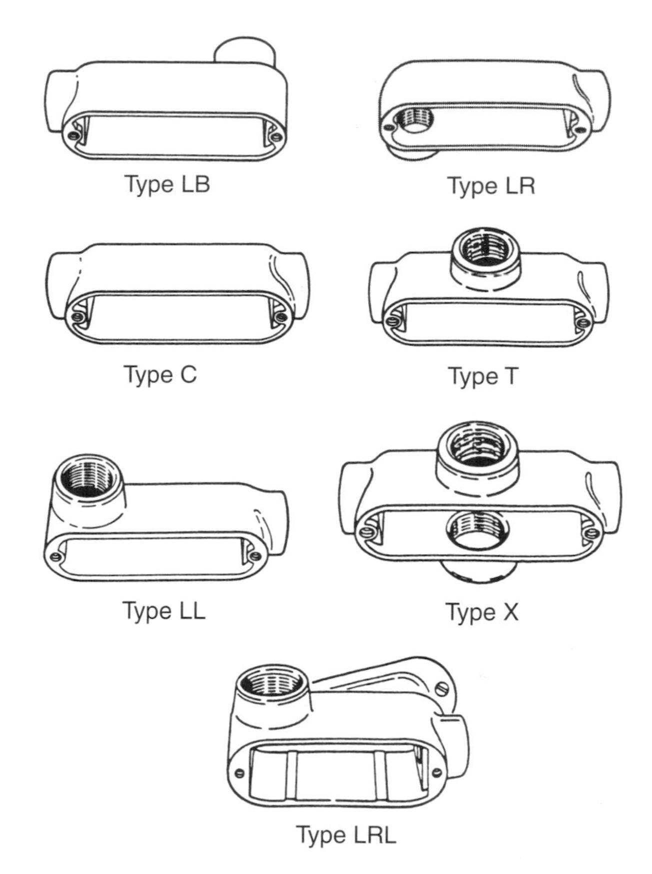

Figure 7-10. Many configurations of conduit bodies are available. (RACO, Inc.)

The *Code* points out that capped elbows, service-entrance elbows, and similar fittings *cannot* contain splices, devices, or taps.

Conduit bodies are subject to the same *Code* rules as boxes with regards to filling, splicing, tapping, and device installation. Very few conduit bodies have adequate volume to accommodate splices, taps, or device installation unless they are oversized by two to three times the largest conduit entering the conduit body. Conduit bodies that are not marked with their capacity cannot contain splices, taps, or devices.

Boxes and conduit bodies must be large enough to provide adequate free space for the conductors entering and exiting. If the enclosure is too small, the ambient temperature around the conductors becomes too hot. This weakens insulation and increases resistance. Carefully read *Section 314.16(C)(2)* to determine if a splice or tap is allowed in a conduit body. *Section 314.5* specifically forbids splices, taps, or devices in short radius conduit bodies (capped elbows and service-entrance elbows).

Enclosure Sizing

Outlet boxes, device boxes, junction boxes, and conduit bodies must be of adequate size to allow for free space for all conductors within the box. Conductors should never be tightly crammed or bunched into a box of inadequate size and volume. Cramming wires into the box can damage insulation and increase the possibility of a ground fault or short circuit.

Pull and Junction Boxes

When 4 AWG or larger conductors are contained in a raceway, the size of the pull and junction boxes is determined by the size of the raceways. The minimum pull box size is determined differently for the two types of pulls—straight and angular.

For ***straight pulls,*** the length of the box must be at least eight times the diameter of the largest raceway entering the box, **Figure 7-11.** For ***angular pulls,*** the box size calculations are a bit more involved. The distance between the raceway entries and the opposite wall of the box is equal to six times the largest raceway diameter on the side *plus* the sum of the diameters of any other conduits on the wall. Also, the distance between raceways having the same conductors must be greater than six times the diameter of the larger raceway. This is illustrated in **Figure 7-12.**

In either case, the depth of the box must be sufficient to allow for the largest conduit, its locknut, bushing, and spacing between adjacent conduits.

Conduit Bodies

For a straight pull through a conduit body, the length of the conduit body must be at least eight times the diameter of the raceway, **Figure 7-13.** For angular pulls through LB, LL, and LR types, the long dimension of the body and the distance between raceway openings must be greater than six times the raceway diameter. However, the

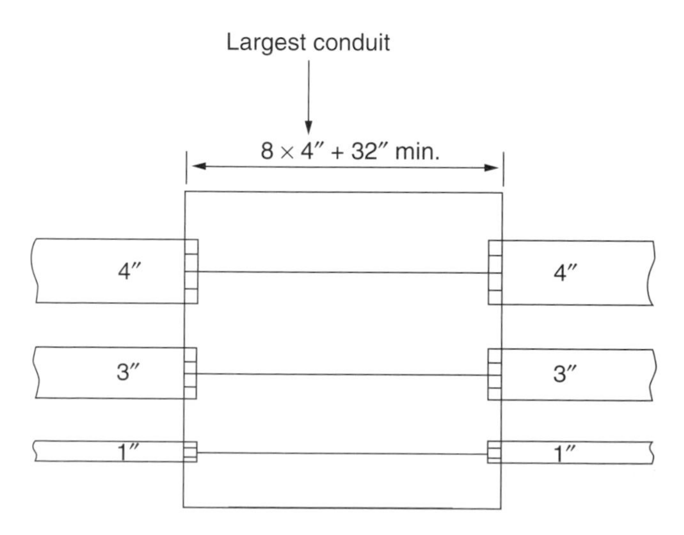

Figure 7-11. The length of the box housing a straight pull must be at least eight times the diameter of the largest conduit entering the box.

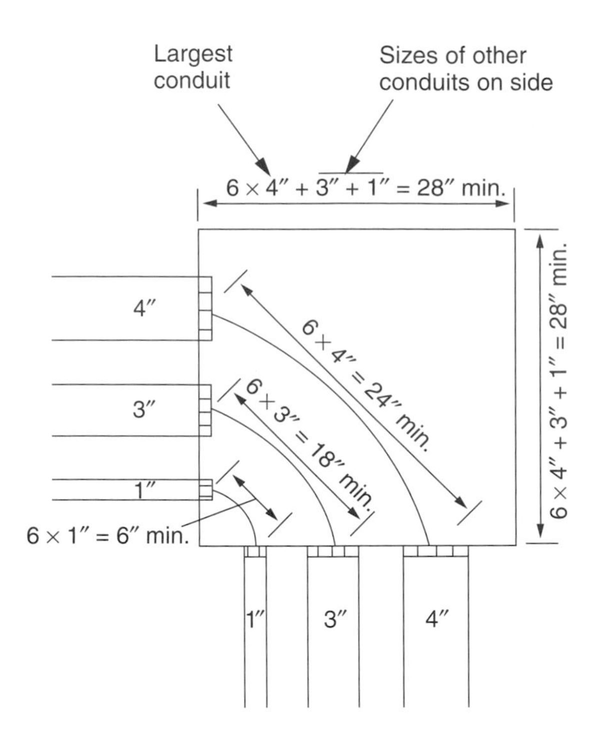

Figure 7-12. Minimum dimensions for a box with angular pulls.

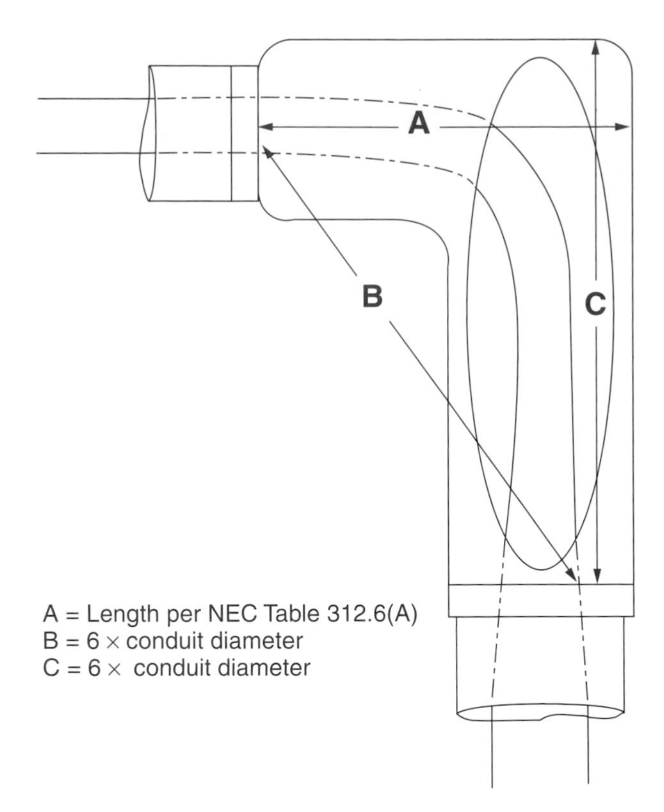

Figure 7-14. Dimension A is found in *Table 312.6(A)* of the *Code.* Dimensions B and C must be at least six times the conduit size.

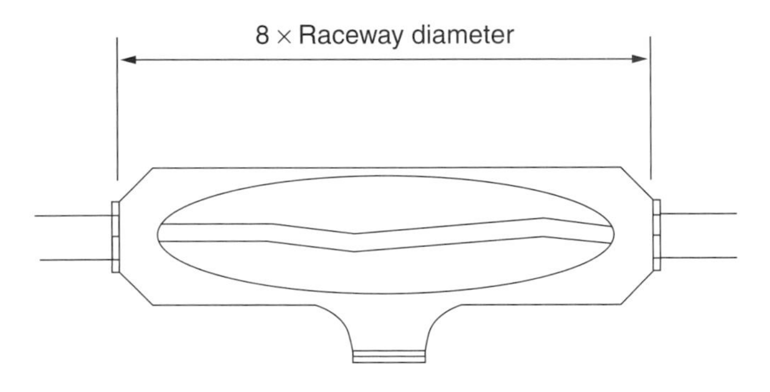

Figure 7-13. Straight pull through a conduit body.

distance between the raceway entry and opposite wall along the short dimension can be less. Minimum distances are shown in *Table 312.6(A)* of the *Code.* See **Figure 7-14.**

Outlet and Device Boxes

Section 314.16 of the *Code* contains requirements needed to adequately size outlet and device boxes containing conductors smaller than 4 AWG.

Figure 7-15 is a reproduction of *Table 314.16(A)* of the *Code.* This table lists the maximum number of conductors permitted in standard box sizes. When conductors of various sizes are contained in a box, the volume of the conductors must be added. **Figure 7-16,** a reproduction of *Table 314.16(B)* of the *Code,* shows the volume assigned to each conductor size. This table is also used for nonmetallic boxes.

When determining the number of conductors contained in the box, other components are also counted as conductors. The following guidelines are illustrated in **Figure 7-17.**

- Each conductor entering the box and being terminated or spliced is counted once. If two conductors enter the box and are spliced together, *each* conductor is counted.
- Conductors that pass through the box without being terminated or spliced are counted once.
- A conductor that does not leave the box, such as a grounding pigtail, is not counted.
- Internal cable clamps are counted as one conductor, regardless of how many there are. They are assumed to have the volume of the largest conductor in the box. If a clamp is located outside of the box, no allowance is needed.
- Fixture studs and hickeys are counted as one conductor, regardless of how many there are. Just as cable clamps, the allowance for the largest conductor is used.
- Each yoke or strap containing a device is counted as *two* conductors. The allowance is based on the largest conductor attached to the device.
- Equipment grounding conductors entering the box are counted *once,* regardless of how many there are. The allowance is based on the largest

Table 314.16(A) Metal Boxes

Box Trade Size			Minimum Volume		Maximum Number of Conductors*						
mm	in.		cm.³	in.³	18	16	14	12	10	8	6
100 × 32	(4 × 1¼)	round/octagonal	205	12.5	8	7	6	5	5	5	2
100 × 38	(4 × 1½)	round/octagonal	254	15.5	10	8	7	6	6	5	3
100 × 54	(4 × 2⅛)	round/octagonal	353	21.5	14	12	10	9	8	7	4
100 × 32	(4 × 1¼)	square	295	18.0	12	10	9	8	7	6	3
100 × 38	(4 × 1½)	square	344	21.0	14	12	10	9	8	7	4
100 × 54	(4 × 2⅛)	square	497	30.3	20	17	15	13	12	10	6
120 × 32	(4¹¹⁄₁₆ × 1¼)	square	418	25.5	17	14	12	11	10	8	5
120 × 38	(4¹¹⁄₁₆ × 1½)	square	484	29.5	19	16	14	13	11	9	5
120 × 54	(4¹¹⁄₁₆ × 2⅛)	square	689	42.0	28	24	21	18	16	14	8
75 × 50 × 38	(3 × 2 × 1½)	device	123	7.5	5	4	3	3	3	2	1
75 × 50 × 50	(3 × 2 × 2)	device	164	10.0	6	5	5	4	4	3	2
75 × 50 × 57	(3 × 2 × 2¼)	device	172	10.5	7	6	5	4	4	3	2
75 × 50 × 65	(3 × 2 × 2½)	device	205	12.5	8	7	6	5	5	4	2
75 × 50 × 70	(3 × 2 × 2¾)	device	230	14.0	9	8	7	6	5	4	2
75 × 50 × 90	(3 × 2 × 3½)	device	295	18.0	12	10	9	8	7	6	3
100 × 54 × 38	(4 × 2⅛ × 1½)	device	169	10.3	6	5	5	4	4	3	2
100 × 54 × 48	(4 × 2⅛ × 1⅞)	device	213	13.0	8	7	6	5	5	4	2
100 × 54 × 54	(4 × 2⅛ × 2⅛)	device	238	14.5	9	8	7	6	5	4	2
95 × 50 × 65	(3¾ × 2 × 2½)	masonry box/gang	230	14.0	9	8	7	6	5	4	2
95 × 50 × 90	(3¾ × 2 × 3½)	masonry box/gang	344	21.0	14	12	10	9	8	7	2
min. 44.5 depth	FS — single cover/gang (1¾)		221	13.5	9	7	6	6	5	4	2
min. 60.3 depth	FD — single cover/gang (2⅜)		295	18.0	12	10	9	8	7	6	3
min. 44.5 depth	FS — multiple cover/gang (1¾)		295	18.0	12	10	9	8	7	6	3
min. 60.3 depth	FD — multiple cover/gang (2⅜)		395	24.0	16	13	12	10	9	8	4

*Where no volume allowances are required by 314.16(B)(2) through 314.16(B)(5).

Figure 7-15. This table can be used to quickly select a standard box size.

Table 314.16(B) Volume Allowance Required per Conductor

Size of Conductor (AWG)	Free Space Within Box for Each Conductor	
	cm³	in.³
18	24.6	1.50
16	28.7	1.75
14	32.8	2.00
12	36.9	2.25
10	41.0	2.50
8	49.2	3.00
6	81.9	5.00

Figure 7-16. When different sizes of conductors are contained in a box, their volumes must be added. This table provides the space required for various sizes of conductors.

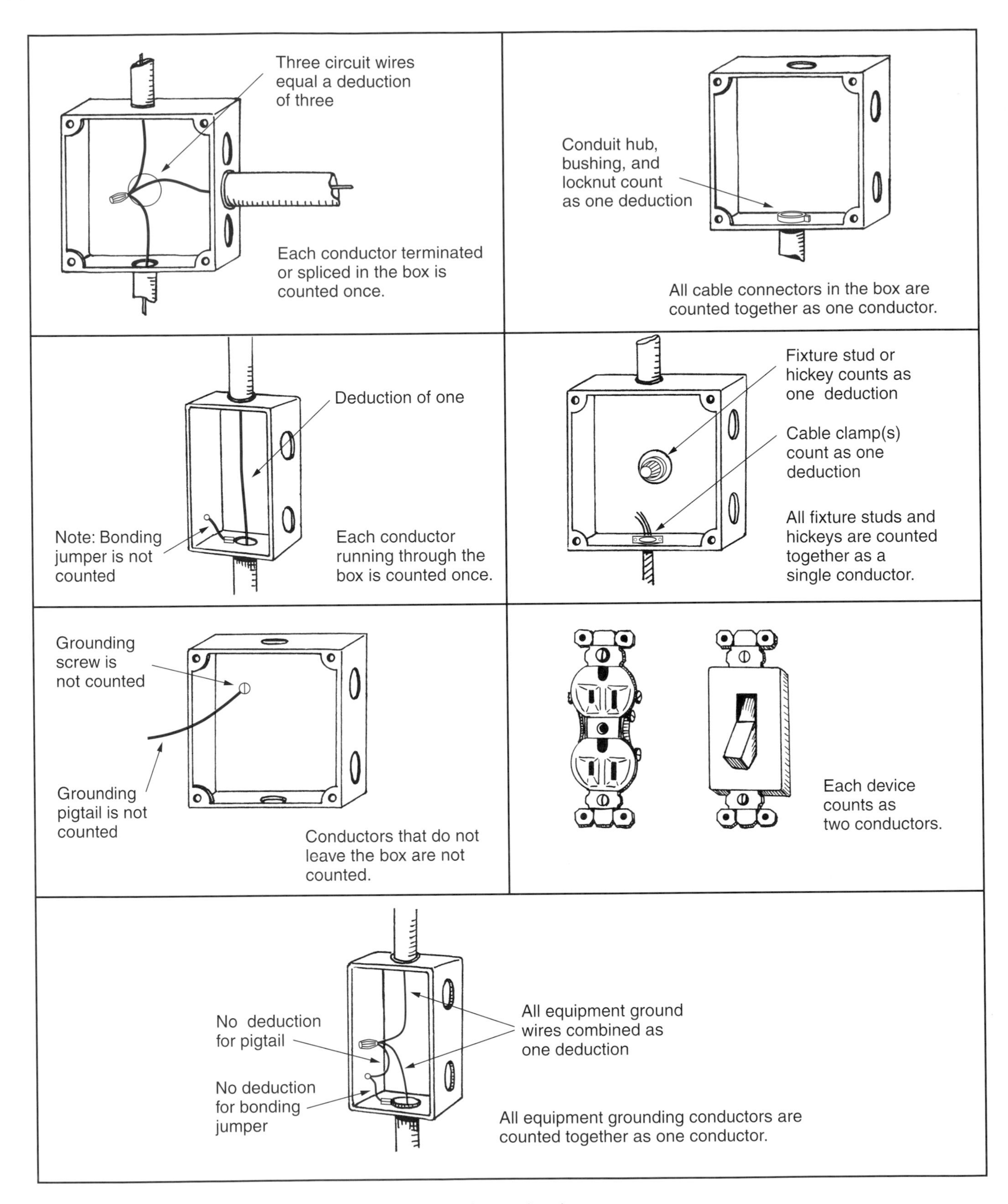

Figure 7-17. Guidelines for determining the number of conductors in a box.

equipment grounding conductor. If an additional set of grounding conductors is present, a second allowance must be included.

Boxes containing conductors 4 AWG and larger must conform to *Section 314.28(A)* and *Section 300.4(F)*, which contain the criteria for pull boxes.

Sample Problem 7-1

Problem: A nonmetallic-sheathed cable (two 12 AWG with one 12 AWG ground) is connected to a receptacle. Another identical cable extends to another outlet. What size box is needed?

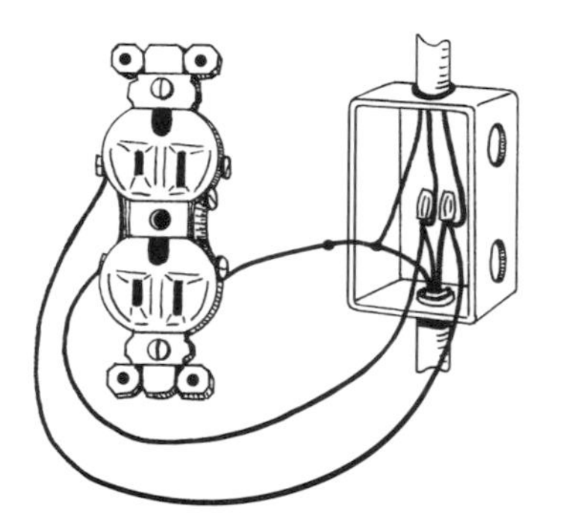

Solution: The conductor count would be as follows:

Circuit conductors:	4
Grounding conductors:	1
All internal clamps:	1
Duplex receptacle:	2
Total:	8

Under the 12 AWG column of the table we find that either a 4 × 1 1/4 square box or 3 × 2 × 3 1/2 device box would be acceptable.

Sample Problem 7-2

Problem: A switch and receptacle are to be installed in a box. The receptacle has 12 AWG conductors and the switch has 14 AWG conductors. Grounds are 12 AWG, and cable clamps are present. What size box is needed?

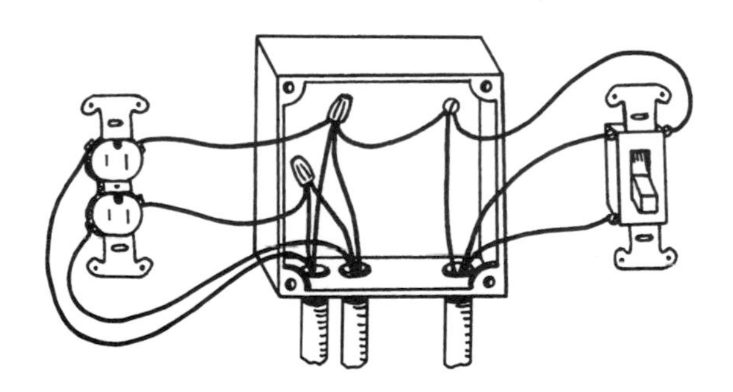

Solution: The conductor volume would be as follows:

Four 12 AWG conductors: $4 \times 2.25\ \text{in}^3 = 9.0\ \text{in}^3$

Two 14 AWG conductors: $2 \times 2.0\ \text{in}^3 = 4.0\ \text{in}^3$

All grounding conductors: $1 \times 2.25\ \text{in}^3 = 2.25\ \text{in}^3$

One receptacle strap: $2 \times 2.25\ \text{in}^3 = 4.50\ \text{in}^3$

One switch strap: $2 \times 2.0\ \text{in}^3 = 4.0\ \text{in}^3$

All clamps: $1 \times 2.25\ \text{in}^3 = 2.25\ \text{in}^3$

Total: $26.0\ \text{in}^3$

Therefore, a box having a minimum capacity of 26.0 in^3 is required.

Referring to *Table 314.16(A)*, a 4 11/16 × 1 1/2 square metal box will be suitable. A nonmetallic box with a minimum of 26.0 in^3 would be equally acceptable.

Table 314.16(A) Metal Boxes

Box Trade Size			Minimum Volume		Maximum Number of Conductors*						
mm	in.		cm.^{3}	in.^{3}	18	16	14	12	10	8	6
100 × 32	(4 × 1¼)	round/octagonal	205	12.5	8	7	6	5	5	5	2
100 × 38	(4 × 1½)	round/octagonal	254	15.5	10	8	7	6	6	5	3
100 × 54	(4 × 2⅛)	round/octagonal	353	21.5	14	12	10	9	8	7	4
100 × 32	(4 × 1¼)	square	295	18.0	12	10	9	8	7	6	3
100 × 38	(4 × 1½)	square	344	21.0	14	12	10	9	8	7	4
100 × 54	(4 × 2⅛)	square	497	30.3	20	17	15	13	12	10	6
120 × 32	(4 11/16 × 1¼)	square	418	25.5	17	14	12	11	10	8	5
120 × 38	(4 11/16 × 1½)	square	484	29.5	19	16	14	13	11	9	5
120 × 54	(4 11/16 × 2⅛)	square	689	42.0	28	24	21	18	16	14	8

Review Questions

Answer the following questions. Do not write in this book.

1. How do you determine the minimum box size for straight pulls?
2. When sizing boxes, describe how spliced conductors are considered when determining the number and size of conductors within the box.
3. What is the primary purpose for the use of conduit bodies?
4. What is the primary use of round boxes?
5. If a box is mounted against a wood surface, is additional support required?
6. What type of box does not need any support in addition to the conduit attached to it?
7. When sizing boxes, describe how equipment grounding conductors are considered when determining the number and size of conductors within the box.
8. What are ganged boxes?
9. How do mounting requirements vary for boxes installed in combustible walls and boxes installed in noncombustible walls?
10. List four wiring methods that can be used with nonmetallic boxes.

USING THE NEC

Refer to the National Electrical Code to answer the following questions. Do not write in this book.

1. What is the capacity (in cubic inches) of a 3 × 2 × 2 metal device box?
2. Drywall surfaces that are damaged around a box must be repaired. What is the maximum allowable gap between the box and the drywall?
3. What is the maximum number of 10 AWG conductors allowed in a 4 × 1 1/2 round metal box?
4. What is the minimum wall thickness for a sheet steel box less than 100 in^3 in size?
5. If the calculated volume of conductors and fittings is 19.2 in^3, what is the smallest square metal box that can be used?
6. Boxes installed in wet locations must be listed for such use. What additional precautions are required for these boxes?
7. How much box volume does a 12 AWG conductor occupy?
8. What is the minimum allowable depth for outlet boxes?
9. Which of the following metal boxes can be used to contain eight 12 AWG conductors, two internal clamps, a 20-amp receptacle, and four 12 AWG equipment grounding conductors? List all acceptable of the following: 4 × 2 1/8 octagonal, 4 × 1 1/2 square, 4 11/16 × 1 1/2 square, 4 × 2 1/8 × 2 1/8 device.
10. What is the minimum allowable length for the box shown in the sketch below?

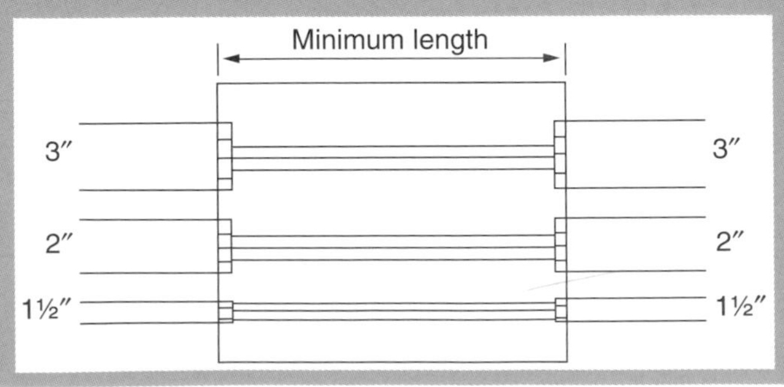

11. Would three ganged 3 × 2 × 1 1/2 device boxes be sufficient to contain the following switches and related wiring? All conductors are 14 AWG.

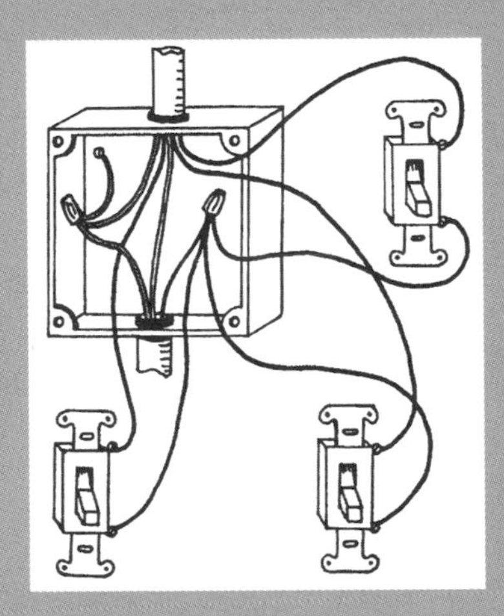

12. In the following sketch, the 1 1/4″ conduit contains five 6 AWG conductors and the 1/2″ conduit contains two 14 AWG conductors. Determine (a) the required box capacity based on conductor volume and (b) the minimum box dimensions based on the angular pull.

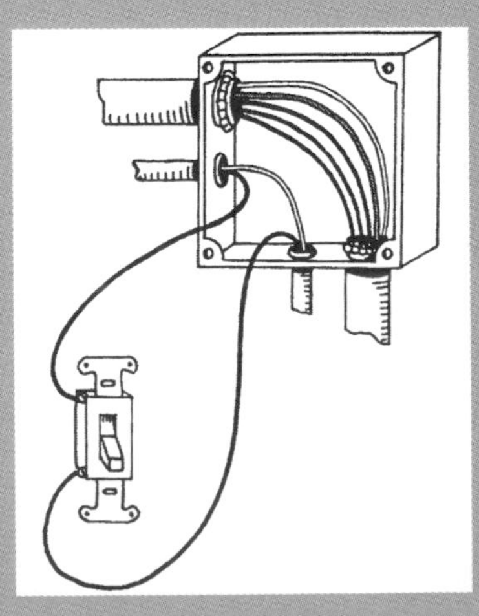

Chapter 8

Overcurrent Protection

Technical Terms

Circuit breaker
Current-limiting fuse
Dual-element fuse
Fuse
Ground fault
Nonrenewable fuse
Overcurrent device
Overload
Renewable fuse
Short circuit
Shunted
Single-element fuse
Tap rules
Tapped conductor
Time-delay fuse

Objectives

After completing this chapter, you will be able to:

- Identify the types, ratings, and characteristics of electrical protective devices.
- Recognize overloads and short circuits.
- List types of fuses.
- Compare fuses and circuit breakers.

Overcurrent protection limits the current passing through a conductor, preventing high temperatures that could damage insulation. Consequently, the equipment connected to the circuit is protected as well. This chapter examines the causes of overcurrent and protection methods.

Overcurrent

There are basically three types of overcurrent:

- **Overload**—When a device is connected to a circuit (such as plugging a lamp into a receptacle), it increases the current in the circuit. If enough electrical equipment is connected to a circuit, the current will increase to the level needed to support all of the equipment. If there is too much current, the conductor could melt its insulation, causing a fire hazard. Overloads generally result from current two to six times the rated current.
- **Short circuit**—The current flowing through circuit conductors is governed by the resistance of the conductor and the resistance of the load connected to the circuit. If the connected load becomes ***shunted*** (paralleled with a very low-resistance path), the normally low resistance of the conductor will allow an extremely high current to flow through it. This condition will generate a large amount of heat, causing the temperature of the conductor to rise to the point where the conductor insulation could melt. See **Figure 8-1.**
- **Ground fault**—A ground fault occurs when a connection is made between a normally ungrounded conductor and a low-resistance ground path. The low resistance of the circuit conductor will allow a high current to flow. See **Figure 8-2.**

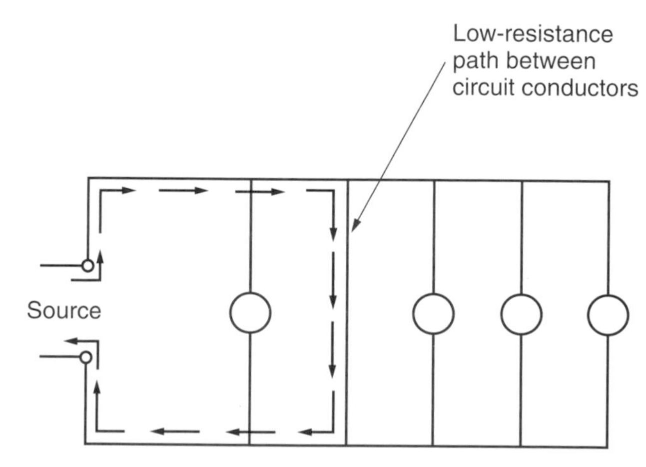

Figure 8-1. A short circuit occurs when there is a low-resistance path between conductors. Current flows through the path of least resistance at a very high current.

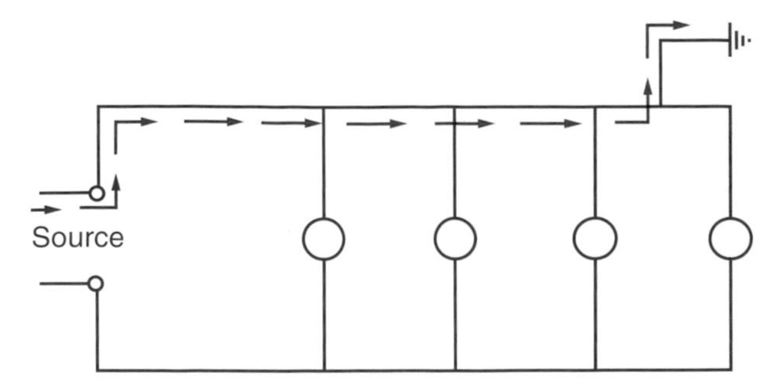

Figure 8-2. A ground fault is a short between a hot conductor and a grounding path. The grounding path can be the equipment grounding system or another path. Very high current results.

> **WARNING**
>
> Both short circuits and ground faults allow very high, uncontrolled current to flow through a conductor and create high temperatures. Both of these forms of overcurrent are fire hazards.

Overcurrent Protection Requirements

All electrical circuits are required to be protected against overcurrent. *Section 240.4* of the *Code* specifically mandates that conductors be protected against overcurrent. The overcurrent device must automatically open the circuit when the current reaches a specified dangerous level.

Fuses and circuit breakers are used as protection devices. The symbols used on schematic drawings to represent fuses and circuit breakers are illustrated in **Figure 8-3.** Other devices, such as thermal relays and thermal cutouts, are sometimes added.

Overcurrent devices must have correct voltage settings for the circuit served and must be marked with their maximum current-interrupting capacity. *Section 240.6* of the *Code* lists standard current ratings for fuses and circuit breakers. These standard ratings are listed in **Figure 8-4.**

> **WARNING**
>
> In order for an overcurrent device to function properly, it must be correctly selected and correctly located in the circuit. The overcurrent device must be operated in a safe manner, both electrically and mechanically, to protect the equipment and prevent harm to personnel operating the equipment.

Overcurrent devices are placed in a system so that only the section of the system where the overcurrent exists will be opened. Other circuits originating from the same power panel do not suffer a power loss from a remote fault.

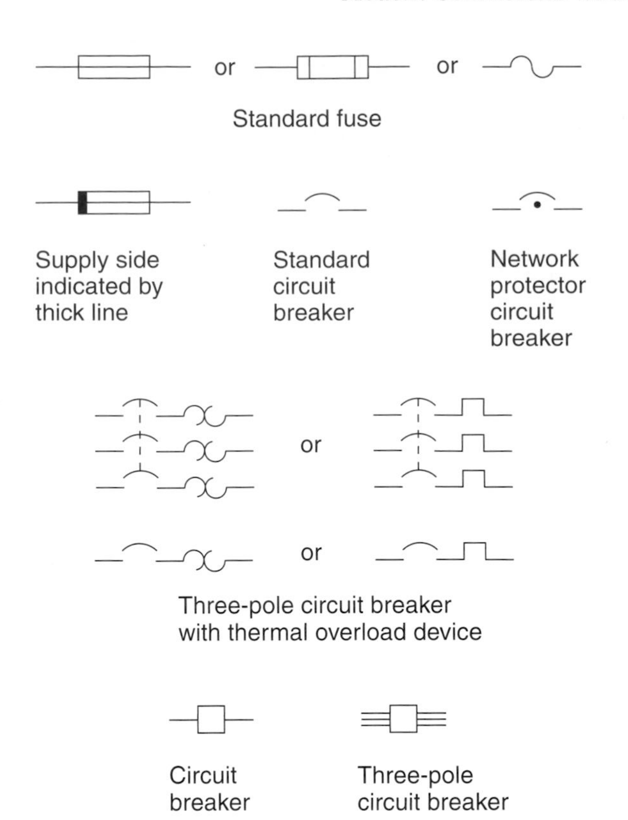

Figure 8-3. A sample of some symbols for overcurrent protective devices.

Ampere Ratings for Fuses and Circuit Breakers

15	20	25	30	35	40
45	50	60	70	80	90
100	110	125	150	175	200
225	250	300	350	400	450
500	600	700	800	1000	1200
1600	2000	2500	3000	4000	5000
6000					

Figure 8-4. Standard ampere ratings for fuses and inverse-time circuit breakers.

Overcurrent Device Sizing Factors

When sizing overcurrent devices, the required size often does not correspond to one of the standard sizes. For circuits less than or equal to 800 amps, the next higher standard size of overcurrent protection can be used if the conditions in *Section 240.4(B)* are met. In circuits exceeding 800 amps, the required size or the next lower standard size of overcurrent protection must be used.

Overcurrent Device Location

The concise rules for locating overcurrent devices, both within the circuits and on the premises, are stated in *Sections 240.20, 240.21,* and *240.24* of the *Code*. Ungrounded (hot) circuit conductors have overcurrent protection at the point of supply.

Overcurrent devices must be readily accessible, free from physical damage, installed in an enclosure, or be otherwise protected and away from flammable material. The movable parts of an overcurrent device, such as handles or power-actuated mechanisms, are guarded to protect personnel.

Typically, branch circuit overcurrent devices are located at the power panel, **Figure 8-5.** There are some exceptions to this practice. Provisions listed in *Section 240.21* of the *Code* and referred to as the ***tap rules*** allow conductors, under certain restricted conditions, to be tapped without overcurrent protection at the connection point. These modifications are discussed later in the chapter.

NOTE

Feeder conductor protection is located before the service drop or service lateral or on the primary side of the transformer.

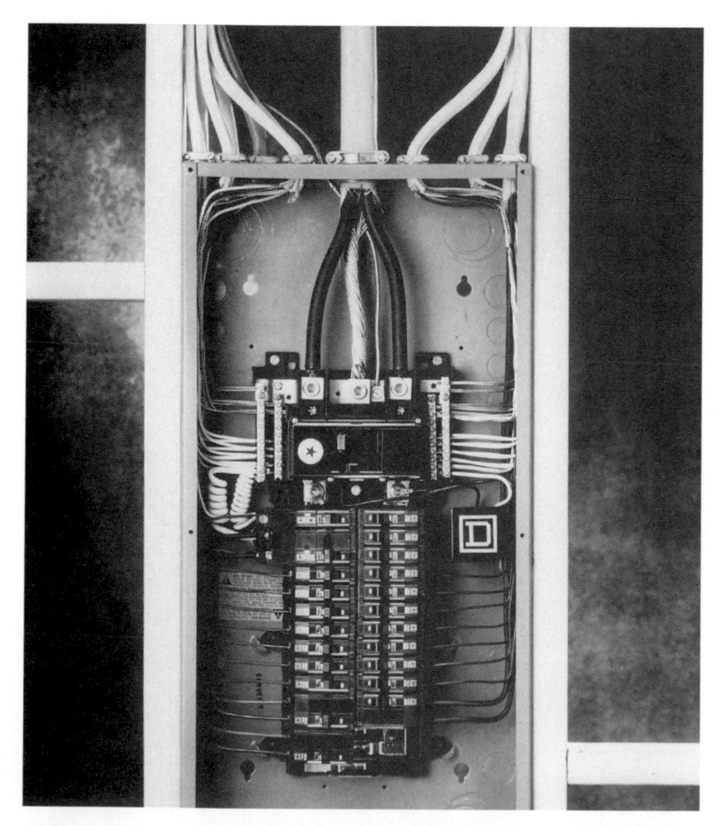

Figure 8-5. Each branch circuit supplied from this panel is protected by a circuit breaker. The circuit breaker is connected to the current-carrying (black) conductors. (Square D Co.)

NEC NOTE 240.8

Individual fuses, circuit breakers, or combinations thereof shall not be connected in parallel unless they are factory assembled in parallel and listed as a unit.

Overcurrent Protective Devices

There are two types of devices used to protect circuits from overcurrent: fuses and circuit breakers. Fuses contain an element that melts under overcurrent load. When the element melts, it breaks the circuit, preventing the flow of electricity. After the cause of the overcurrent is located and corrected, the "blown" fuse is replaced with a new one.

Circuit breakers are mechanical devices that open the circuit when an overcurrent is detected. These devices can be reset after being "tripped," making them more convenient than fuses.

Voltage Rating

Overcurrent protective devices are rated for voltage and should be used in appropriate circuits. That is, the voltage rating of a fuse should be equal to or slightly higher than the circuit rating, but never lower. The table in **Figure 8-6** lists the maximum voltage ratings of various types of overcurrent protective devices.

Ampere Rating

Every overcurrent protective device has a specific ampere rating. Usually the ampere rating of a device is chosen based on the type of circuit and loads on the circuit, as well as *Code* requirements. The ampere rating should be the same as the current-carrying capacity of the circuit conductor, but there are exceptions (as with motors) where it is necessary to use higher ampere-rated fuses.

Overcurrent protective devices also have an interrupting rating. This rating is the highest fault current the device is intended to interrupt at its rated voltage. Fuses have an interrupting rating of 10,000 amps unless otherwise marked. Fuses with higher interrupting ratings are called high-interrupting-rated fuses. Circuit breakers have an interrupting rating of 5000 amps unless otherwise marked.

Fuses

Fuses are highly reliable and safe overcurrent devices. They consist of a resistance-sensitive link encapsulated in a tube. The ends of the fuse are connected to contact terminals of the conductor. The link has very low

Overcurrent Protection Device Ratings

OPD Classification	Voltage Rating (V)	Maximum Current (A)
Plug Fuses		
Edison-base	Up to 125	30
Type S	Up to 125	30
Cartridge Fuses		
Class G	300	60
Class H	250 and 600	600
Class J	600	600
Class K	250 and 600	600
Class L	600	601 to 6000
Class R	250 and 600	600
Class T	300	1200
	600	800
Class CC	600	30

Figure 8-6. Maximum voltage ratings of overcurrent protective devices.

resistance. If high current occurs, the link melts and opens the circuit.

Two general types of fuses are available: plug and cartridge. Within these two types, fuses can be further classified as the following:

- Single-element or dual-element.
- Fast-acting or time-delay.
- Renewable or nonrenewable.

Single-element fuses have a single element of one or more links that quickly melts in response to an overcurrent. The link melts in approximately three seconds at a 200% overload current and usually within thousandths of a second (less than a half-cycle of alternating current) for short circuits and ground faults. Since these fuses act rapidly, they are only used in circuits with no transient surge currents or temporary overloads. **Figure 8-7** illustrates a fuse melting and opening a circuit.

Dual-element fuses have two elements arranged in series. One element is similar to the element in a single-element fuse, and performs the same function. The other element is designed to allow for low-level overloads of about five times the ampere rating for over ten seconds. This type of fuse is particularly useful in protecting a circuit where temporary surges are normal. A dual-element fuse is shown in **Figure 8-8.** These fuses are marked with a "D."

A time-delay feature is inherent in dual-element fuses. However, not all time-delay fuses are dual-element. The main feature of ***time-delay fuses,*** regardless of number of elements, is that they allow for momentary overloads.

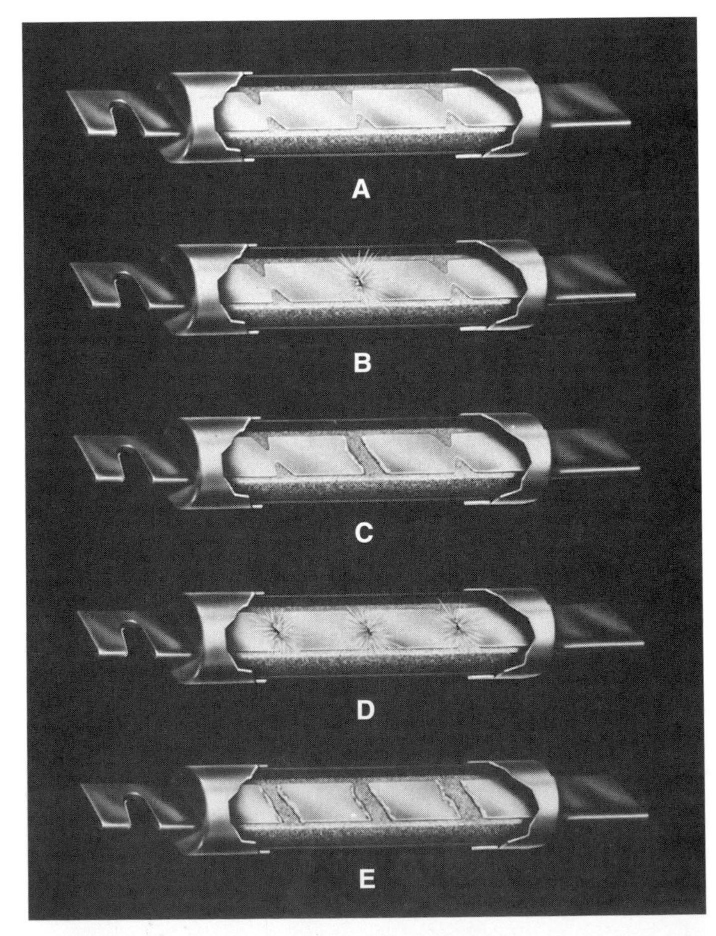

Figure 8-7. A single-element cartridge fuse. A—Fuse in normal working condition. B—An overload melts the fusible element in one location. C—After an overload. D—A short melts all fusible elements. E—After a short circuit. (Bussmann Division of Cooper Industries)

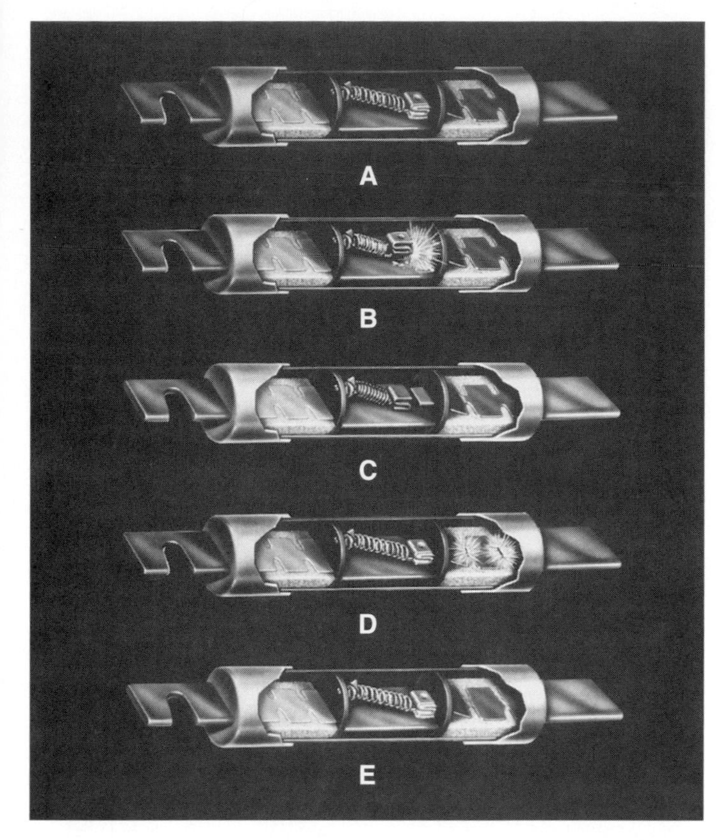

Figure 8-8. A dual-element fuse. A—Normal condition. B—Fuse opens from overload after a brief time delay. C—After an overload. D—Short melts fusible element. E—After a short circuit. (Bussmann Division of Cooper Industries)

In ***renewable fuses,*** the fusible link can be replaced. This allows the fuse holder to be reused, reducing cost. A ***nonrenewable fuse*** (***one-time fuse***) must be discarded after use and replaced with a new fuse.

Current-limiting fuses quickly open the circuit when a specified overcurrent is reached. These fuses clear the circuit of current in a fraction of a second.

Plug fuses

Plug fuses are available in two varieties: Edison-base and Type S, **Figure 8-9.** Plug fuses have a maximum rating of 30 amps. Fuses rated as 15 amps or less have a hexagonal window, while higher-rated fuses have a circular window, **Figure 8-10.**

Plug fuses are limited by the following voltage restrictions:

- Maximum 125 volts between conductors
- Maximum 150 volts between ungrounded conductors and the ground in grounded-neutral systems

Edison-base fuses are interchangeable—fuses of greater current rating can be used in circuits of lesser capacities. This potentially dangerous oversizing problem is eliminated by using a Type S fuse, which uses a screwshell adapter. Type S fuses must be used on all

Figure 8-9. Type S plug fuses (left) have a thinner stem than Edison-base type fuses (right).

> **WARNING**
>
> Edison-base plug fuses have no inherent limit to the use of higher ampacities. For example, nothing prevents a 30-amp fuse from being installed in a 15-amp circuit. Using a fuse with an ampere rating higher than that of the circuit is a fire hazard.

Figure 8-10. Fuses rated for 15 amps or less have a hexagonal window.

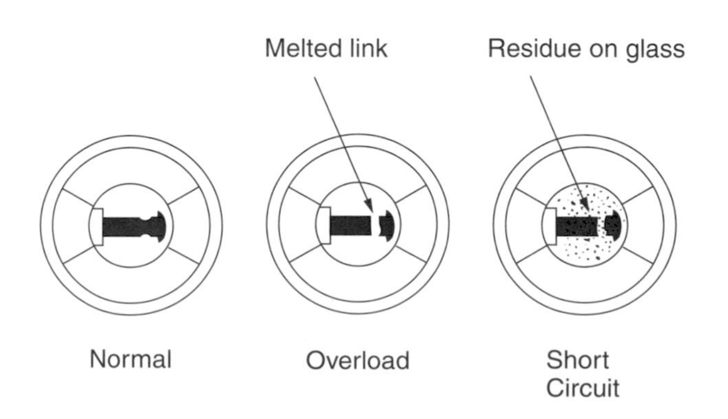

Figure 8-11. A blown fuse identifies the type of overcurrent.

new installations and for replacement installations. There are three classifications of Type S fuses with specific adapters based on their ampere ratings: 0–15 amps, 16–20 amps, and 21–30 amps.

The appearance of a blown fuse identifies the type of overcurrent that occurred. An overload causes the fusible link to neatly melt, while the high currents associated with a short circuit vaporize the link and leave residue on the glass. See **Figure 8-11.**

Cartridge fuses

Cartridge fuses are used in fused systems over 120 volts or 30 amps. There are two types of cartridge fuse connections: ferrule type and knife-blade type. See **Figure 8-12.**

Underwriters Laboratories classifies different types of cartridge fuses. Most types have unique dimensions. This prevents one class from being replaced by another class in a circuit. Some of the most common types include the following:

- **Class H**—These fuses can be renewable or nonrenewable. The nonrenewable types can be time-delay. They have a low interrupting rating of only 10,000 amperes and are not current limiting. The low interrupting rating restricts their use in new construction.
- **Class J**—Class J fuses are nonrenewable and current limiting with an interrupting rating of 200,000 amperes. They are smaller than Class H fuses and are not interchangeable with any other class. They are rated for 600 volts and up to 600 amperes. These fuses have no time delay. A new "cube" style of Class J fuses is also available, **Figure 8-13.**
- **Class R**—These are nonrenewable, current-limiting fuses with an interrupting rating of 200,000 amperes. These fuses have a rejection feature so no other type of fuse can be installed in the same fuse holder. They may be used as replacements for Class H fuses and are rated for 250 volts and 600 volts and up to 600 amperes.
- **Class T**—A nonrenewable, current-limiting fuse with an interrupting rating of 200,000 amperes. There is no time delay with Class T fuses. They are not interchangeable with any other fuse and are rated at 300 volts and 600 volts and up to 1200 amperes.

Circuit Breakers

Circuit breakers are permitted to serve not only as the overcurrent protective device, but also as a disconnecting switch. They are often operated electrically or pneumatically, but must also be capable of manual operation.

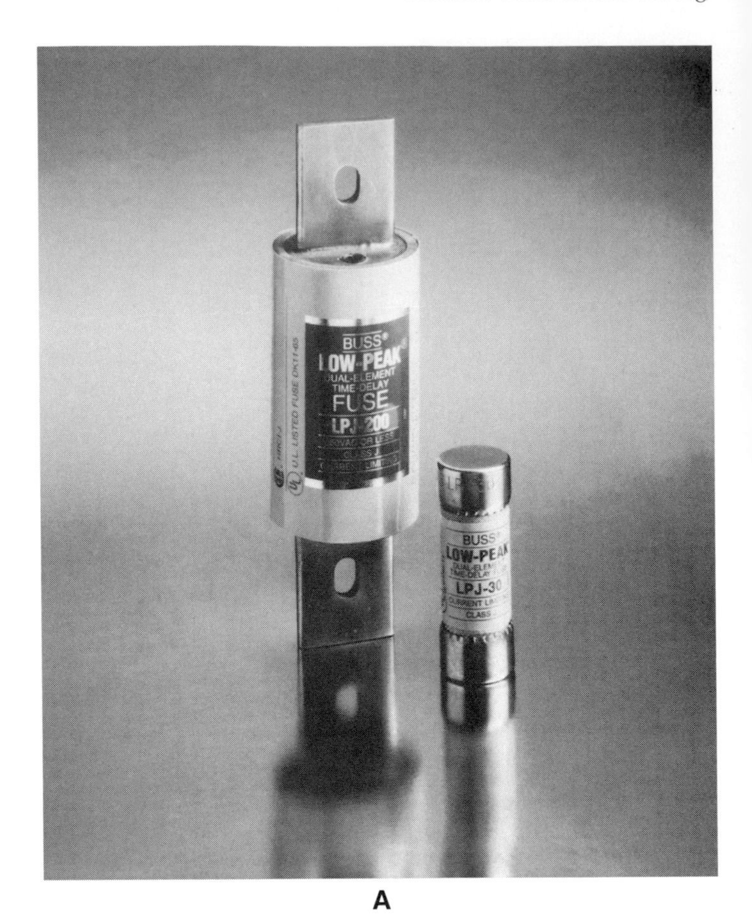

A

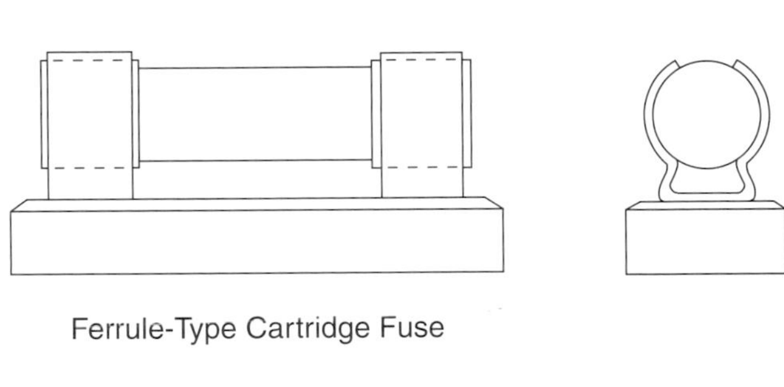

Ferrule-Type Cartridge Fuse

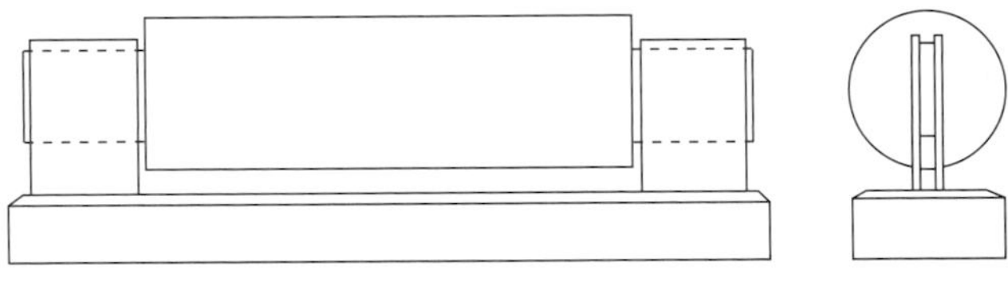

Knife-Blade Cartridge Fuse

B

Figure 8-12. Knife-blade and ferrule-type cartridge fuses. (Bussmann Division of Cooper Industries)

NEC NOTE 240.6(B)

The rating of adjustable-trip circuit breakers having external means for adjusting the current setting (long-time pickup setting), not meeting the requirements of 240.6(C), shall be the maximum setting possible.

Figure 8-13. The CUBEFuse™ is a new style of Class J fuse. It is more compact than other similar fuses. (Bussmann Division of Cooper Industries)

Circuit breakers must be marked to show if they are "ON" or "OFF". This indicates if the circuit is closed or open, respectively. There are two types of circuit breakers:

- **Instantaneous-trip circuit breaker**—As its name indicates, this circuit breaker opens the circuit instantly when its interrupting rating is reached. It serves the same purpose as a nondelay, current-limiting fuse. See **Figure 8-14.**
- **Inverse-time circuit breaker**—Like a dual-element fuse, this circuit breaker has two mechanisms for sensing overloads. One mechanism allows for a certain amount of overload to occur, the other opens the circuit at the fault-level current. An inverse-time circuit breaker will allow as much as ten times the overload current prior to opening the circuit.

Figure 8-15 illustrates the internal features common to most circuit breakers.

Circuit breaker ratings

Circuit breakers, like fuses, have standard interrupting ratings. These interrupting ratings must be clearly marked, along with the intended voltage rating, on each circuit breaker. Breakers of different interrupting ratings and voltages are constructed with different mounting features so they are not interchangeable.

NEC NOTE **240.83(C)**

Every circuit breaker having an interrupting rating of other than 5000 amperes shall have its interrupting rating shown on the circuit breaker.

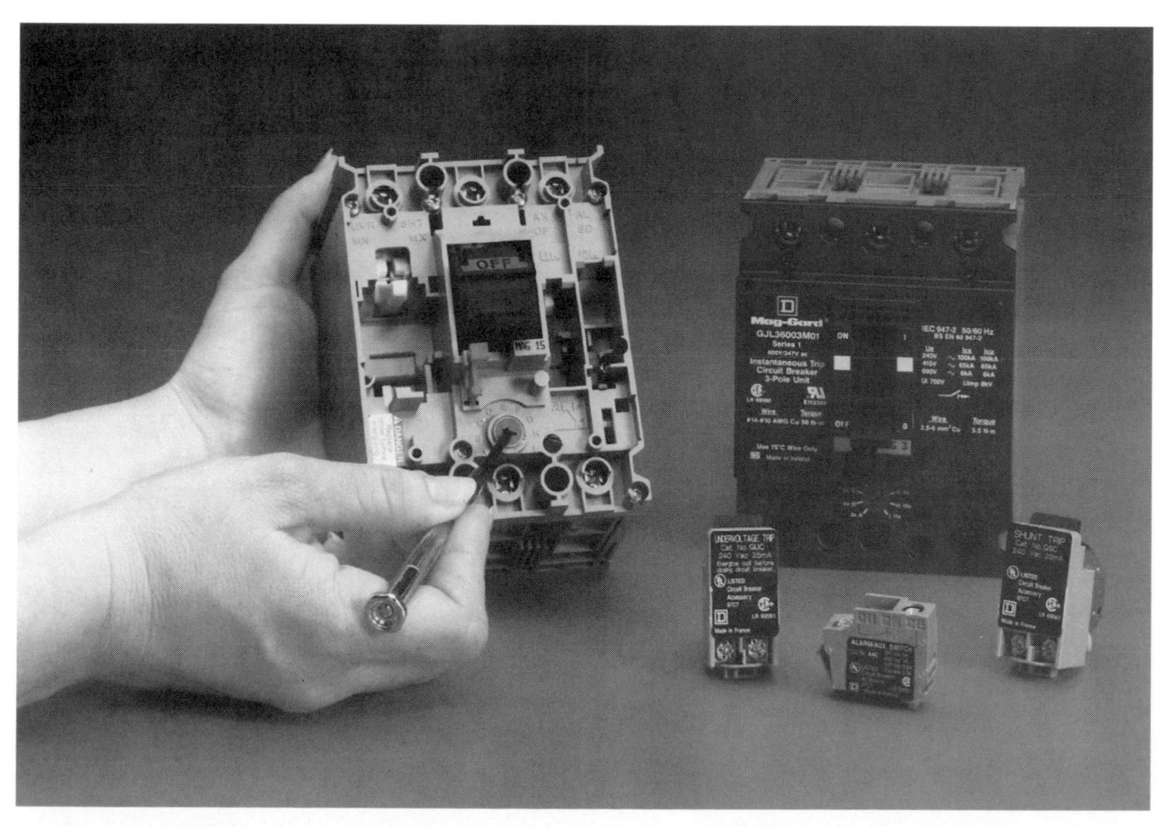

Figure 8-14. These circuit breakers have an adjustable instantaneous-trip range. Circuit breaker accessories can be added as needed. (Square D Co.)

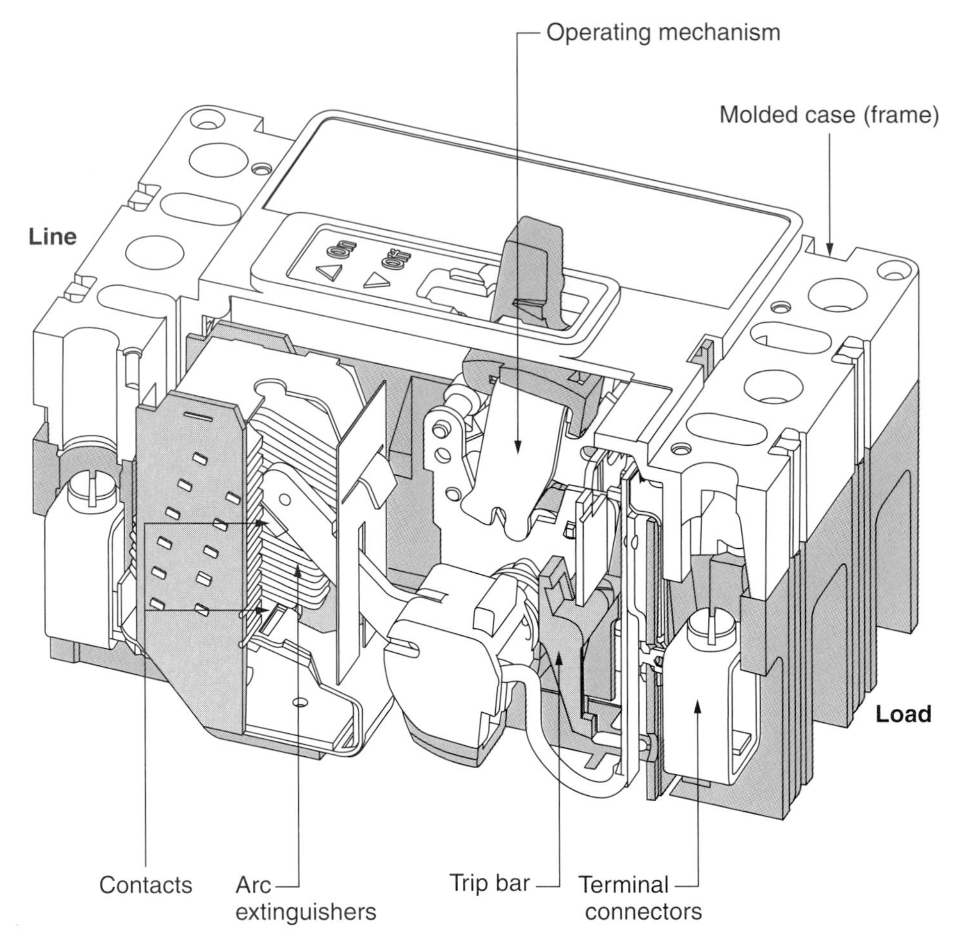

Figure 8-15. Internal mechanisms common in circuit breakers.

Conductor Taps

Conductors normally have overcurrent protection at their point of supply. In the case of ***tapped conductors*** (conductors that are connected to another conductor), this additional protection is not required as long as certain conditions are met. These conditions are listed in *Section 240.21* of the *Code.*

Section 240.21(B)(1), referred to as the "10-foot tap rule," allows a feeder conductor of smaller ampacity to be tapped, without protection at the point of tap, to the secondary side of a transformer *provided* the tap does not exceed 10′ in length and also meets the following conditions (**Figure 8-16**):

- The tap conductors must have a sufficient current rating to supply its loads. The current rating also must be greater than the rating of the overcurrent device at its termination point.
- The tap conductors must terminate at the device, panelboard, transformer, or switchboard that it supplies.
- The tap conductors must be enclosed in a raceway between the point of tap and its destination point.

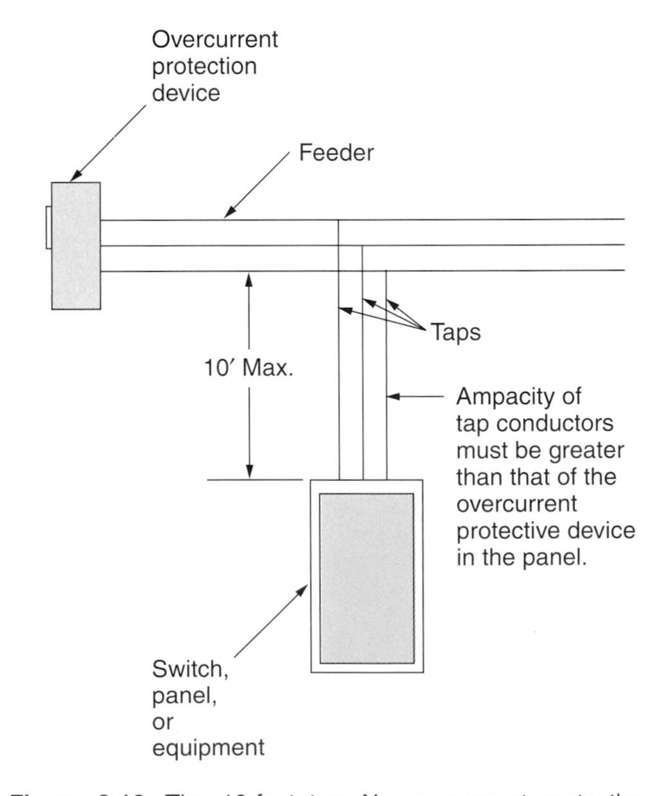

Figure 8-16. The 10-foot tap. No overcurrent protection is needed at the tap point as long as other conditions are met.

Section 240.21(B)(2) (the "25-foot tap rule") permits taps less than 25′ long without overcurrent protection at the point of supply as long as the following conditions are met (**Figure 8-17**):

- The tap conductors and the overcurrent protective device located at their terminations are rated at a minimum of 1/3 of the ampacity rating of the feeder's overcurrent protective device.
- The tap conductors are enclosed in a raceway for protection.
- The tap conductors must terminate at an overcurrent device that is rated to the ampacity of the tap conductors.

Section 240.21(B)(4) allows for an unprotected tap up to 100′ long in high bay manufacturing buildings with walls over 35′ high. These systems must be maintained and supervised by qualified individuals, **Figure 8-18.**

In addition, the following rules apply:

- The tap conductors must not travel over 25′ horizontally from the point at which the tap is made and the total tap conductor length must not exceed 100′.
- The tap conductor ampacity must be at least 1/3 of the ampacity rating of the feeder's overcurrent protective device.
- The tap conductor must terminate at an overcurrent device that is rated to the proper ampacity.
- The tap must be physically protected (installed in raceway).
- The tap conductors may not be spliced.

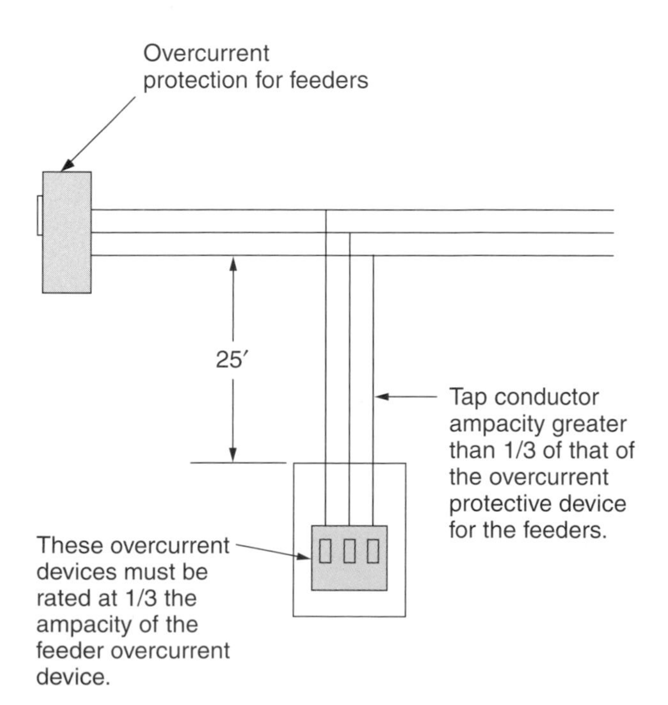

Figure 8-17. The 25-foot tap. The tap conductors must be rated for at least 1/3 of the rating of the overcurrent protective device protecting the feeder.

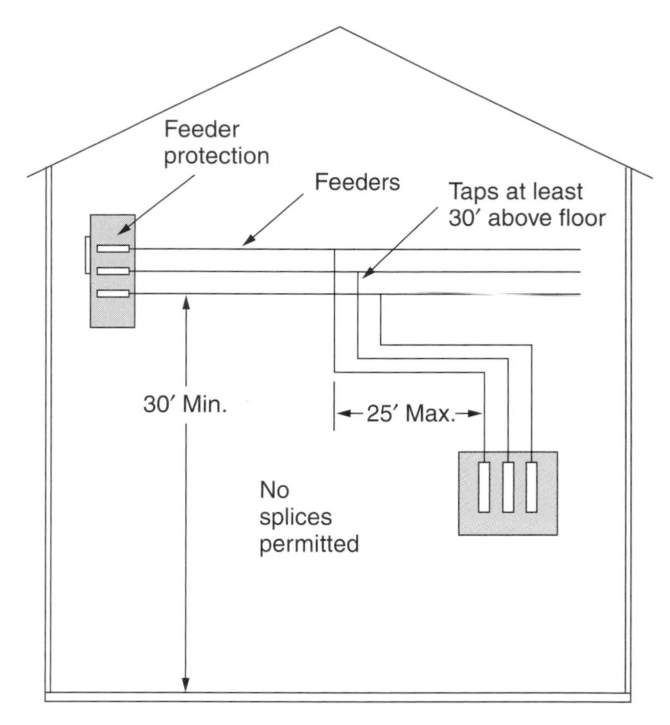

Figure 8-18. The 100-foot tap rules require the tap to be located at least 30′ above finished floor level.

- The tap conductors must be 6 AWG copper (4 AWG aluminum) or larger.
- The tap conductors may not pass through ceilings, floors, or walls.
- The point at which the tap is made must be at least 30′ above the finished floor of the building.

Review Questions

Answer the following questions. Do not write in this book.

1. Describe the three types of overcurrent.
2. What are the two main types of overcurrent protective devices used to protect circuit conductors and the equipment they serve.
3. List the two main types of circuit breakers. Which type is similar to a single-element fuse and which type is similar to a dual-element fuse?
4. Explain what is meant by the phrase "circuit breakers must be indicating."
5. At what point in a circuit must a breaker or fuse be placed?
6. What is the maximum ampere rating for a Class J cartridge fuse?
7. What is the interrupting rating of a fuse that does not have a rating marked on it?
8. What feature is unique to time-delay fuses?
9. Why are Type S fuses safer than Edison-base fuses?
10. The glass of a blown fuse is coated with a dark residue on the inside. What type of overcurrent caused the fuse to blow?

USING THE NEC

Refer to the National Electrical Code to answer the following questions. Do not write in this book.

1. Where in *Article 240* does it state that conductor overload protection shall not be required where the interruption of the circuit would create a hazard?
2. To which article should you refer to find overcurrent protection requirements for air-conditioning and refrigerating equipment?
3. What five items should be marked on a cartridge fuse?
4. How must circuit breakers used as switches in 120-volt and 277-volt fluorescent lighting circuits be marked?
5. Overcurrent protective devices cannot be located in bathrooms. What *Code* section states this?
6. Can a Type S adapter be removed after being inserted in a fuse holder?
7. What is the definition of "current-limiting overcurrent protective device"?
8. In what two cases can an overcurrent protective device be connected in series to a grounded conductor?

Chapter 9

Service and Distribution

Technical Terms

Drip loop
Service
Service disconnect
Service drop
Service lateral
Service mast

Objectives

After completing this chapter, you will be able to:

- Describe the two basic types of service.
- Explain service terminology.
- Find service drop clearance in the *Code*.
- Compare the arrangement and construction of service drops and service laterals.
- Identify the required working clearances at the service equipment.
- Explain the various supply voltages available in the United States and their common applications.

Power provided by the utility company to a building is called the *service.* The service is then distributed throughout the facility. Regardless of the size and electrical requirements of a facility, the service and distribution methods are similar.

Service

The requirements of the electrical service are provided in the *Code* by *Article 230—Services*. This article is divided into the following parts:

- *Part I. General*
- *Part II. Overhead Service-Drop Conductors*
- *Part III. Underground Service-Lateral Conductors*
- *Part IV. Service-Entrance Conductors*
- *Part V. Service Equipment—General*
- *Part VI. Service Equipment—Disconnecting Means*
- *Part VII. Service Equipment—Overcurrent Protection*
- *Part VIII. Services Exceeding 600 Volts, Nominal*

The ***service*** is defined as the conductors extending from the utility pole or equipment (such as a transformer) to the facilities service equipment. This equipment is also considered as part of the service.

The service conductors must be run either overhead or underground. They are connected to the ***service disconnect,*** **Figure 9-1,** a piece of equipment where the power can be disconnected.

NEC NOTE 230.2

A building or other structure served shall be supplied by only one service, unless as permitted in 230.2(A) through (D).

Figure 9-1. The main service panel (left) receives the utility service. The power is then distributed to various transformers and panels (including the one shown here).

There are two methods of bringing the service into a facility. The service can be run overhead through exposed wiring or routed below the ground. When run overhead, the service is referred to as the ***service drop.*** When routed underground, the service is called the ***service lateral.***

NEC NOTE 100

Service Drop. The overhead service conductors from the last pole or other aerial support to and including the splices, if any, connecting to the service-entrance conductors at the building or other structure.

Service Drop

Overhead services are discussed in *Part II* of *Article 230.* Because service drops are exposed, their accessibility must be limited. A ***service mast,*** **Figure 9-2,** supports the conductors. Specific clearances must be observed.

NEC NOTE 230.28

Where a service mast is used for the support of service-drop conductors, it shall be of adequate strength or be supported by braces or guys to withstand safely the strain imposed by the service drop. Where raceway-type service masts are used, all raceway fittings shall be identified for use with service masts. Only power service-drop conductors shall be permitted to be attached to a service mast.

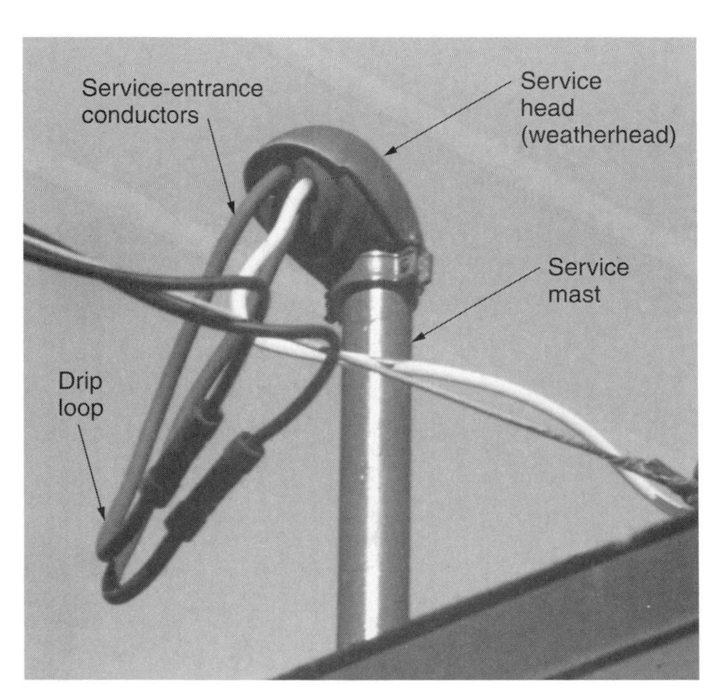

Figure 9-2. The service mast limits accessibility. The drip loop prevents water from running inside the mast. (Jack Klasey)

Vertical clearances

The minimum vertical clearance of service-drop conductors above finished grade ranges from 10′ to 18′, depending on the service voltage and what is below the conductors. **Figure 9-3** illustrates the minimum vertical clearances above finished grade and the controlling parameters.

Service drops routed over flat roofs must have a minimum vertical clearance of 8′, unless that roof is subject to vehicular traffic, in which case the minimum clearance becomes 10′. See **Figure 9-4.**

This rule is further modified for sloped roofs. Where the slope is at least 4 in 12, the minimum vertical clearance is only 3′. See **Figure 9-5.**

If the service mast passes through the roof within 4′ of the roof edge, the vertical clearance is reduced to 18″ regardless of pitch, as long as the voltage-to-ground does not exceed 300 volts. The service-drop conductors can only pass above the roof between the service mast and the roof edge, **Figure 9-6.**

NEC NOTE 230.9(C)

Overhead service conductors shall not be installed beneath openings through which materials may be moved... and shall not be installed where they obstruct entrance to these building openings.

Miscellaneous clearance conditions

Service-drop conductors are not permitted to pass over swimming pools. The service drop cannot be placed within 3′ below or to the side of a window, but it can be attached above a window. See *Section 230.9(A).*

Service Lateral

When the service conductors are routed underground, it is called a ***service lateral.*** The service feeders extend from a utility pole or pad-mounted transformer and can be directly buried (using USE cable) at a minimum depth of 2′. See **Figure 9-7.** Rigid nonmetallic conduit, rigid metal conduit, or intermediate metal conduit can be used. *Section 300.5* and *Table 300.5* contain burial and trenching requirements. In this text, buried conductors are discussed in Chapter 5, Wiring Methods.

NEC NOTE 230.7

Conductors other than service conductors shall not be installed in the same service raceway or service cable. See exceptions.

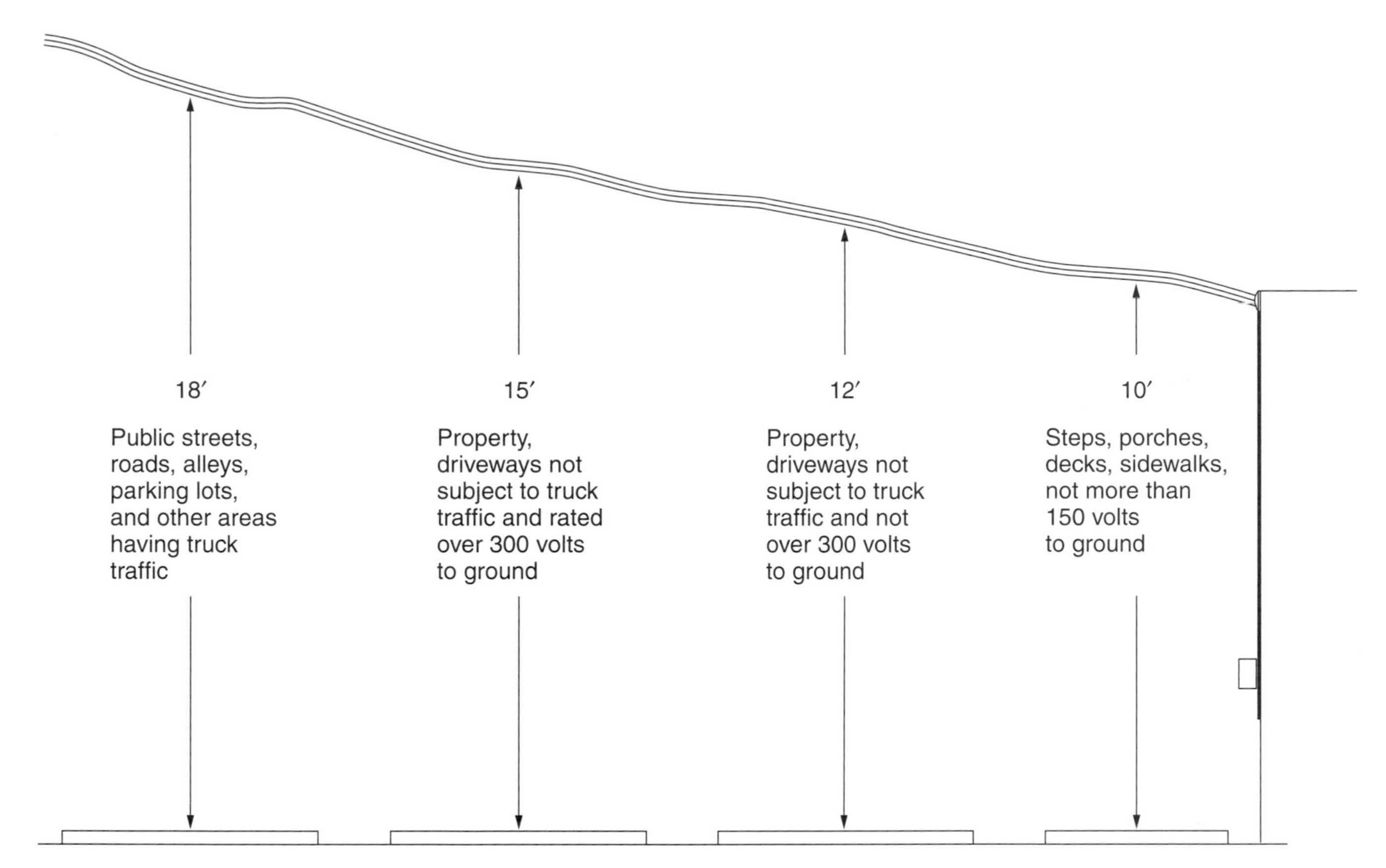

Figure 9-3. Minimum clearance of service-drop conductors depends on the service voltage and what is below the conductors. Check with the local authority and utility providers for any further restrictions.

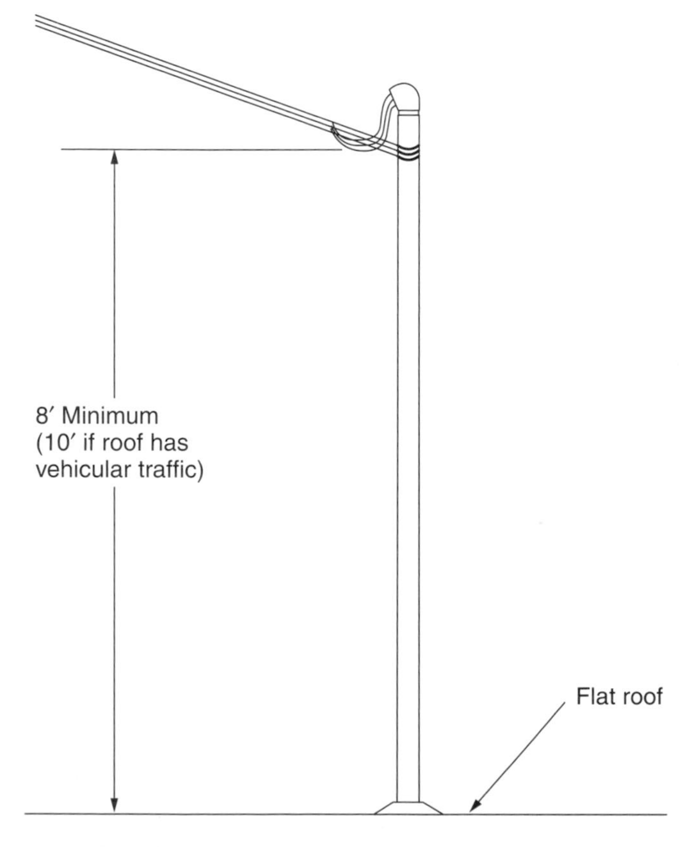

Figure 9-4. An 8′ minimum clearance is required for service drops on flat roofs. This clearance increases to 10′ if the roof is subject to vehicular traffic.

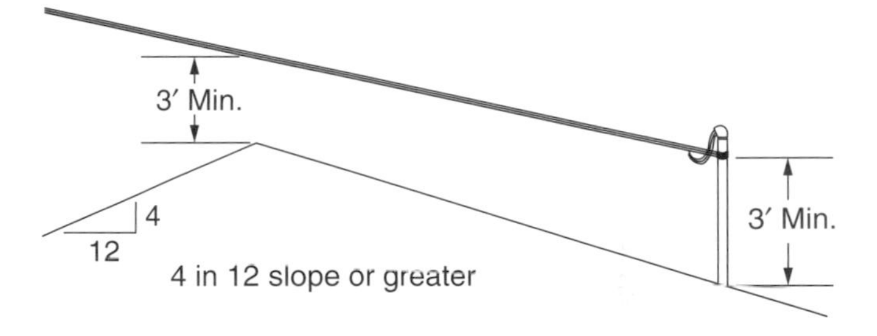

Figure 9-5. On a sloped roof with at least a 4 in 12 slope, the minimum clearance for the service drop is 3′. See *Section 230.24.*

Service Conductors

Whether run overhead or underground, service conductors must have an ampacity rating equal to or greater than the intended load. At the point of entry to a building, the service conductors become the service-entrance (SE) conductors. The service-entrance conductors can be insulated conductors or a multiconductor cable.

NEC NOTE 230.23(B)

Service-drop conductors shall not be smaller than 8 AWG copper or 6 AWG aluminum or copper-clad aluminum.

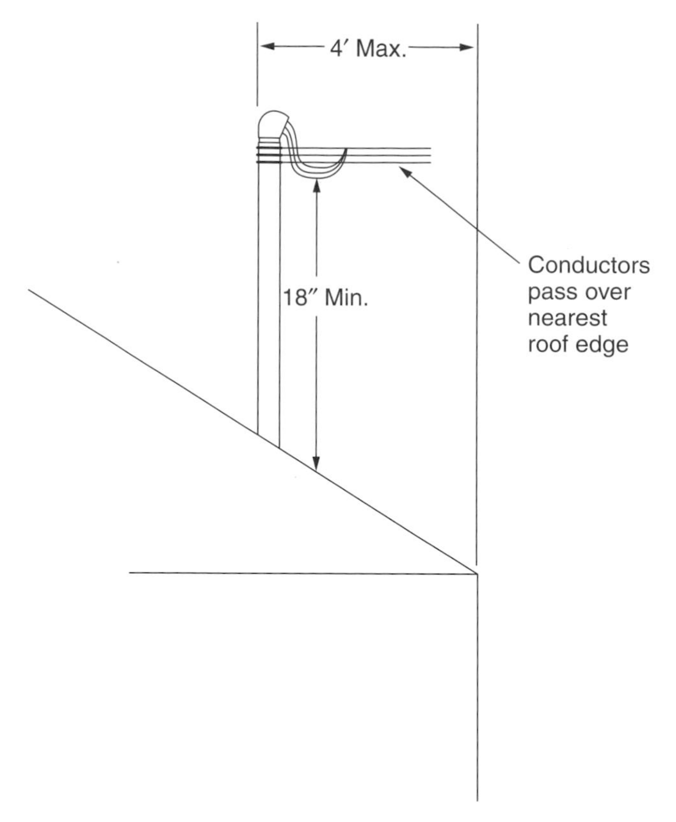

Figure 9-6. If the service mast is located within 4′ of the edge of a roof, vertical clearance is reduced and the service-drop conductors must pass over the nearest roof edge.

At the service head, the service conductors are connected to the service-entrance (SE) conductors. The SE cable can be run down the side of the building. The cable must be supported within 12″ of the service head, 12″ above the meter enclosure, and at intervals not exceeding 30″ along its length. See **Figure 9-8.** If this cable might be exposed to physical damage, then it must be installed in conduit or have some other protection.

Where connected to the service drop, service-entrance conductors must be provided with a raintight service head or gooseneck, **Figure 9-9.** In addition, the service-entrance conductors should be left long enough (24″ to 36″) to form a drip loop where they exit the service head fitting. The ***drip loop*** prevents water from running into the service head and down the service mast. **Figure 9-10** illustrates the conditions necessary for properly installing service-entrance conductors at the service drop. Regardless of the manner of service-drop connection, it must be at least 10′ above grade at its point of attachment. See **Figure 9-11.**

Normally, a single set of service conductors supply a single service-entrance. In multiple occupancies, up to six disconnects may be supplied by a single set of service-entrance conductors. Individual disconnects are mounted to a gutter or wireway in which the feeder taps to other disconnects are housed, **Figure 9-12.**

Figure 9-7. Digging a trench for service-lateral conductors.

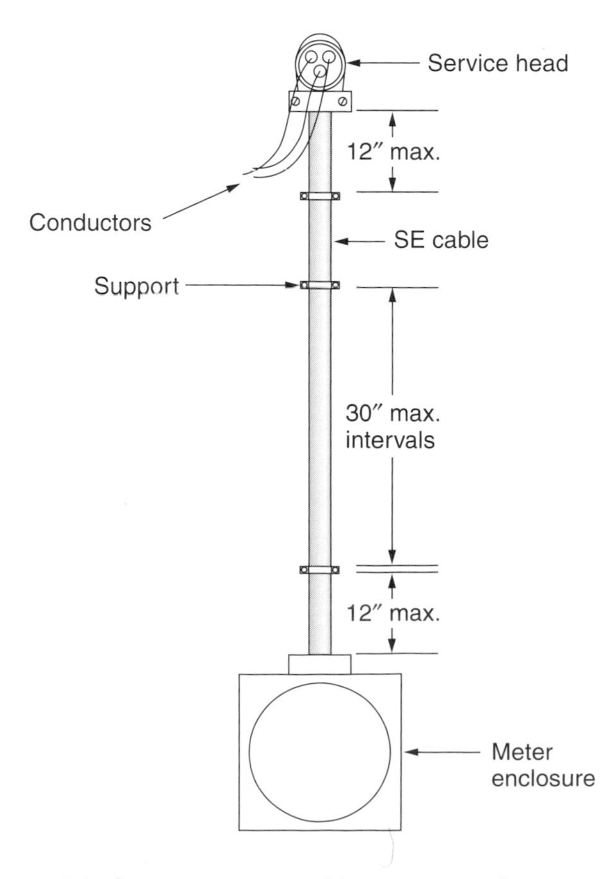

Figure 9-8. Service-entrance cable support requirements.

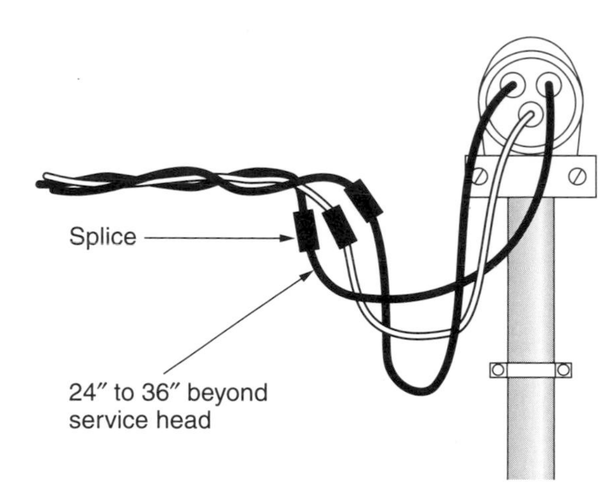

Figure 9-9. Service-entrance conductors extend 24″ to 36″ beyond the service head to allow formation of a drip loop. The service-entrance conductors are spliced to the service-drop conductors.

All service-entrance conductors must carry enough ampacity to safely supply the facility's current requirement. Weather-resistant insulation should always be used on service-entrance conductors.

NEC NOTE 230.3

Service conductors supplying a building or other structure shall not pass through the interior of another building or other structure.

NEC NOTE 230.22

Individual service conductors shall be insulated or covered. Exception: The grounded conductor of a multiconductor cable shall be permitted to be bare.

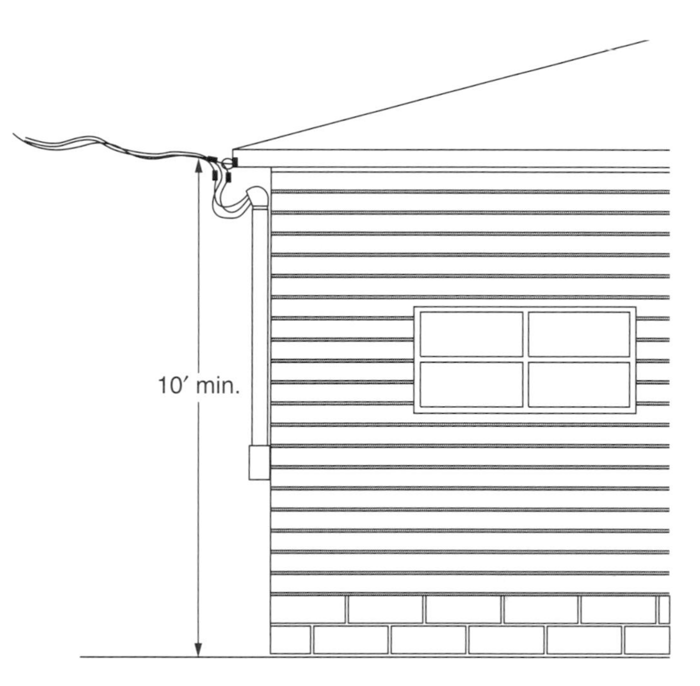

Figure 9-11. Minimum height requirement for service drop installation. See *Figure 9-3* as well.

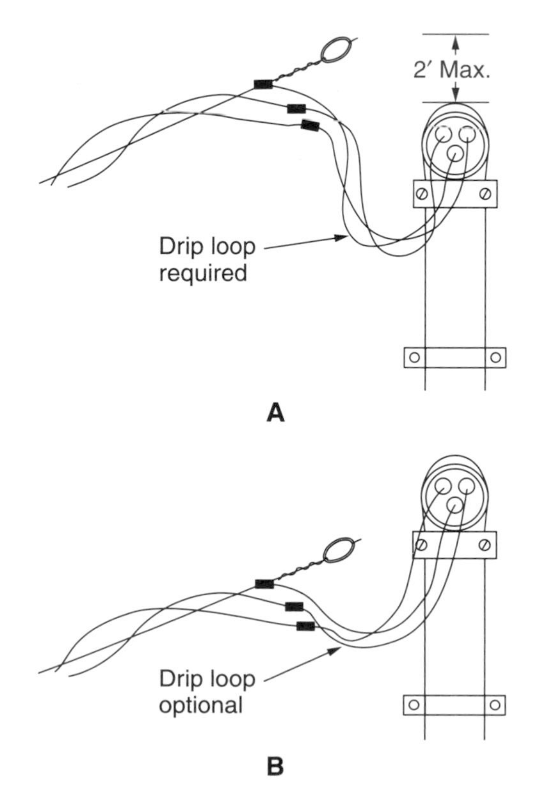

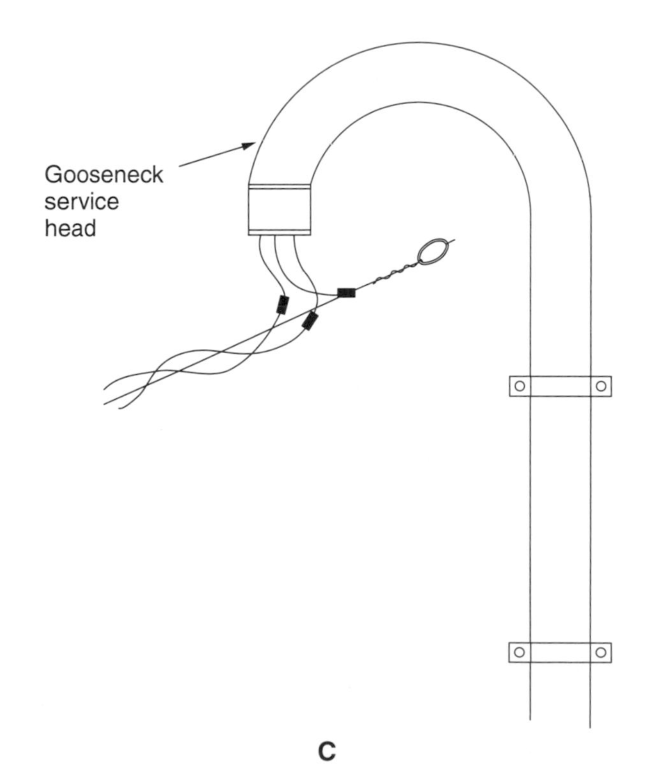

Figure 9-10. Service head installation. A—If the point of attachment is above the service head, a drip loop is required. B—If the point of attachment is below the service head, a drip loop is optional. C—Gooseneck service head.

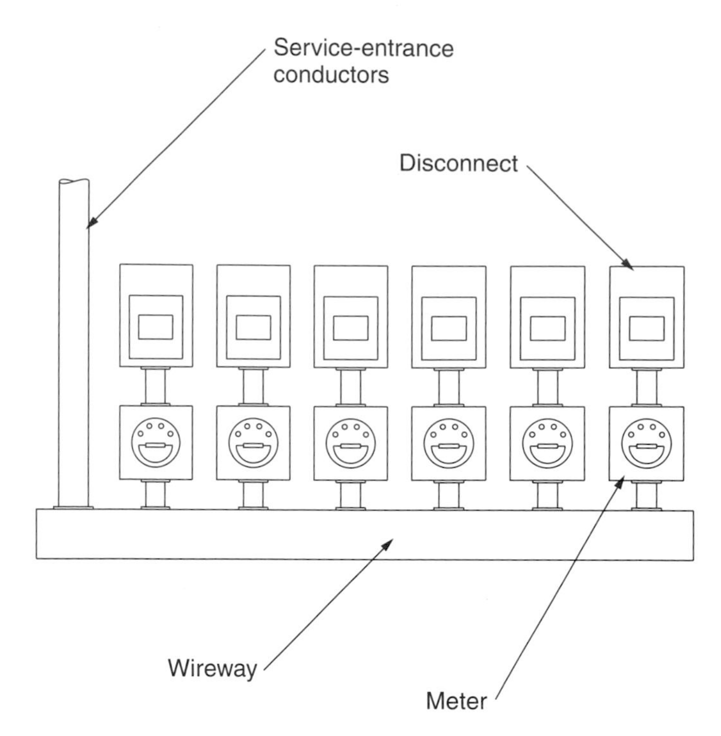

Figure 9-12. As many as six service disconnects may be supplied by a single set of service conductors for multiple-occupancy buildings.

Service Disconnects

Buildings must have a means of disconnecting all power feeders from the service-entrance conductors. This disconnecting means can be a fused disconnect switch, a breaker in a service enclosure, or the main circuit breaker in the service panel. It may be inside or outside the structure, but must be readily accessible and clearly identified as the service disconnect. The handle of the disconnect switch or main breaker must be mounted in an unobstructed location no higher than 6′-6″ above the floor, **Figure 9-13.**

The entry point of the service conductors and the disconnect should be as close as possible. Local codes may have specific limits.

For safe operation and maintenance, electrical service equipment rated up to 600 volts requires at least 30″ of clear horizontal workspace in front of it. The panel, disconnect switch, and enclosure can be mounted anywhere within the 30″ area, but no other obstructions or equipment are allowed in this space.

For industrial/commercial equipment with larger services (over 600 volts), individual equipment rooms are normally provided and additional work space requirements (contained in *Article 110—Requirements for Electrical Installations*) must be observed. These work space clearance requirements vary, depending on nominal voltage and surrounding conditions. See *Sections 110.33* and *110.34(A)* for more information.

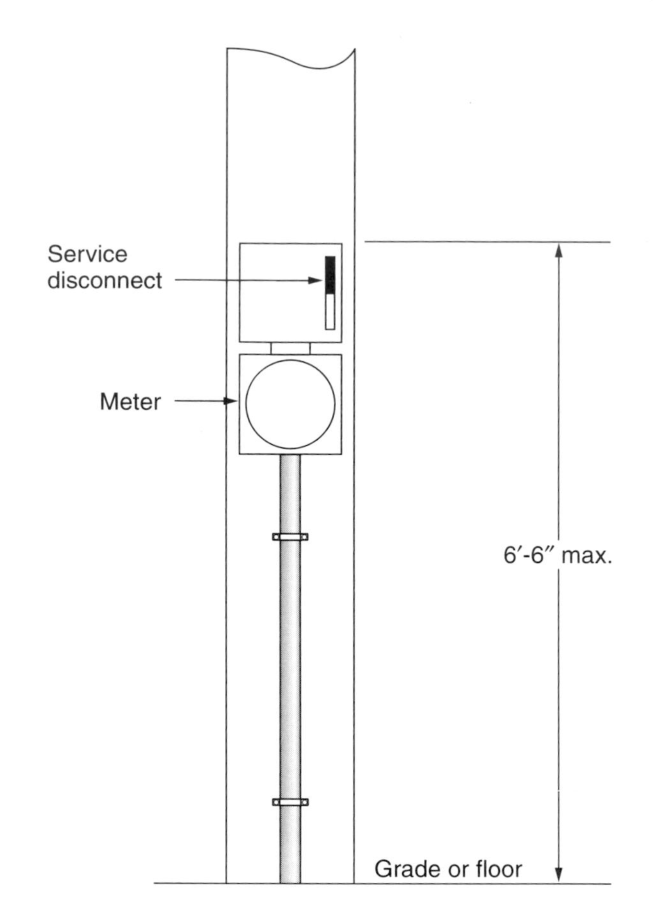

Figure 9-13. The service disconnect must be within 6′-6″ of the floor.

Distribution Systems

The power utility company provides numerous voltage levels for its customers. In addition, the voltage received at the service can be changed by transformers to meet certain specific processing or manufacturing needs.

Supply Voltage Levels

There are several common ac voltage levels available in the United States. The voltage is selected based on the type of equipment being supplied.

120/240-volt, single-phase, three-wire system

This type of system is used primarily for lighting, small appliance, and motor circuits. It is used in homes, small stores, and light commercial facilities where the total load is relatively minimal. See **Figure 9-14.**

120/208-volt, three-phase, four-wire system

Many small commercial establishments require this type of distribution system. It provides a highly efficient, economical, and flexible supply. Both single- and three-phase loads can be supplied, such as lighting and motors, respectively. See **Figure 9-15.**

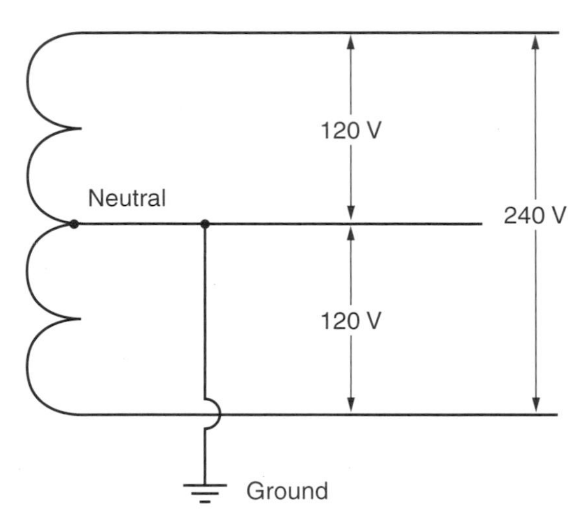

Figure 9-14. 120/240-volt, single-phase distribution system.

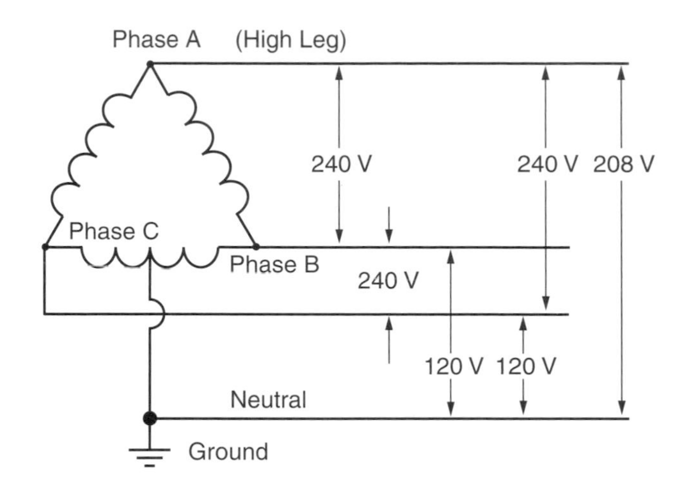

Figure 9-16. 240-volt, three-wire, three-phase system changed to a 240/120 volt system by adding a fourth wire.

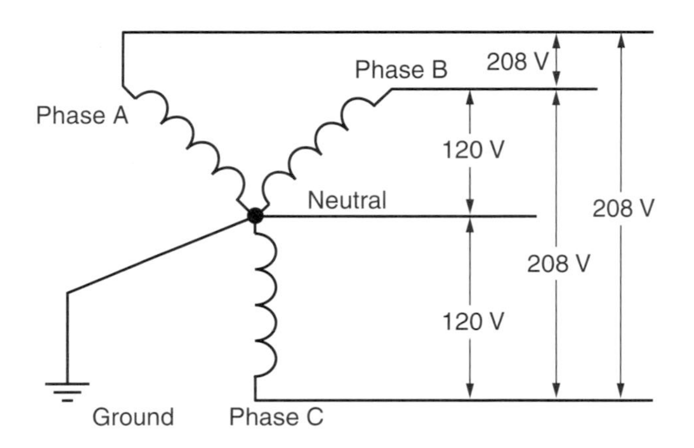

Figure 9-15. 208/120-volt, three-phase distribution system.

240-volt, three-phase, three-wire system

Used in commercial and industrial facilities, this system provides power to panelboards, busways, and wireways supplying motor branch circuits. As shown in **Figure 9-16,** adding a grounded center tap to one of the phase legs creates a neutral for circuits requiring 120 volts, such as lighting. When modified in such a manner, the distribution system is made into a 240/120-volt, three-phase, four-wire system.

480-volt, three-phase, three-wire system

When commercial and industrial facilities have fairly large motor power requirements, this system provides a highly efficient and economical supply alternative. The 480-volt power can be reduced with a step-down transformer for lighting and receptacle loads.

480/277-volt, three-phase, four-wire system

This type of distribution provides not only power for facilities having substantial motor loads, but also power for large fluorescent lighting loads commonly found in large commercial and industrial complexes. It is one of the most widely used distribution systems. Step-down transformers can be used for circuits requiring 120 volts, such as lighting and receptacles. See **Figure 9-17.**

4160/2400-volt, three-phase, four-wire system

This distribution system is used in industrial plants. A grounded neutral and step-down transformers are used to supply a wide variety of voltages for motors, lighting, and substations. A 2400-volt, three-phase, three-wire system can also be used for the same facilities, but it is not as efficient.

4800-volt, three-phase, three-wire system

These systems are primarily used in industrial complexes to feed 480-volt substations for motors and lighting. The 7200-volt supply is used in a similar manner, being stepped down at substations and transformers in order to supply lower-voltage requirements.

13,800-volt, three-phase, four-wire system

For large industrial plants, the 13.8-kilovolt, three-phase, four-wire system is often used because of its

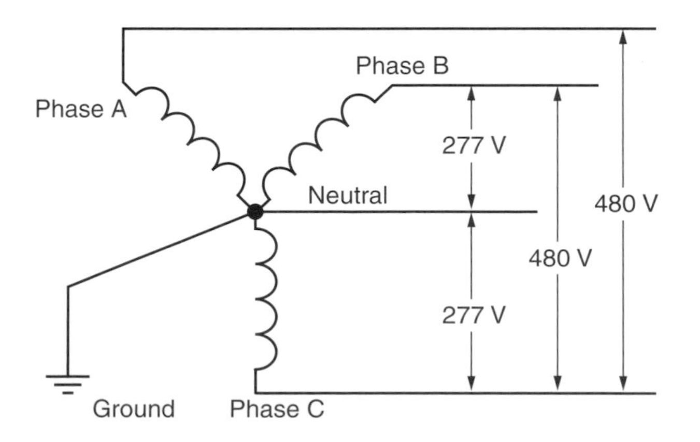

Figure 9-17. 480/277-volt distribution system.

tremendous versatility. This voltage supply is fed to substations within the plant that step-down the voltage to 480 volts, 240 volts, 277 volts, 208 volts, or 120 volts, depending on the load requirements.

Review Questions

Answer the following questions. Do not write in this book.

1. Briefly explain what is meant by the term *service*.
2. What is a service disconnect?
3. What is the minimum vertical clearance below service-drop conductors above a flat roof that is not subject to vehicular traffic?
4. Explain the difference between a service drop and a service lateral.
5. What is the importance of a drip loop?
6. What is the maximum height above grade for the handle for a service disconnect?
7. What are the support requirements for service-entrance cable between the service head and the meter?
8. What is the maximum number of disconnects permitted to be installed for each service?
9. What is the minimum vertical clearance below service-drop conductors passing over a parking lot?
10. Which article in the *Code* contains the requirements for workspace clearances around large (over 600-volt) services?

USING THE NEC

Refer to the National Electrical Code to answer the following questions. Do not write in this book.

1. What is the minimum size for service-drop conductors and service-lateral conductors?
2. Which *Code* section specifies minimum burial depth of service-lateral conductors?
3. Which *Code* section prohibits service conductors passing through one building to serve another building or structure?
4. Ungrounded service conductors must have overcurrent protection. Where is this overcurrent protection located?
5. For aboveground service-entrance conductors exceeding 600 volts, which *Code* section lists the approved wiring methods?

Chapter 10 Transformers

Technical Terms

Askarel
Autotransformer
Constant-current transformer
Control transformer
Copper loss
Core
Core loss
Current transformer
Dry-type transformer
Heat loss
I^2R loss
Induction
Isolation transformer
Less-flammable liquid-insulated transformer
Liquid-insulated transformer
Magnetic field
Potential transformer
Primary winding
Resistance loss
Secondary winding
Step-down transformer
Step-up transformer
Transformer efficiency
Transformer impedance
Transformer windings
Turns ratio
Voltage ratio
Windings

Objectives

After completing this chapter, you will be able to:

- Define the purpose and uses of transformers.
- Identify the basic components and construction of a transformer.
- Explain how a transformer works.
- List the types of transformers.
- Size overcurrent protective devices for transformer primaries and secondaries.
- Perform transformer calculations and solve practical transformer problems.

For economic reasons, power is transmitted from the utility generating station to the consumer at high voltages and low currents. This permits the use of smaller wire sizes. A ***transformer*** is a device that converts electrical power from one voltage and current to another voltage and current.

Transformer Theory

A transformer is made up of three parts: a primary coil, a core, and a secondary coil. All three parts are stationary. The operation of a transformer depends on magnetic fields, rather than mechanical movement.

The two separate coils, or ***windings,*** are wrapped around a core. The magnetic core is normally made of iron or bound steel strips. When current is applied to one of the windings, an alternating magnetic field is created in the core. This magnetic field causes current to flow in the other winding. This principle, called ***induction,*** is the basis of transformer operation.

The ***primary winding*** is energized with current and produces the magnetic field. The ***secondary winding*** produces current induced by the magnetic field. See **Figure 10-1.**

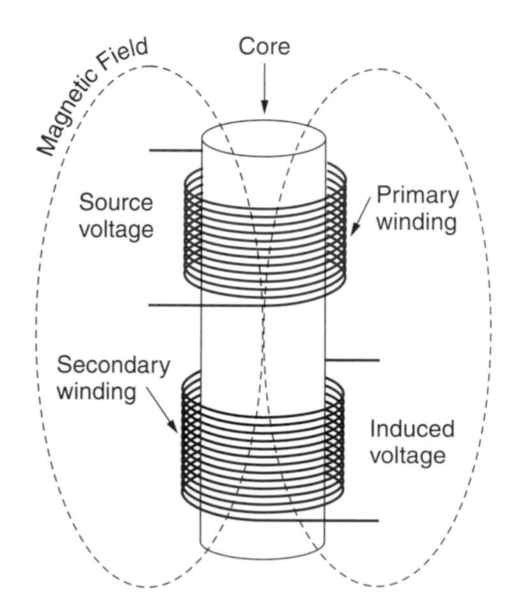

Figure 10-1. Basic parts of a transformer. The current in the primary winding produces a magnetic field, which induces current in the secondary winding.

Windings are normally referred to as high-voltage or low-voltage, rather than primary or secondary. High-voltage terminals are marked as H, H1, H2, etc., while low-voltage terminals are marked as X, X1, X2, etc.

There is a direct relationship (the ***voltage ratio***) between the voltage in the primary winding and the voltage in the secondary winding. The number of turns of wire in each coil determines the ratio. For example, if a transformer has a 10:1 voltage ratio, one coil has ten times as many turns of wire as the other. This winding will produce ten times the voltage of the other winding. The voltage ratio is also referred to as the ***turns ratio.***

If the primary coil has more turns than the secondary coil, the primary voltage will be greater than the secondary voltage. This type of transformer is called a ***step-down transformer*** because the voltage leaving the transformer is less than the voltage supplied. A ***step-up transformer*** has more turns in the secondary coil and produces a voltage higher than the input voltage.

Power Rating

Transformer power is designated in units of volt-amperes (VA), or more commonly, kilovolt-amperes (kVA). The mathematical formula for power was presented in Chapter 1 of this text.

$$\text{Power} = \text{Voltage} \times \text{Current}$$

$$P = E \times I$$

In a transformer, the current and voltage are different for each winding. However, the power is identical for both windings. To compute the power of a transformer, multiply the voltage and current of the primary *or* secondary winding. For single-phase transformers, the following formulas can be used:

$$P = V_P \times I_P$$

or

$$P = V_S \times I_S$$

where

V_P = Voltage in primary coil
I_P = Current in primary coil
V_S = Voltage in secondary coil
I_S = Current in secondary coil

In the case of three-phase transformers, multiply the voltage by 1.732 before calculating the power. The power rating of a transformer indicates the maximum allowable power. **Figure 10-2** contains tables of full-load currents for various transformers and voltages.

Single-Phase Transformer Full-Load Current

Power Rating (kVA)	Voltage					
	120 V	208 V	240 V	480 V	600 V	2400 V
1	8.3	4.8	4.2	2.1	1.7	0.42
1.5	12.5	7.2	6.2	3.1	2.5	0.62
2	16.7	9.6	8.3	4.2	3.3	0.83
3	25	14.4	12.5	6.2	5.0	1.25
5	41.7	24	20.8	10.4	8.3	2.1
7.5	62.5	36.1	31.2	15.6	12.5	3.1
10	83.4	48	41.6	20.8	16.7	4.16
15	125	72	62.5	31.2	25	6.25
25	208	120	104	52	41.7	10.4
37.5	312	180	156	78	62.5	15.6
50	417	240	208	104	83.5	20.8
75	625	361	312	156	125	31.2
100	834	480	416	208	167	41.6
125	1042	600	520	260	208	52.0
167.5	1396	805	698	349	279	70.0
200	1666	960	833	416	333	83.3
250	2080	1200	1040	520	417	104
333	2776	1600	1388	694	555	139
500	4170	2400	2080	1040	835	208

Three-Phase Transformer Full-Load Current

Power Rating (kVA)	Voltage					
	208 V	240 V	480 V	600 V	2400 V	4160 V
3	8.3	7.2	3.6	2.9	0.72	0.42
6	16.6	14.4	7.2	5.8	1.44	0.83
9	25	21.6	10.8	8.7	2.16	1.25
15	41.6	36	18	14.4	3.6	2.1
22.5	62	54	27	21.6	5.4	3.1
25	69	60	30	24	6	3.5
30	83	72	36	29	7.2	4.15
37.5	104	90	45	36	9	5.2
45	125	108	54	43	10.8	6.25
50	139	120	60	48	12	7
75	208	180	90	72	18	10.4
112.5	312	270	135	108	27	15.6
150	416	360	180	144	36	20.8
225	625	542	271	217	54	31.2
300	830	720	360	290	72	41.5
500	1390	1200	600	480	120	69.4
600	1665	1443	722	577	144	83
750	2080	1800	900	720	180	104
1000	2775	2400	1200	960	240	139
1500	4150	3600	1800	1440	360	208
2000	5550	4800	2400	1930	480	277
2500	6950	6000	3000	2400	600	346
5000	13900	12000	6000	4800	1200	694
7500	20800	18000	9000	7200	1800	1040
10000	27750	24000	12000	9600	2400	1386

Figure 10-2. These tables show full-load current for single-phase and three-phase transformers with various voltage ratings.

Sample Problem 10-1

Problem: What is the full-load current limit for a 5-kVA transformer supplying a 120-volt secondary?

Formula: Rearrange $P = E \times I$ to solve for I:

$$I = \frac{P}{E}$$

Solution: Solve for current using the values of voltage and power. However, the power and voltage values are expressed in different units, so they must be converted to similar units:

$$1 \text{ kVA} = 1000 \text{ VA}$$
$$5 \text{ kVA} = 5000 \text{ VA}$$

Now that the units agree, the numbers can be put into the formula:

$$\begin{aligned} I &= \frac{P}{E} \\ &= \frac{5000 \text{ VA}}{120 \text{ V}} \\ &= 41.67 \text{ A} \end{aligned}$$

The full-load current for this transformer is 41.67 amps.

Impedance

The impedance of a transformer is an indication of the short-circuit current delivered when a direct short is placed across the secondary terminals. The impedance factor is listed as a percentage on the transformer nameplate. This factor is used in sizing overcurrent protective devices.

The transformer's ***available secondary short-circuit current (ASC)*** is computed by dividing the ***secondary full-load current (FLC)*** by the impedance (Z), and then multiplying by 100%. In this formula, impedance is expressed as a percentage.

$$ASC = \frac{FLC \times 100\%}{Z}$$

Sample Problem 10-2

Problem: A 50-kVA transformer has a single-phase secondary voltage of 240 volts and an impedance of 5%. What is the available secondary short-circuit current?

Formulas:

$$I = \frac{P}{E}$$

$$ASC = \frac{FLC \times 100\%}{Z}$$

Solution: First, the secondary full-load current must be determined. This is computed using the power and the voltage:

$$\begin{aligned} FLC &= \frac{P}{E} \\ &= \frac{50{,}000 \text{ VA}}{240 \text{ V}} \\ &= 208.3 \text{ A} \end{aligned}$$

The available short-circuit secondary current is calculated using the following formula:

$$\begin{aligned} ASC &= \frac{FLC \times 100\%}{Z} \\ &= \frac{208.3 \text{ A} \times 100\%}{5\%} \\ &= 4166 \text{ A} \end{aligned}$$

The available secondary short-circuit current is 4166 amps. This means that the overcurrent device, breaker, or fuse (which must be sized and rated to protect the circuit adequately) should be 225 A with a short-circuit withstand of at least 5000 A.

Voltage, Current, and Impedance Calculations

The table in **Figure 10-3** shows the algebraic equations relating voltage, current, and impedance in a transformer. Notice how the voltage, current, and power are related:

- Voltage is directly proportional to the number of turns.
- Current is inversely proportional to the voltage (and number of turns).
- Power is equal in each of the windings.

Sample Problem 10-3

Problem: A transformer is provided with a 480-volt source and must supply 120 volts and 30 amps. What is the required primary current?

Solution: We know the primary voltage (E_P), secondary voltage (E_S), and secondary current (I_S), and we want to determine the primary current (I_P). In the "Current" column of Figure 10-3, find an equation for I_P using the known variables:

$$I_P = \frac{E_S \times I_S}{E_P}$$

$$= \frac{120 \text{ V} \times 30 \text{ A}}{480 \text{ V}}$$

$$= 7.5 \text{ A}$$

The primary current is 7.5 A.

Sample Problem 10-4

Problem: A transformer with a primary winding with 500 turns and a secondary winding with 250 turns is connected to a 480-volt source. What voltage is produced on the secondary side?

Solution: We know N_P, N_S, and E_P, and want to determine E_S. Using an equation from the table in Figure 10-3:

$$E_S = \frac{E_P \times N_S}{N_P}$$

$$= \frac{480 \text{ V} \times 250}{500}$$

$$= 240 \text{ V}$$

The secondary side would supply 240 volts.

Sample Problem 10-5

Problem: A transformer has a primary coil with 2500 turns and a secondary coil with 500 turns. It must deliver 25 amps to its load side. What current should be supplied to the transformer?

Solution: Using an equation from the table in Figure 10-3:

$$I_P = \frac{I_S \times N_S}{N_P}$$

$$= \frac{25 \text{ A} \times 500}{2500}$$

$$= 5 \text{ A}$$

The transformer should be supplied with a current of 5 amps.

A transformer adjusts to meet the requirement of the load-side current. When no current is being used on the load side, no current flows in the primary, except a very small amount needed to maintain the magnetic field.

Transformer Formulas

Voltage (E)	Current (I)	Impedance (Z)
$\frac{E_S}{E_P} = \frac{N_S}{N_P}$	$\frac{I_P}{I_S} = \frac{N_S}{N_P}$	$\frac{Z_S}{Z_P} = \frac{N_S^2}{N_P^2}$
$E_P \times N_S = E_S \times N_P$	$I_P \times N_P = I_S \times N_S$	
$E_P = I_S \times \frac{E_S}{I_P}$	$I_P = E_S \times \frac{I_S}{E_P}$	$Z_P \times N_S^2 = Z_S \times N_P^2$
$E_P = N_P \times \frac{E_S}{N_S}$	$I_S = I_P \times \frac{E_P}{E_S}$	
$E_S = E_P \times \frac{N_S}{N_P}$	$I_P = N_S \times \frac{I_S}{N_P}$	$\frac{Z_P}{Z_S} = \frac{N_P^2}{N_S^2}$
$E_S = I_P \times \frac{E_P}{I_S}$	$I_S = I_P \times \frac{N_P}{N_S}$	
$N_P = E_P \times \frac{N_S}{E_S}$	$N_P = I_S \times \frac{N_S}{I_P}$	
$N_S = N_P \times \frac{E_S}{E_P}$	$N_S = N_P \times \frac{I_P}{E_S}$	

E_P = Primary voltage
N_P = Primary turns
I_P = Primary current
Z_P = Primary impedance
E_s = Secondary voltage
N_S = Secondary turns
I_S = Secondary current
Z_S = Secondary impedance

Figure 10-3. This table shows useful algebraic relationships for transformer voltage, current, and impedance.

Transformer Efficiency

Transformers are highly efficient, but they do have some losses. These losses are due to heating and power consumption within the components. Two kinds of losses are inherent with transformers. Both are small, but always present.

- **Core loss**—This type of loss is independent of the load supplied by the transformer. It occurs due to the power consumed in creating the magnetic flux. This power loss is about 1% of the transformer power rating.
- **Copper loss**—This type of loss is also referred to as ***heat loss***, ***resistance loss***, and ***I^2R loss***. It is caused by resistance in the windings and is numerically equal to $I^2 \times R$. This loss varies with the square of the current and is normally less than 2% of the transformer power rating.

Sample Problem 10-6

Problem: If a transformer has a winding resistance of 4 ohms and a load current of 6 amperes, what is the copper loss?

Solution: Using the formula presented in the previous discussion:

$$\begin{aligned}\text{Copper loss} &= I^2R \\ &= (6\text{ A})^2 \times 4\ \Omega \\ &= 144\text{ VA}\end{aligned}$$

The copper loss is 144 volt-amperes.

Copper loss and core loss are combined and compared to the overall transformer power rating to determine the efficiency:

$$\text{Efficiency} = \frac{\text{Output}}{\text{Input} \times 100\%}$$

The efficiency is normally listed on the transformer nameplate.

Transformer Varieties

Transformers are identified and classified in many ways. The construction, type of cooling system, winding connections, and usage may be used to describe a given transformer and to differentiate it from other transformers.

Transformers can be classified as dry-type or liquid-insulated. This classification is based on whether or not the transformer core and winding are submersed in an insulating liquid.

In ***dry-type transformers,*** air surrounds the core and windings. Ventilation openings in the transformer enclosure allow air to flow through the transformer, dissipating the heat produced by the windings. Larger transformers use a blower to circulate more air through the enclosure.

The core and windings of a ***liquid-insulated transformer*** are enclosed in a sealed tank filled with liquid. The liquid serves as an insulator between the components and also provides a medium for heat to be dissipated. Heat is transferred from the windings, through the liquid, to the tank. The air in contact with the outer tank surface then removes the heat.

Small liquid-insulated transformers have a smooth tank. Tanks on larger transformers have fins or external radiators to provide a larger surface in contact with the air. The larger contact surface allows heat to dissipate more quickly. A blower can also be used to circulate air around the tank. A sensor measuring oil temperature operates a thermostat in the blower circuit.

NEC NOTE 450.9

The ventilation shall be adequate to dispose of the transformer full-load losses without creating temperature rise that is in excess of the transformer rating.

Sample Problem 10-7

Problem: A 50-kVA transformer has a copper loss of 0.45 kVA and a core loss of 0.35 kVA. Determine the efficiency of the transformer.

Solution: First, the core and copper losses are combined to determine the total losses:

$$\begin{aligned}\text{Total losses} &= 0.45\text{ kVA} + 0.35\text{ kVA} \\ &= 0.80\text{ kVA}\end{aligned}$$

The output power is the nominal power rating of the transformer. The input power is the output power plus the total losses. From these values, efficiency is computed:

$$\text{Output power} = 50\text{ kVA}$$

$$\begin{aligned}\text{Input power} &= 50\text{ kVA} + 0.80\text{ kVA} \\ &= 50.8\text{ kVA}\end{aligned}$$

$$\begin{aligned}\text{Efficiency} &= \frac{\text{Output}}{\text{Input} \times 100\%} \\ &= \frac{50\text{ kVA}}{50.8\text{ kVA} \times 100\%} \\ &= 98.4\%\end{aligned}$$

The transformer efficiency is 98.4%.

The liquid adds additional weight to liquid-insulated transformers, making them heavier than comparable dry-type transformers. Also, extra precautions must be taken to check for leaks when installing liquid-insulated transformers.

In the past, liquid-insulated transformers contained either oil or ***askarel,*** a type of nonflammable oil. However, askarel-filled transformers are no longer used due to environmental hazards. They have been replaced with ***less-flammable liquid-insulated transformers.*** The insulating liquid in these transformers may burn, but flames will not spread from the point of ignition.

Transformers can also be classified by their windings. In an ***isolation transformer,*** the primary and secondary winding are physically separate and independent of one another. In an ***autotransformer,*** there is only one winding, which is shared by the primary and secondary circuits. See **Figure 10-4.** This type of transformer is used when voltage reduction is needed (for example, starting ac motors). Though economical to manufacture and very efficient, they are limited in application due to their low voltage ratios.

There are several other types of transformers that are more specialized in their design and application than basic isolation transformers and autotransformers. These include potential transformers, current transformers, constant-current transformers, and control transformers.

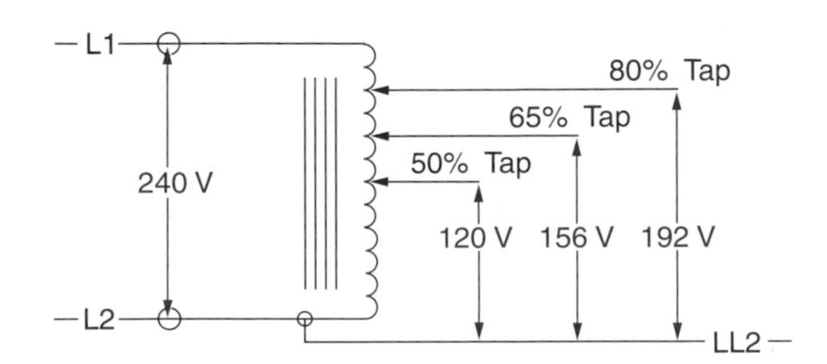

Figure 10-4. The secondary voltage in this step-down autotransformer varies, depending on the location of the tap on the windings.

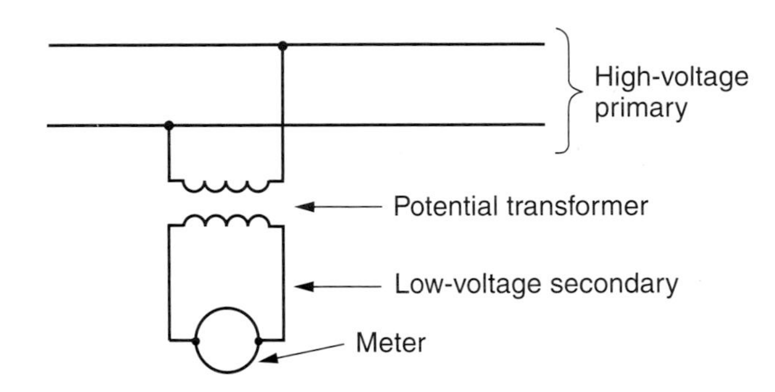

Figure 10-5. A potential transformer reduces high voltage to protect measuring devices.

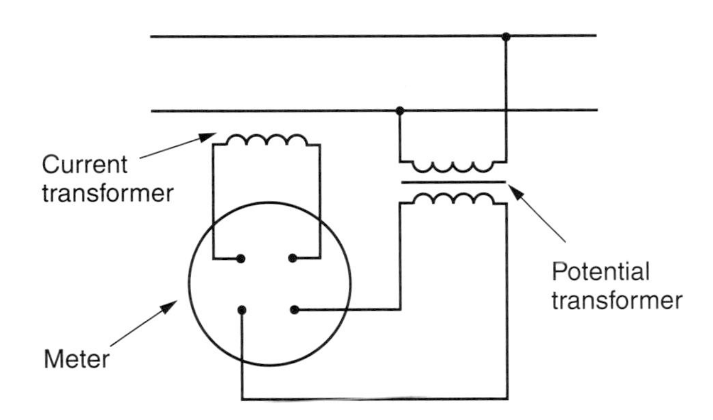

Figure 10-6. Potential transformers and current transformers are often paired to effectively step-down the voltage and current to an instrument. The primary of a current transformer is always connected in series with the high voltage line.

Potential Transformers (PT)

These step-down transformers are used to operate measuring instruments, such as voltmeters and wattmeters. The secondary voltage is a percentage of the primary voltage, and is normally 120 volts. This type of transformer is illustrated in **Figure 10-5.**

Current Transformers (CT)

Current transformers are used for reducing high current. Like potential transformers, they are used with instruments. The primary of a current transformer is connected in series with the high-voltage line. The secondary current is proportional to the primary current and is usually designed to be 5 amps. Current and potential transformers are often used together, **Figure 10-6.**

Constant-Current Transformers

Constant-current transformers maintain an unchanging current on the secondary side, even if the load demand varies. Regardless of what changes take place among the operating devices on the secondary side, the constant-current transformer will maintain the same delivered current (within a 1% tolerance) at all times.

Control Transformers

Control transformers reduce the primary voltage to 120 volts or less on the secondary. The secondary is used to operate solenoids, relays, and contactors of a step-down transformer. Sometimes these transformers are incorrectly referred to as signal or instrument transformers.

Transformer Installation

Most of the *Code* requirements for transformer installation are found in *Part II* of *Article 450.* These requirements are mainly concerned with preventing a transformer fire from spreading to a building or other combustible material. Provisions vary depending on the transformer type, its voltage rating, and whether it is installed indoors or outdoors.

Transformers should be located in such a manner that they will not be physically damaged, exposed to severe weather, or liable to having foreign objects inserted into them. Ventilation openings must not be blocked by walls or other obstructions.

Transformers must be located where they are easily accessible. If the transformer is concealed or obstructed, maintenance will be difficult.

NEC NOTE 450.21(B)

Unless otherwise specified in this article, the term *fire-resistant* means a construction having a minimum fire rating of 1 hour.

Dry-Type Transformer Location

Dry-type transformers located outdoors must have a weatherproof enclosure. In addition, transformers rated above 112 1/2 kVA must be located at least 12″ away from buildings and combustible materials. This clearance may be reduced if the requirements of *Section 450.22* of the *Code* are met.

When installed indoors, dry-type transformers over 35,000 volts must be installed in a transformer vault. Transformers rated 112 1/2 kVA or higher must be installed in a transformer room of fire-resistant construction. Transformers with a lower rating must have 12″ clearance from any part of the building or combustible materials, unless rated for 600 volts or less and completely enclosed (except for ventilation openings).

In accordance with *Section 450.13,* dry-type transformers with a voltage rating less than 600 volts do *not* need to be readily accessible in the following two situations:

- The transformer is located in the open on a wall or column.
- The power rating does not exceed 50 kVA and the transformer is located with proper ventilation in a fire-resistant hollow space that is not permanently enclosed in the structure.

Liquid-Insulated Transformer Locations

Special precautions are needed when an oil-insulated transformer is installed outdoors near a building or on a roof. Any nearby buildings and combustible materials must be protected in case of fire. This protection can be any or a combination of the following:

- Sufficient distance from transformer.
- Automatic fire protection system, such as sprinklers.
- Fire-resistant barriers.
- Enclosure to contain spilled oil, such as a curbed area or trench.

Normally, oil-insulated transformers installed indoors must be located in a transformer vault. There are exceptions listed in *Section 450.26,* including the following:

- No vault is needed for transformers rated less than 600 volts if the total transformer capacity in one location is less than 75 kVA (for fire-resistant areas) or 10 kVA (for combustible areas), and if proper measures are taken to ensure that an oil fire cannot ignite other materials.
- No vault is needed if the transformer is located in a detached building used only for supplying electrical service, accessible to qualified personnel only, with contents that would not create a hazard in the event of fire.

Less-flammable liquid-insulated transformers rated less than 35,000 volts do *not* require a transformer vault if protected by an automatic fire extinguishing system, provided with a liquid-containment area, and located away from combustible materials. If the transformer is rated over 35,000 volts and is installed indoors, it must be located in a transformer vault.

Transformer Vaults

Transformer vaults are designed to contain transformer oil spills and prevent fire from escaping a confined area. When installed indoors, all transformers rated over 35,000 volts and many rated under 35,000 volts must be installed in a vault. The *Code* addresses transformer vaults in *Part III* of *Article 450.*

The walls, ceiling, and floor of a transformer vault must be constructed of material with a minimum three-hour fire rating. If the floor rests on earth, it must be at least 4″ thick.

Interior doors must be tight-fitting with a three-hour fire-resistance. The sill below the door must be high enough to contain the amount of oil in the largest transformer, with a 4″ minimum. Doors must swing out from the vault and must be locked, with only qualified personnel having access.

NOTE

Where protected by an acceptable automatic fire extinguishing system, walls, ceilings, floors, and doors with one-hour fire-resistant construction are allowed.

Sufficient ventilation must be provided to keep transformer temperatures at an acceptable level. Vents should be located as far as possible from doors, windows, and combustible materials. Whenever possible, vaults should be located so they can be ventilated with fresh air without the use of ducts and flues. Ventilation openings between the vault and the building must have 1 1/2-hour fire-resistant dampers that automatically close in the event of fire.

Materials should not be stored in transformer vaults. Pipes and ducts that are not related to the vault should not pass through the vault.

Grounding

All noncurrent-carrying metal parts of the transformer, including the enclosure, must be bonded together and properly grounded. *Section 250.110* provides the requirements for grounding fastened equipment. Grounding points on the secondary windings are shown in **Figure 10-7.** The winding cannot be grounded by connecting the neutral to the enclosure.

NEC NOTE 450.10

Exposed, noncurrent-carrying metal parts of transformer installations, including fences, guards, etc., shall be grounded where required under the conditions and in the manner specified for electric equipment and other exposed metal parts in *Article 250.*

NOTE

Grounding is discussed in Chapter 11 of this text.

Overcurrent Protection

Transformers require overcurrent protection. General requirements for transformer overcurrent protection are specified in *Section 450.3* of the *Code.*

Normally, separate overcurrent protective devices are required for the transformer primary and secondary. The device rating is based on the current rating of the primary or secondary. The sizing requirements are different for transformers rated over 600 volts and for transformers rated 600 volts or less.

NOTE

These overcurrent protection requirements are for the transformer only. The conductors connected to the transformer primary and secondary must have overcurrent protection as specified in *Article 240* and *Article 310.*

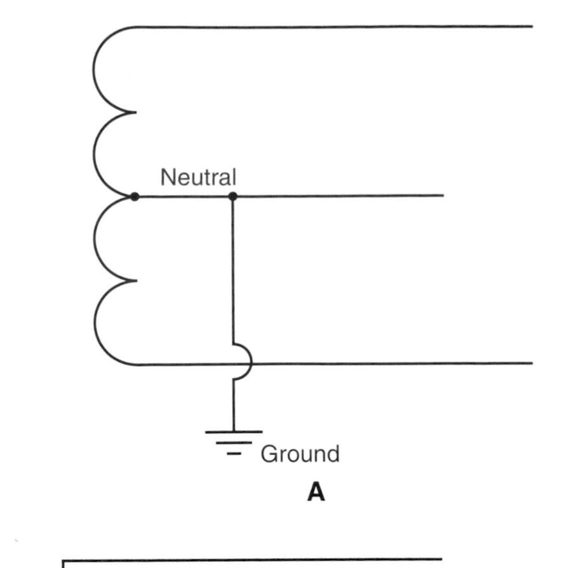

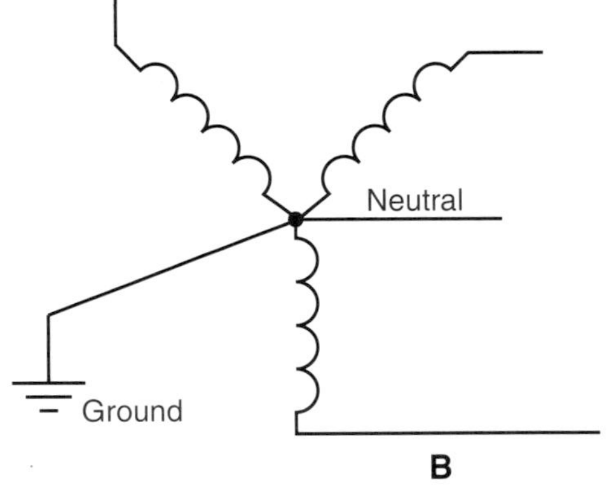

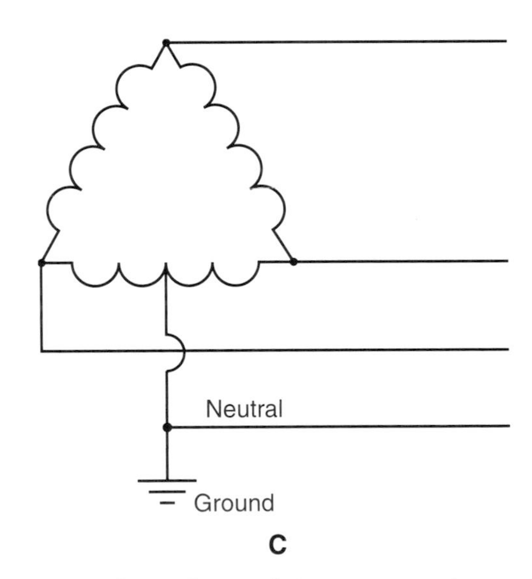

Figure 10-7. Grounding points on secondary windings. A—Single-phase, three-wire system. B—Three-phase, four-wire wye system. C—Three-phase, four-wire delta system.

Transformers Rated over 600 Volts

Both primary and secondary overcurrent protective devices are normally required for transformers rated over 600 volts. The maximum allowable settings for the overcurrent devices depend on the rated transformer impedance, the type of device used, and the rated voltage of the secondary. See **Figure 10-8,** which is a reproduction of *Table 450.3(A)* of the *Code*. If the required rating does not correspond to a standard device rating, the next higher device rating may be used.

Normally, overcurrent protection is required for both the transformer primary and secondary. The required overcurrent protective device rating depends on the rated impedance of the transformer. If the rated impedance is 6% or less, higher circuit breaker or fuse ratings are allowed.

If the transformer is located in a supervised location, secondary overcurrent protection may not be necessary if the primary overcurrent protection has a maximum setting of 300% for circuit breakers or 250% for fuses. Also, secondary overcurrent protection may be omitted if the transformer is supervised and equipped with coordinated thermal overload protection. *Note 3* in *Table 450.3(A)* defines *supervised location.*

NOTE

Check with the authority having jurisdiction regarding supervised transformer installations. Their definition of "supervised" may differ from yours, and *their decision* determines the requirements.

Transformers Rated 600 Volts or Less

Section 450.3(B) of the *Code* addresses overcurrent protection for transformers rated 600 volts or less. Carefully review this section of the *Code*; there are different protection arrangements depending on conditions and transformer ratings. The protective device ratings depend on two factors: the protection method (primary only or primary and secondary) and the current rating of the transformer. See **Figure 10-9.**

- **Primary protection only**—The overcurrent protective device for the transformer primary

Table 450.3(A) Maximum Rating or Setting of Overcurrent Protection for Transformers Over 600 Volts (as a Percentage of Transformer-Related Current)

				Secondary Protection (see Note 2)		
		Primary Protection Over 600 Volts		Over 600 Volts		600 Volts or Less
Location Limitations	Transformer Rated Impedance	Circuit breaker (see Note 4)	Fuse Rating	Circuit Breaker (see Note 4)	Fuse Rating	Circuit Breaker or Fuse Rating
Any location	Not more than 6%	600% (see Note 1)	300% (see Note 1)	300% (see Note 1)	250% (see Note 1)	125% (see Note 1)
	More than 6% and not more than 10%	400% (see Note 1)	300% (see Note 1)	250% (see Note 1)	225% (see Note 1)	125% (see Note 1)
Supervised locations only (see Note 3)	Any	300% (see Note 1)	250% (see Note 1)	Not required	Not required	Not required
	Not more than 6%	600%	300%	300% (see Note 5)	250% (see Note 5)	250% (see Note 5)
	More than 6% and not more than 10%	400%	300%	250% (see Note 5)	225% (see Note 5)	250% (see Note 5)

Notes:
1. Where the required fuse rating or circuit breaker setting does not correspond to a standard rating or setting, a higher rating or setting that does not exceed the next higher standard rating or setting shall be permitted.
2. Where secondary overcurrent protection is required, the secondary overcurrent device shall be permitted to consist of not more than six circuit breakers or six sets of fuses grouped in one location. Where multiple overcurrent devices are utilized, the total of all the device ratings shall not exceed the allowed value of a single overcurrent device. If both breakers and fuses are utilized as the overcurrent device, the total of the device ratings shall not exceed that allowed for fuses.
3. A supervised location where conditions of maintenance and supervision ensure that only qualified persons will monitor and service the transformer installation.
4. Electronically actuated fuses that may be set to open at a specific current shall be set in accordance with settings for circuit breakers.
5. A transformer equipped with a coordinated thermal overload protection by the manufacturer shall be permitted to have separate secondary protection omitted.

Figure 10-8. Overcurrent protection requirements for transformers over 600 volts. If the required rating does not correspond to a standard rating, the next higher device rating may be used.

Table 450.3(B) Maximum Rating or Setting of Overcurrent Protection for Transformers 600 Volts and Less (as a Percentage of Transformer-Rated Current)

	Primary Protection			Secondary Protection (see Note 2)	
Protection Method	**Currents of 9 Amperes or More**	**Currents Less than 9 Amperes**	**Currents Less than 2 Amperes**	**Currents of 9 Amperes or More**	**Currents Less than 9 Amperes**
Primary only protection	125% (see Note 1)	167%	300%	Not required	Not required
Primary and secondary protection	250% (see Note 3)	250% (see Note 3)	250% (see Note 3)	125% (see Note 1)	167%

Notes:
1. Where 125 percent of this current does not correspond to a standard rating of a fuse or nonadjustable circuit breaker, a higher rating that does not exceed the next higher standard rating shall be permitted.
2. Where secondary overcurrent protection is required, the secondary overcurrent device shall be permitted to consist of not more than six circuit breakers or six sets of fuses grouped in one location. Where multiple overcurrent devices are utilized, the total of all the device ratings shall not exceed that allowed value of a single overcurrent device. If both breakers and fuses are utilized as the overcurrent device, the total of the device ratings shall not exceed that allowed for fuses.
3. A transformer equipped with coordinated thermal overload protection by the manufacturer and arranged to interrupt the primary current, shall be permitted to have primary overcurrent protection rated or set at a current value that is not more than six times the rated current of the transformer for transformers having not more than 6 percent impedance, and not more than four times the rated current of the transformer for transformers having more than 6 percent but not more than 10 percent impedance.

Figure 10-9. Overcurrent protection requirements for transformer primaries 600 volts and less.

must be rated at the value shown in the first row of *Table 450.3(B)*. If the required rating does not correspond to a standard device rating, the next higher rating may be used. If the overcurrent protection for the primary feeder circuit satisfies this requirement, individual circuit protection for the transformer is not required.

- **Primary and secondary protection**—The bottom row of *Table 450.3(B)* lists overcurrent protective device settings for transformers with primary and secondary protection. The primary devices can have higher ratings than allowed for primary-only protection.

NOTE

Requirements for overcurrent protection of autotransformers can be found in *Section 450.4* and *Section 450.5* of the *Code.* Overcurrent protection requirements for secondary ties are found in *Section 450.6(B).*

Sample Problem 10-8

Problem: What size fuses are required to protect the primary side of a single-phase, 4160-volt, 120-kVA transformer in a supervised location if no individual secondary overcurrent protection is provided?

Solution: First, convert the power from kVA to VA:

$$120 \text{ kVA} = 120{,}000 \text{ VA}$$

Solving the power formula ($P = E \times I$) for I:

$$I = \frac{P}{E} = \frac{120{,}000 \text{ VA}}{4160 \text{ V}} = 28.8 \text{ A}$$

From *Table 450.3(A)*, the fuse rating for the primary is 250% (as shown on the next page).
The required fuse rating can be found by multiplying the voltage times 2.5 (250%).

$$\text{Fuse rating} = 28.8 \text{ A} \times 2.5 = 72 \text{ A}$$

Because this is not a standard fuse size (see *Section 240.6*), the next highest standard fuse (90 A) is used.

(Continued on the following page.)

Sample Problem 10-8 *Continued*

Table 450.3(A) Maximum Rating or Setting of Overcurrent Protection for Transformers Over 600 Volts (as a Percentage of Transformer-Related Current)

		Primary Protection Over 600 Volts		Secondary Protection (see Note 2) Over 600 Volts		Secondary Protection (see Note 2) 600 Volts or Less
Location Limitations	**Transformer Rated Impedance**	**Circuit breaker (see Note 4)**	**Fuse Rating**	**Circuit Breaker (see Note 4)**	**Fuse Rating**	**Circuit Breaker or Fuse Rating**
Any location	Not more than 6%	600% (see Note 1)	300% (see Note 1)	300% (see Note 1)	250% (see Note 1)	125% (see Note 1)
	More than 6% and not more than 10%	400% (see Note 1)	300% (see Note 1)	250% (see Note 1)	225% (see Note 1)	125% (see Note 1)
Supervised locations only (see Note 3)	Any	300% (see Note 1)	250% (see Note 1)	Not required	Not required	Not required
	Not more than 6%	600%	300%	300% (see Note 5)	250% (see Note 5)	250% (see Note 5)
	More than 6% and not more than 10%	400%	300%	250% (see Note 5)	225% (see Note 5)	250% (see Note 5)

Sample Problem 10-9

Problem: What size fuses are needed to protect a single-phase, 2-kVA transformer with a rated primary voltage of 240 volts? No additional overcurrent protection is provided for the transformer secondary.

Solution: First determine the current:

$$I = \frac{P}{E} = \frac{2000\text{ VA}}{240\text{ V}} = 8.3\text{ A}$$

The transformer voltage is less than 600 volts and the current is less than 9 amps. Use *Table 450.3(B)* to determine the fuse rating.

The fuse can be no larger than 167% of the primary current:

$$\text{Fuse size} = 8.3\text{ A} \times 1.67 = 13.9\text{ A}$$

This is not a standard size, so the next higher standard size (15 A) is used.

Table 450.3(B) Maximum Rating or Setting of Overcurrent Protection for Transformers 600 Volts and Less (as a Percentage of Transformer-Rated Current)

	Primary Protection			Secondary Protection (see Note 2)	
Protection Method	**Currents of 9 Amperes or More**	**Currents Less than 9 Amperes**	**Currents Less than 2 Amperes**	**Currents of 9 Amperes or More**	**Currents Less than 9 Amperes**
Primary only protection	125% (see Note 1)	167%	300%	Not required	Not required
Primary and secondary protection	250% (see Note 3)	250% (see Note 3)	250% (see Note 3)	125% (see Note 1)	167%

Transformer Connections and Wiring Diagrams

A single-phase, three-wire system is used for most residential and many commercial installations. This type of service has several advantages:

- There are two voltages available: 120 volts for lighting and small appliance loads, 240 volts for heavy appliance and motor loads.
- The system uses 35% less copper than a two-wire, 120-volt system having the same rating and transforming efficiency.
- Current can be balanced between the neutral and either of the two hot conductors. Thus, voltage drop is minimized and the voltage is more constant at the load end of the circuit.

Most electrical energy is transmitted by three-phase, ac generating devices. Three single-phase transformers can be used to transform the voltages of three-phase systems to single-phase current. There are several types of these transformers (**Figure 10-10**):

- Delta-delta
- Wye-wye
- Delta-wye
- Wye-delta
- Open delta

Sample Problem 10-10

Problem: A three-phase transformer is rated as 15 kVA, supplied with 480 volts, and supplies power for a 120/208-volt lighting panel with a continuous load of 30 amps. What size overcurrent protective device is needed for primary-only protection?

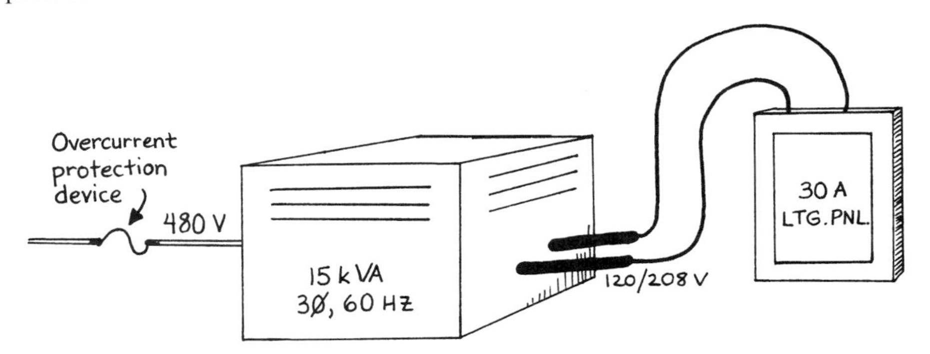

Solution: First, convert the power from kilovolt-amperes to volt-amperes:

$$15 \text{ kVA} = 15{,}000 \text{ VA}$$

Because this is three-phase current, the actual voltage is 1.732 times the nominal voltage:

$$P = 1.732 \times 480 \text{ V} = 831 \text{ V}$$

Then determine the primary current:

$$I = \frac{P}{E} = \frac{15{,}000 \text{ VA}}{831 \text{ V}} = 18 \text{ A}$$

Table 450.3(B) states that primary overcurrent protection for the transformer is to be rated at not more than 125% of the rated primary current for current greater than 9 amps:

$$\text{Fuse rating} = 18 \text{ A} \times 1.25 = 22.5 \text{ A}$$

The next highest standard fuse size (25 A) is used.

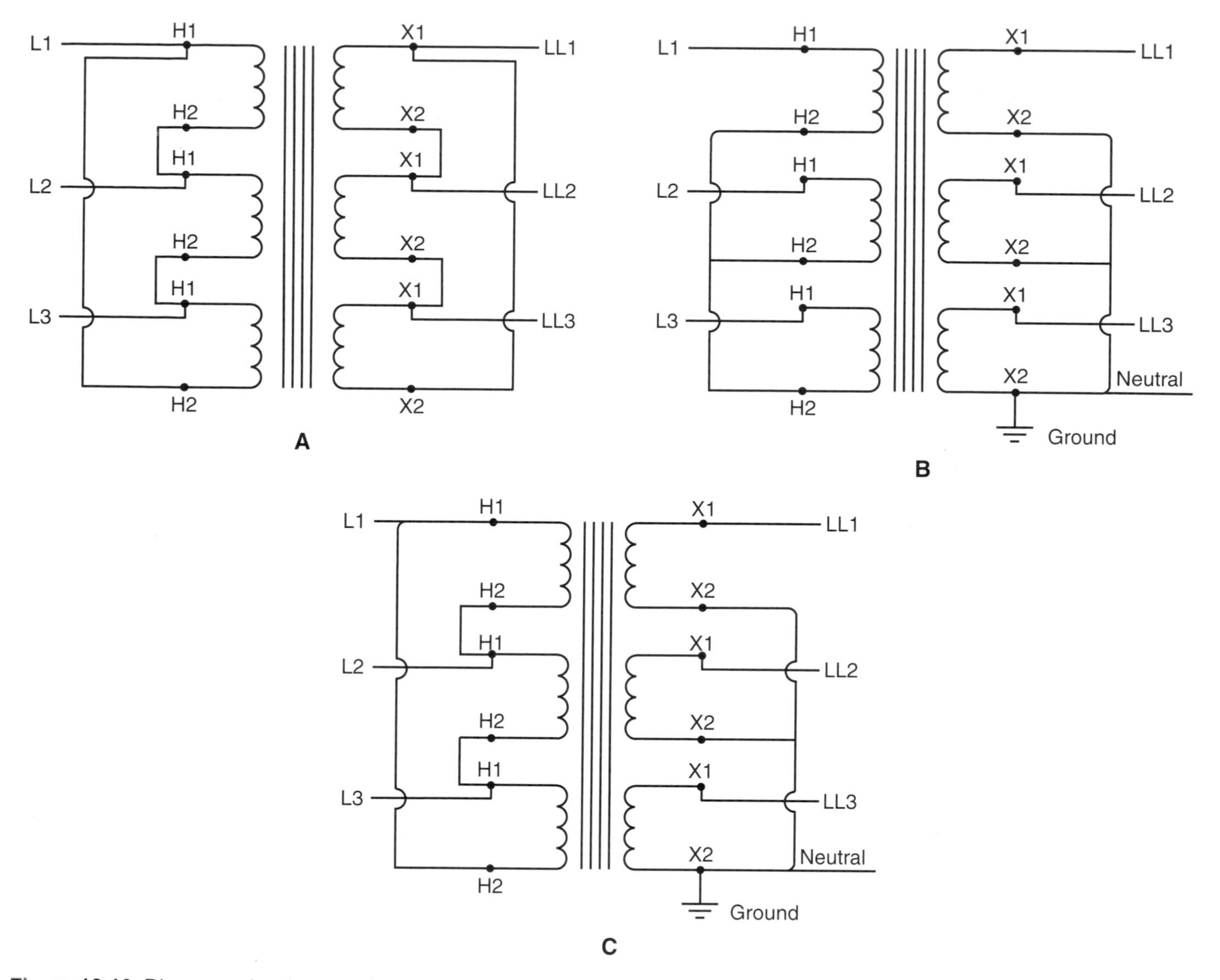

Figure 10-10. Diagrams showing transformer connections. A—Delta-delta transformer. B—Wye-wye transformer. C—Delta-wye transformer.

Sample Problem 10-11

Problem: What are the primary and secondary current ratings of a 500-kVA, three-phase, step-down transformer reducing a 480-volt delta to 120/208-volt wye?

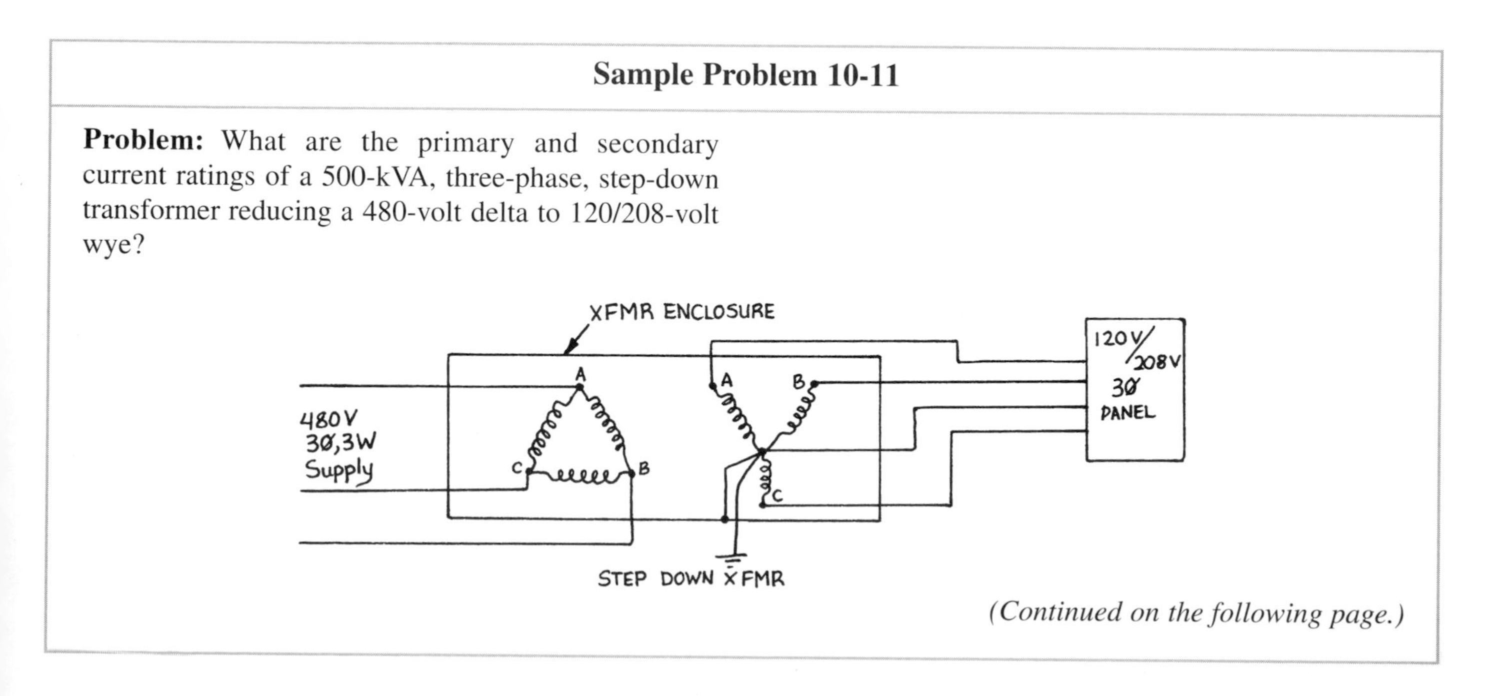

(Continued on the following page.)

Sample Problem 10-11 *Continued*

Solution: Determine the actual voltage (three-phase):

$$E = 480 \text{ V} \times 1.732$$
$$= 831 \text{ V}$$

Convert the power to units of volt-amperes:

$$500 \text{ kVA} = 500{,}000 \text{ VA}$$

Find the current on the delta primary side:

$$I = \frac{P}{E}$$
$$= \frac{500{,}000 \text{ VA}}{831 \text{ V}}$$
$$= 602 \text{ A}$$

Then, determine the current for the wye secondary. First, calculate the three-phase voltage:

$$208 \text{ V} \times 1.732 = 360 \text{ V}$$

Then the current is calculated:

$$I = \frac{P}{E}$$
$$= \frac{500{,}000 \text{ VA}}{360 \text{ V}}$$
$$= 1389 \text{ A}$$

The primary current is 602 A and the secondary current is 1389 A.

Sample Problem 10-12

Problem: A single-phase, 50-kVA transformer has a 480-volt primary and a 240-volt secondary. What are the primary and secondary current ratings of the transformer?

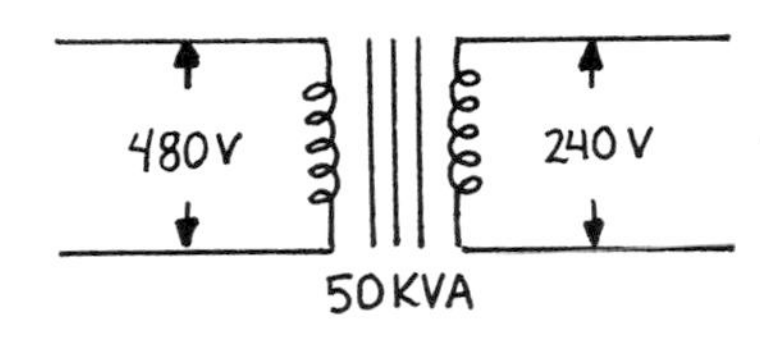

Solution: Convert power from kilovolt-amperes to volt-amperes:

$$50 \text{ kVA} = 50{,}000 \text{ VA}$$

Calculate the primary and secondary currents:

$$I_P = \frac{P}{E_P}$$
$$= \frac{50{,}000 \text{ VA}}{480 \text{ V}}$$
$$= 104 \text{ A}$$

$$I_S = \frac{P}{E_S}$$
$$= \frac{50{,}000 \text{ VA}}{240 \text{ V}}$$
$$= 208 \text{ A}$$

The primary current is 104 amps and the secondary current is 208 amps.

Review Questions

Answer the following questions. Do not write in this book.

1. Describe the main purpose of a transformer.
2. In a transformer vault, why is a sill below the door required and what is the minimum sill height?
3. Describe the construction of a transformer.
4. What does the kVA rating of a transformer represent?
5. Briefly define core loss and copper loss.
6. How many windings does an autotransformer have?
7. Explain what is meant by the term *step-up transformer*.
8. What two purposes does the oil in a liquid-insulated transformer serve?
9. List two precautions taken to protect nearby buildings when installing a liquid-insulated transformer outdoors.
10. Why would a liquid-insulated transformer have fins on its tank?

For Questions 11–18, determine the quantity indicated with a question mark using the values provided in the row.

	Primary				Secondary			
	Voltage	Current	Impedance	Turns	Voltage	Current	Impedance	Turns
11.	480 V	?			120 V	50 A		
12.		100 A		2500		?		500
13.			?	500			2.5%	250
14.	480 V			1000	?			250
15.	?	40 A			240 V	100 A		
16.			8 %	2500			2%	?
17.	120 V			?	240 V			500
18.		150 A	8%	400		?	4.5%	

19. A 15-kVA transformer supplies a single-phase, 120-volt secondary. What is the full-load current of the secondary?
20. If the transformer in Question #19 has an impedance of 4%, what is the available secondary short-circuit current?
21. What size overcurrent protective device should be used on the primary side of a 1.5-kVA, 240-volt, single-phase transformer? No additional secondary overcurrent protection is provided.
22. What size overcurrent protective device should be used on the primary side of a 25-kVA, 480-volt, single-phase transformer? No additional secondary overcurrent protection is provided.
23. What size primary and secondary fuses should be used for a 750-kVA, three-phase transformer with a 4160-volt primary, 208-volt secondary, and 2% impedance?

USING THE NEC

Refer to the National Electrical Code to answer the following questions. Do not write in this book.

1. What is the maximum number of fuses or breakers that can be grouped together as the secondary overcurrent protective device?
2. What information must be included on the transformer nameplate?
3. List three types of transformers that are *not* included in the requirements contained in *Article 450.*
4. When installing a transformer in hazardous (classified) locations, which *Code* articles in addition to *Article 450* must be satisfied?
5. What is a *secondary tie*?
6. For a transformer vault ventilated by natural circulation, what is the minimum combined net area of all ventilating openings if the vault houses a 75-kVA transformer?
7. When a combination of fuses and circuit breakers are used as the transformer secondary overcurrent protection, is their rating based on the maximum rating for a single fuse or the maximum rating for a single circuit breaker?

Transformers are available in a wide range of sizes. This huge high-voltage transformer would be used for very large commercial or industrial installations. (Cooper Industries, Cooper Power Systems)

Chapter 11 Grounding

Technical Terms

Bonding
Bonding jumper
Equipment grounding
Ground
Ground-fault circuit-interrupter (GFCI)
Grounded conductor
Grounding conductor
Grounding electrode
Grounding electrode conductor
Ground ring
Made electrodes
System grounding
Triad

Objectives

After completing this chapter, you will be able to:

- State important grounding concepts.
- Explain the difference between system grounding and equipment grounding.
- Compare the characteristics of grounded and ungrounded systems.
- List the primary reasons for grounding equipment.
- Select the appropriate size system and equipment grounding conductors using the *Code*.
- Identify the requirements for a service grounding system.
- Explain the purpose, operation, and installation of ground-fault circuit-interrupters.
- Describe various grounding electrode systems and their *Code* requirements.
- Size grounding electrode conductors using the *Code*.

Grounding is one of the most important facets of electrical safety for both personnel and equipment protection. In this chapter, the principles and applications of grounding are covered in detail. In the *Code, Article 250—Grounding* addresses the requirements for grounding and bonding. The article is divided into several parts:

- *Part I—General*
- *Part II—Circuit and System Grounding*
- *Part III—Grounding Electrode System and Grounding Electrode Conductor*
- *Part IV—Enclosure, Raceway, and Service Cable Grounding*
- *Part V—Bonding*
- *Part VI—Equipment Grounding and Equipment Grounding Conductors*
- *Part VII—Methods of Equipment Grounding*
- *Part VIII—Direct-Current Systems*
- *Part IX—Instruments, Meters, and Relays*
- *Part X—Grounding of Systems and Circuits of 1 kV and Over (High Voltage)*

When you are looking for a specific *Code* rule, first determine which part would most likely contain the rule. Then find the appropriate part in *Article 250* and begin your search there.

WARNING

Improper grounding can be as dangerous as no grounding. Grounding is the most important and least understood area of the *Code*. Become very familiar with *Article 250*.

Grounding Theory

Grounding terms with similar definitions are often confused. Their distinct differences should be carefully noted.

- **Grounding**—This is a broad term referring to the intentional act of making a permanent low-impedance electrically conductive path to the earth. The term is used in a general way to encompass the overall subject.

- **Ground**—This is a path from a circuit or equipment to the earth. It can be an intentional or accidental connection.
- **Grounded**—Connected to the earth through a conductor, such as a driven pipe or metal plate.
- **Grounded conductor**—A circuit conductor that serves as part of the system grounding path. Often called the neutral.
- **Grounding conductor**—Also called the equipment grounding conductor. A conductor used to connect noncurrent-carrying metallic parts of equipment to the system grounding busbar. The grounding conductor is part of the equipment grounding path.
- **Grounding electrode**—A device that establishes an electrical connection to the earth.
- **Grounding electrode conductor**—The conductor used to connect the grounding electrode(s) to the equipment grounding conductor.

Grounding eliminates currents from traveling through noncurrent-carrying components of electrical systems, **Figure 11-1.** A low-impedance path is provided by connecting the metal components of an electrical system to a ***ground,*** through which current can safely flow.

When a conductor contacts grounded metal equipment, the low-resistance ground path results in large currents. These currents cause the overcurrent protection device to trip.

If there is no ground path, the metal equipment becomes energized and acts simply as another resistor in the circuit. This resistance may be high enough to limit the current to a level that will not trip the overcurrent protective device. The equipment remains energized and will heat up. This creates a shock and fire hazard, and can also damage the equipment.

There are two types of grounding: system grounding and equipment grounding. System grounding is the connection of a circuit to the ground. Equipment grounding is the connection of noncurrent-carrying metal items to the ground. See **Figure 11-2.**

Bonding

Bonding is the mechanical joining of metallic parts to form an electrically conductive path. This path should have as little resistance as possible, and the joints should be mechanically sound and permanent. This gives the bonded parts continuity and conductivity (low resistance).

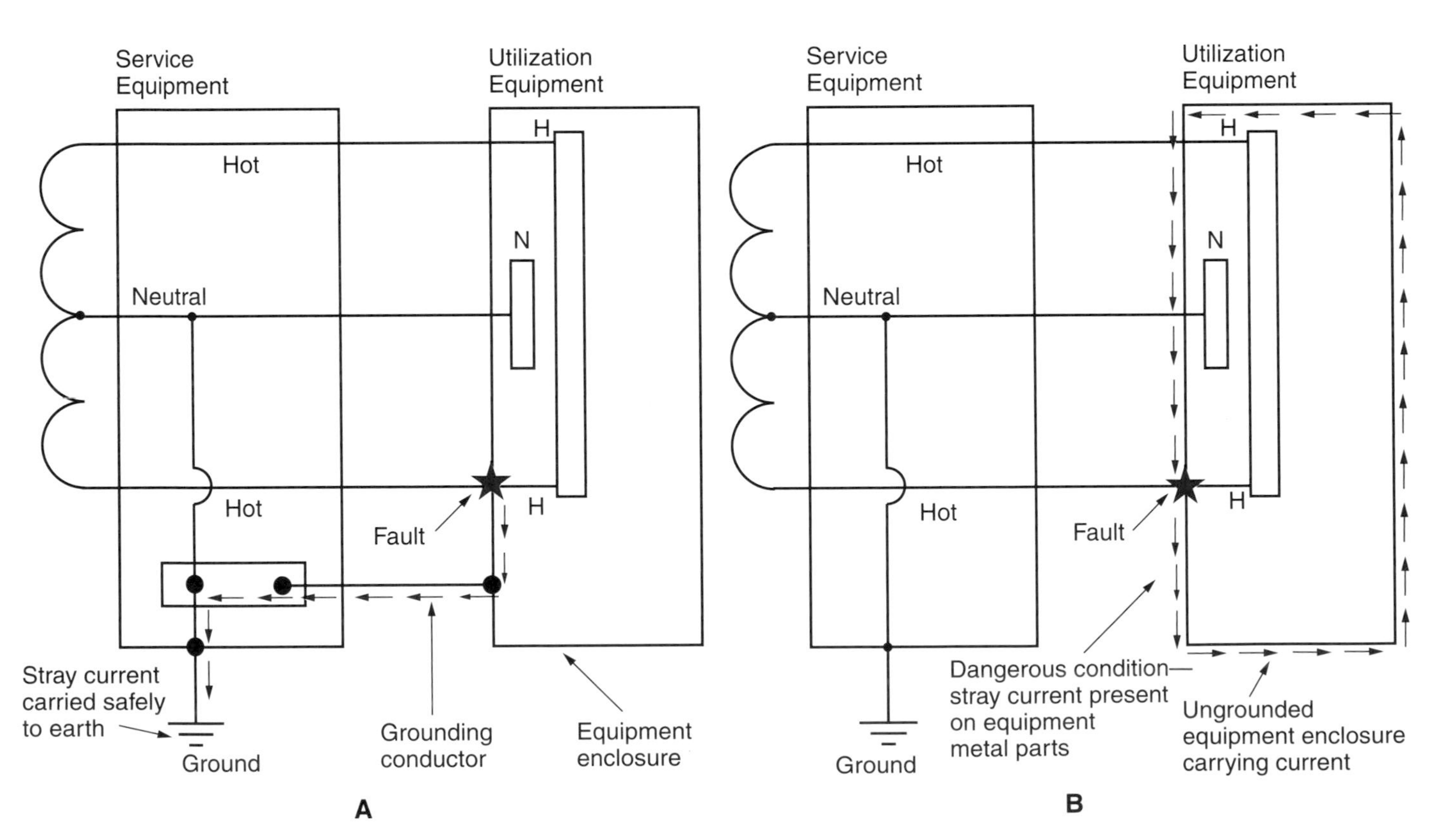

Figure 11-1. A fault occurs when the current-carrying conductor contacts the metal enclosure. A—If the enclosure is grounded, the current follows the path of least resistance to ground. Electricity flows at a very high current, tripping the overcurrent protective device. B—Dangerous stray currents can be present in an equipment enclosure if there is no ground. The resistance of the energized part may prevent the current from increasing enough to trip the overcurrent protective device.

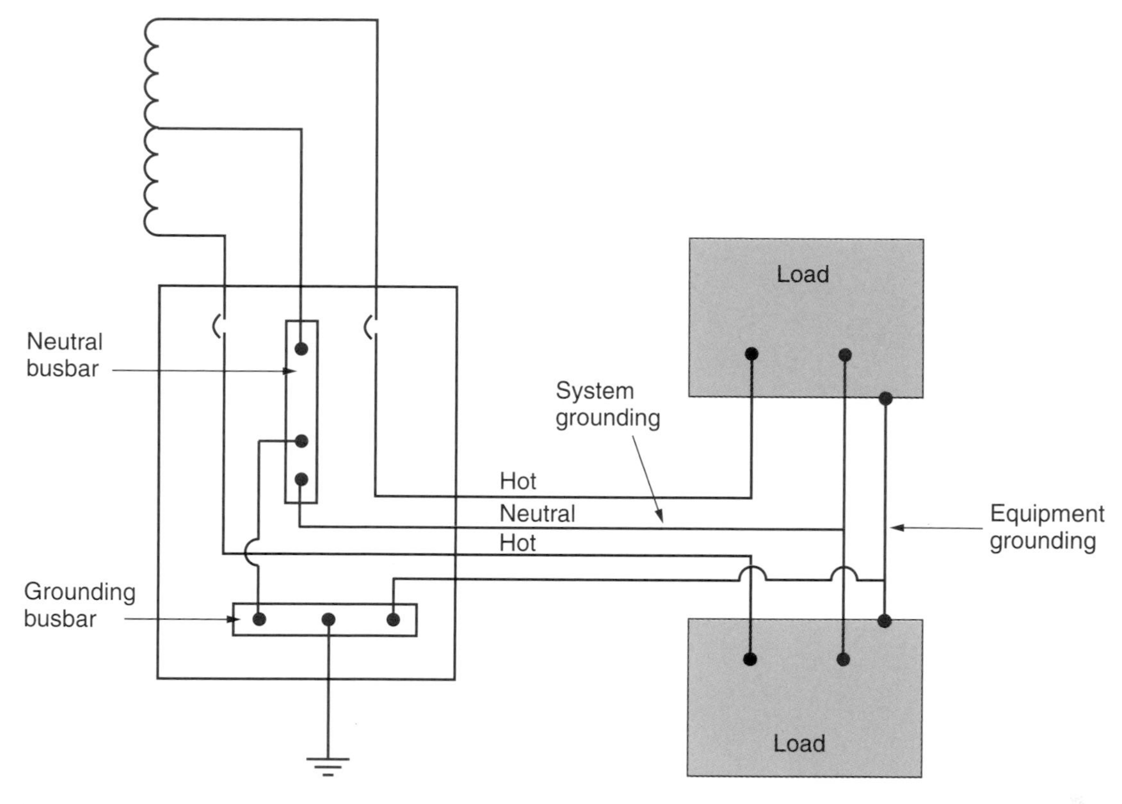

Figure 11-2. The system ground is connected to the neutral conductor. Equipment grounds are connected to equipment enclosures.

NEC NOTE 100

Bonding (Bonded): The permanent joining of metallic parts to form an electrically conductive path that ensures electrical continuity and the capacity to conduct safely any current likely to be imposed.

Conductors used for bonding are called ***bonding jumpers.*** There are several types (see **Figure 11-3**):

- **Equipment bonding jumper**—The connection between two or more portions of the equipment grounding conductor.
- **Main bonding jumper**—The connection between the grounded circuit conductor and the equipment grounding conductor at the service.
- **System bonding jumper**—The connection between the grounded circuit conductor and the equipment grounding conductor at a separately derived system.

NEC NOTE 250.28

For a grounded system, an unspliced main bonding jumper shall be used to connect the equipment grounding conductor(s) and the service-disconnect enclosure to the grounded conductor of the system within the enclosure for each service disconnect.

Due to the high currents available on the supply side of the service equipment, the bonding jumpers must be the same size or larger than the grounding electrode conductor. *Section 250.28(D)* discusses bonding jumper size in detail.

If a bonding jumper is impractical, another means of bonding, such as a grounding wedge, grounding bushing, or ground clamp, is used. See **Figure 11-4.**

NEC NOTE 250.8

Grounding conductors and bonding jumpers shall be connected by exothermic welding, listed pressure connectors, listed clamps, or other listed means. Connection devices or fittings that depend solely on solder shall not be used. Sheet metal screws shall not be used to connect grounding conductors to enclosures.

System Grounding

In ***system grounding,*** one circuit conductor of each circuit is connected to the ground. This connection is made at the service-entrance equipment. The grounded conductors are connected to the neutral busbar, which is connected by the main bonding jumper to the grounding busbar. A grounding electrode conductor is connected between the grounding busbar and the grounding electrode.

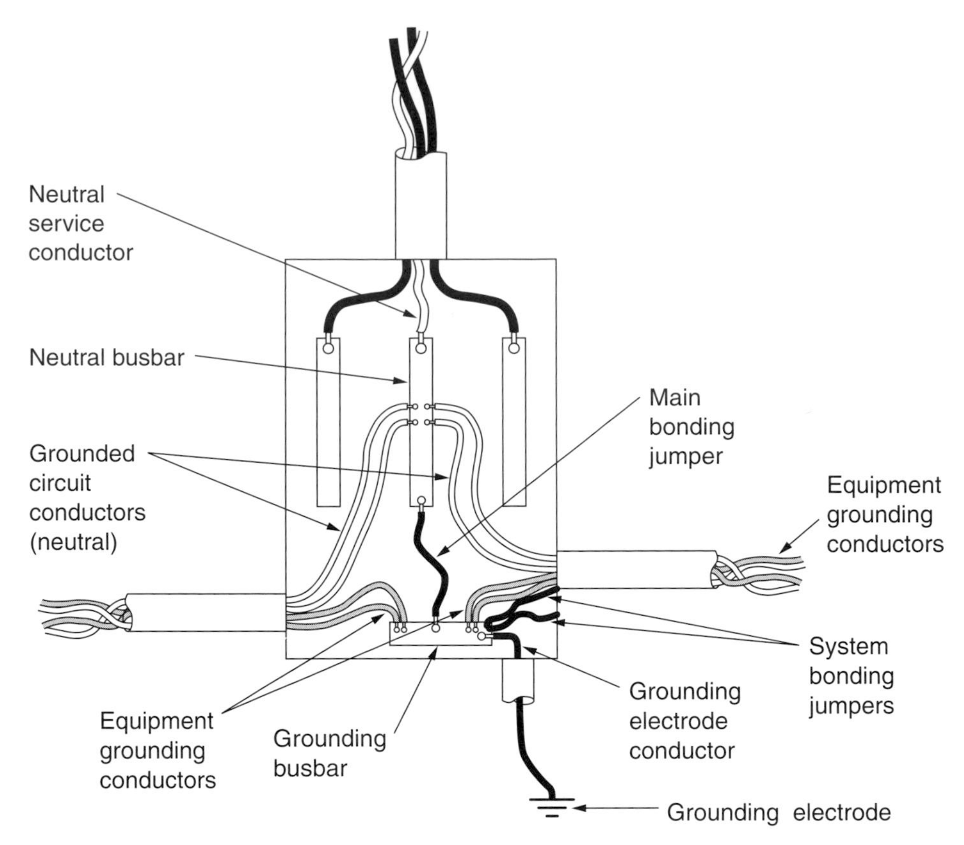

Figure 11-3. This sketch illustrates some basic components of a grounding system.

Figure 11-4. Bonding clamps ensure electrical continuity between connected items. (Arlington Industries, Inc.)

Grounded Circuit Conductor

Wiring System	Grounded Conductor
Single-phase, two-wire	Grounded circuit conductor
Single-phase, three-wire	Neutral conductor
Multiphase with one wire common to all phases	Common conductor
Multiphase with one phase grounded	One phase conductor
Multiphase with one phase used as a single-phase, three-wire system	Neutral conductor

Figure 11-5. The conductor used as the grounded conductor varies among different wiring systems.

The grounded circuit conductor serves to limit the voltage at which the system operates. Further, this ground restricts the amount of voltage imposed on the system by accidental shorts between conductors or lightning strikes. The grounded circuit conductor can be the neutral conductor or one of the phase conductors, depending on the wiring system being used. See **Figure 11-5.** Additional information on system grounding is found in *Part II* of *Article 250* in the *Code.*

The electrical system grounding is made continuous from the supply transformer to the service equipment. The following connections, which are illustrated in **Figure 11-6,** accomplish the grounding:

- At the service equipment, the neutral conductor is grounded (in accordance with *Section 250.24*) by connection with a grounding electrode.
- The service-entrance neutral is connected to the grounded supply conductor and also connected to the grounding busbar at the service disconnect.
- The grounding busbar is connected to a grounding electrode conductor, which is connected to the grounding electrode. If more than one electrode is used, they must be bonded together. Also, the

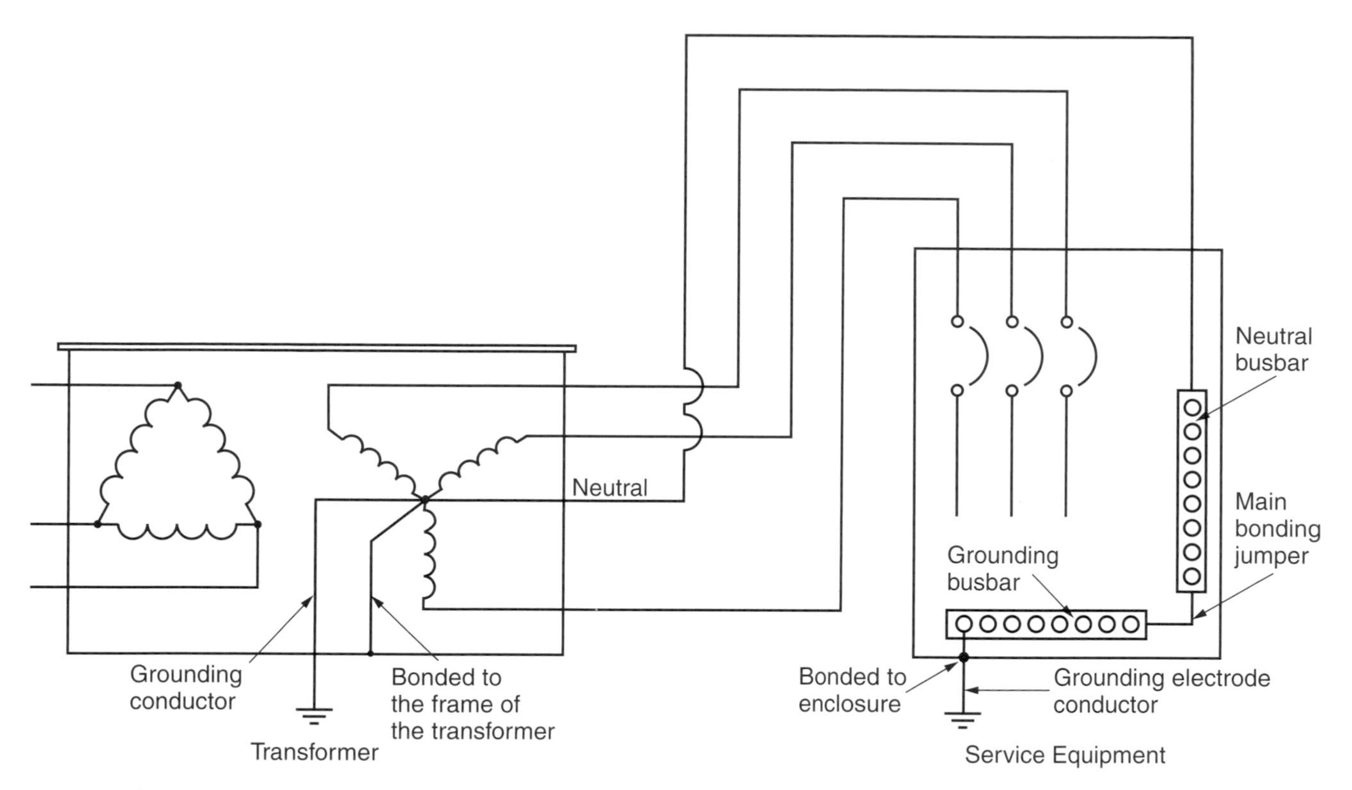

Figure 11-6. System grounding is continuous from the supply transformer to the service equipment. The neutral conductor is grounded at the supply transformer. The service neutral is connected to grounding busbar at the service equipment. The neutral busbar, grounding busbar, and service equipment enclosure are all connected by the main bonding jumper.

grounding busbar is bonded to the metal enclosure of the service disconnect.

Circuits Requiring Grounding

Most circuits require grounding. *Section 250.20* lists the specific types of circuits for which grounding is mandated.

For ac circuits less than 50 volts, grounding is required in the following conditions:

- The transformer supplying the voltage has a source that exceeds 150 volts to ground.
- The overhead conductors are installed outdoors.
- The transformer supplying the voltage receives its supply from an ungrounded source.

For ac circuits between 50 volts and 1000 volts, grounding is required in the following conditions:

- The system can be grounded so that the maximum voltage to ground is not greater than 150 volts.
- A three-phase, four-wire wye system that uses the neutral as a circuit conductor.
- A three-phase, four-wire delta system that uses the midpoint of one phase winding as a circuit conductor. See **Figure 11-7.**

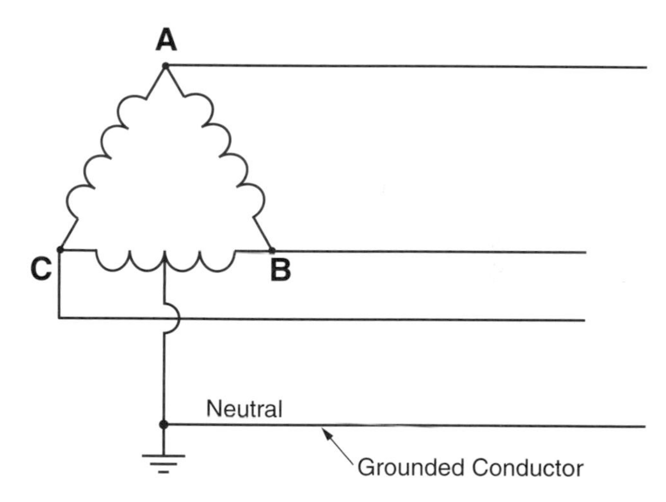

Figure 11-7. A three-phase, four-wire delta system has a neutral conductor at the midpoint of one winding. This neutral conductor is used as the grounded conductor.

Ground-Fault Circuit-Interrupters

A ***ground-fault circuit-interrupter (GFCI)*** is a device designed to open an electrical circuit when a ground fault occurs. The GFCI opens the circuit at a current that is less than the setting of the overcurrent protective device. GFCIs open a circuit if the current-to-ground exceeds 6 milliamperes (0.006 amps). Current above 0.006 amps is sufficient to cause injury.

A sensor in the GFCI monitors the balance between the hot and neutral conductors. If a fault occurs, some current will return to the source by a path other than the neutral. This will create an imbalance between the two conductors. When this happens, the sensor signals the internal circuitry, which "breaks" the hot connection and opens the circuit. See **Figure 11-8.**

A GFCI must be self-testing so that it can be checked periodically. There are numerous types of GFCI devices available, such as circuit breakers, direct-wired receptacles, and plug-in receptacles.

Arc-Fault Circuit-Interrupters

The *Code* requires that any 125 V, 15 or 20 A branch circuit that supplies devices in bedrooms must be protected by an arc-fault circuit-interrupter (AFCI). The AFCI opens the circuit when an arc fault is detected. By detecting arc faults, AFCIs will help prevent fires from starting. The breaker form of the AFCI is similar in appearance to a GFCI breaker. See **Figure 11-9.**

The AFCI detects arcing faults that are not detected by standard breakers or GFCIs. Typical conditions that may generate arc faults include the following:

- Damaged wires
- Frayed conductor insulation
- Loose electrical connections
- Overheated or otherwise stressed electrical cords
- Damaged electrical appliances

Equipment Grounding

In addition to circuit grounding, metal enclosures and parts that could be energized must also be grounded. This process, called ***equipment grounding,*** is needed for the following reasons:

- The grounded conductor (neutral) of a circuit will, at times, carry an unbalanced load and will therefore become a current-carrying conductor.
- By grounding equipment and conductive enclosures, the voltage-to-ground potential is limited.
- Equipment grounding provides a current pathway during a fault condition so the overcurrent protection device operates properly.

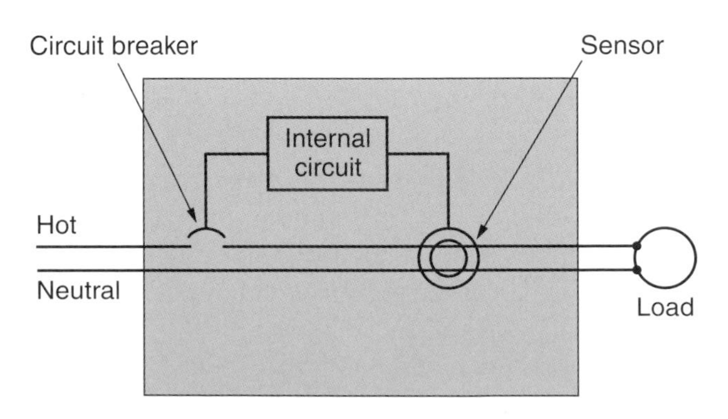

Figure 11-8. This schematic illustrates the components of a ground-fault circuit-interrupter.

Figure 11-9. AFCI breakers are required in bedroom circuits. (Cutler-Hammer)

- Stray currents (leakage) and static electricity can drain to ground, preventing injury to people and damage to equipment.

Two sections of the *Code* discuss enclosure grounding: *250.80* and *250.86*. These sections require that nearly all metal enclosures be effectively grounded. *Part VII* of *Article 250* lists grounding methods for specific equipment.

Wiring Method

If equipment is supplied by a metallic wiring method, it must be grounded. This applies to the following wiring methods:

- Metal conduit (including flexible)
- Metal tubing (including flexible)
- Metal raceway
- Wireway
- Underfloor duct
- Bus duct
- Armored cable
- Mineral-insulated cable (metal-sheathed)
- Metal-clad cable
- Metal-jacketed cable

> **WARNING**
>
> Equipment must *never* be grounded to the grounded conductor (neutral) on the load side of the service. This would create a very dangerous situation.

Location

If a person contacts a grounded enclosure and an ungrounded, energized enclosure simultaneously, a ground path through the person's body could occur. This could cause severe injury or death. The *Code* contains requirements to prevent this dangerous situation. All electrical equipment within 5′ horizontally and 8′ vertically of any grounded surface device must be grounded. This prevents two pieces of equipment from having a potential difference.

Water increases the possibility of electrical shorts and faults. If it contacts a current-carrying conductor, water can conduct the current to other items with which it is in contact, such as the equipment enclosure. Also, if a short energized an ungrounded enclosure and a person touched the enclosure while contacting water, the current would flow through the person. Therefore, all electrical equipment installed in a damp or wet location must be grounded.

Any nonelectrical equipment that is in contact or likely to come in contact with electrical equipment requires grounding.

Some equipment and enclosures do not require grounding. As a general rule, devices that are double-insulated, guarded, or isolated are exempt from grounding.

When entering junction boxes, equipment grounding conductors should be connected to each other and bonded to the metal box. This maintains the continuity of the grounding system.

Sizing the Equipment Grounding Conductor

The size of the equipment grounding conductor is based on the size of the overcurrent protective device protecting the circuit conductors. *Table 250.122* is used to size the grounding conductor.

Sample Problem 11-1

Problem: A piece of equipment is supplied by a circuit protected with a 100-amp circuit breaker. What size equipment grounding conductor is needed?

Table 250.122 Minimum Size Equipment Grounding Conductors for Grounding Raceway and Equipment

Rating or Setting of Automatic Overcurrent Device in Circuit Ahead of Equipment, Conduit, etc., Not Exceeding (Amperes)	Size (AWG or kcmil)	
	Copper	Aluminum or Copper-Clad Aluminum*
15	14	12
20	12	10
30	10	8
40	10	8
60	10	8
100	8	6
200	6	4
300	4	2

Solution: Using *Table 250.122*, for a 100-amp overcurrent protective device, an 8 AWG copper conductor or 6 AWG aluminum conductor can be used for the equipment grounding conductor.

NEC NOTE 250.4(A)(5)

The earth shall not be considered as an effective ground–fault current path.

Ungrounded Systems

In an ungrounded system, the neutral conductor is not grounded. The equipment grounding conductor, metal conduit, armored cable, or BX cable is used to bond all metal, noncurrent-carrying parts of the circuit to a grounding electrode.

Grounding Electrode System

Currents carried by system grounds and equipment grounds travel back to the service equipment, then through the grounding electrode conductor to the grounding electrode. From the grounding electrode, the current is dissipated into the soil.

There are many types of grounding electrodes permitted by the *Code*. Sometimes, a portion of the construction can be used as the grounding electrode. The following are some grounding electrode systems:

- **Underground water piping**—An external metal water pipe that extends for at least 10′ underground can be used as the grounding electrode. Any metering in the pipe must be bypassed with bonding jumpers. This type of electrode must be supplemented with an additional grounding electrode, such as a driven ground rod. The two electrodes are then bonded to each other.

NEC NOTE 250.52(B)

Aluminum electrodes or a metal underground gas piping system shall not be used as a grounding electrode.

- **Building structural steel**—The metal framing of a structure can serve as all or part of the grounding electrode system. In most instances, the building steel is connected to three driven ground rods placed near the building in a triangular pattern. The ground rod grouping, often referred to as a ***triad,*** ensures grounding integrity.
- **Concrete-encased rebar**—Concrete reinforcing bars that are 1/2″ in diameter (#4 rebar) or larger and at least 20′ long can be used as a grounding electrode. No other supplemental electrodes are needed. 4 AWG (or larger) copper conductor at least 20′ long embedded in concrete is also an acceptable grounding electrode.

- **Ground ring**—A 20′ or greater length of bare 2 AWG copper conductor buried at least 2′-6″ below grade and encircling the structure is frequently used as the grounding electrode.

If none of these electrodes are available, one of the following systems can be used. Electrodes that serve no other purpose are called ***made electrodes.***

- **Buried tanks**—Metal underground piping and structures such as buried storage tanks can serve as a suitable grounding electrode.
- **Metal plate**—A metal plate consisting of at least 2 ft^2 of exposed area and buried at least 2′-6″ below grade can serve as the grounding electrode.
- **Ground rod**—A driven ground rod at least 8′ long placed vertically or at an angle no greater than 45° can serve as the grounding electrode. In areas where subsurface rock prevents driving the ground rod, it can be buried horizontally at a minimum depth of 2′-6″.

Table 250.66 Grounding Electrode Conductor for Alternating-Current Systems

Size of Largest Service-Entrance Conductor or Equivalent Area for Parallel Conductors[1]		Size of Grounding Electrode Conductor	
Copper	Aluminum or Copper-Clad Aluminum	Copper	Aluminum or Copper-Clad Aluminum[2]
2 or smaller	1/0 or smaller	8	6
1 or 1/0	2/0 or 3/0	6	4
2/0 or 3/0	4/0 or 250 kcmil	4	2
Over 3/0 through 350 kcmil	Over 250 kcmil through 500 kcmil	2	1/0
Over 350 kcmil through 600 kcmil	Over 500 kcmil through 900 kcmil	1/0	3/0
Over 600 kcmil through 1100 kcmil	Over 900 kcmil through 1750 kcmil	2/0	4/0
Over 1100 kcmil	Over 1750 kcmil	3/0	250 kcmil

Notes:
1. Where multiple sets of service-entrance conductors are used as permitted in Section 230.40, Exception No. 2, the equivalent size of the largest service-entrance conductor shall be determined by the largest sum of the areas of the corresponding conductors of each set.
2. Where there are no service-entrance conductors, the grounding electrode conductor size shall be determined by the equivalent size of the largest service-entrance conductor required for the load to be served.
[a]This table also applies to the derived conductors of separately derived ac systems.
[b]See installation restrictions in Section 250.64(A).

Figure 11-10. *Table 250.66* in the *Code* specifies the minimum size for grounding electrode conductors.

Sizing the Grounding Electrode Conductor

Regardless of the type of grounding electrode used, there must be a conductor connecting this electrode to the grounded busbar at the service equipment.

According to *Section 250.66*, the size of the grounding electrode conductor is based on the size of the ungrounded service-entrance conductors. *Table 250.66* is used to determine minimum sizes. See **Figure 11-10.**

NEC NOTE 250.68(A)

The connection of a grounding electrode conductor or bonding jumper to a grounding electrode shall be accessible.

Sample Problem 11-2

Problem: A service is fed by 250 kcmil copper THW conductors. What is the minimum size grounding electrode conductor for the system?

Table 250.66 Grounding Electrode Conductor for Alternating-Current Systems

Size of Largest Service-Entrance Conductor or Equivalent Area for Parallel Conductors[1]		Size of Grounding Electrode Conductor	
Copper	Aluminum or Copper-Clad Aluminum	Copper	Aluminum or Copper-Clad Aluminum[2]
2 or smaller	1/0 or smaller	8	6
1 or 1/0	2/0 or 3/0	6	4
2/0 or 3/0	4/0 or 250 kcmil	4	2
Over 3/0 through 350 kcmil	Over 250 kcmil through 500 kcmil	2	1/0
Over 350 kcmil through 600 kcmil	Over 500 kcmil through 900 kcmil	1/0	3/0
Over 600 kcmil through 1100 kcmil	Over 900 kcmil through 1750 kcmil	2/0	4/0
Over 1100 kcmil	Over 1750 kcmil	3/0	250 kcmil

Solution: Using *Table 250.66*, either a 2 AWG copper conductor or a 2 AWG aluminum or copper-clad aluminum conductor is required.

Grounding Methods

Article 250 contains rules for constructing a grounding circuit. The numerous equipment grounding conductors must be connected and sized correctly to ensure high conductivity, low impedance, and absolute continuity.

Panelboards

The basic arrangement for grounding distribution panels is shown in **Figure 11-11.** There are two separate busbars: one for equipment grounds and one for neutrals (system grounds). The neutral bar is not bonded to the enclosure and is insulated by a nonconductive shim. The grounding bar is bonded to the enclosure to ensure electrical continuity.

The neutral is the return path for current flow. If the neutral bar was bonded to the enclosure, this would create an alternate route for current, which is not desired. Therefore, the neutral bar is insulated.

When grounding panelboards, follow these guidelines:

- Install an approved grounding bar.
- The grounding bar is securely fastened to the inside of the enclosure. The cabinet and bar must be bonded.
- Equipment grounding conductors are connected to the grounding bar. Neutral conductors are not.
- Never connect the neutral bar to the grounding bar. This is permitted only at the service equipment.

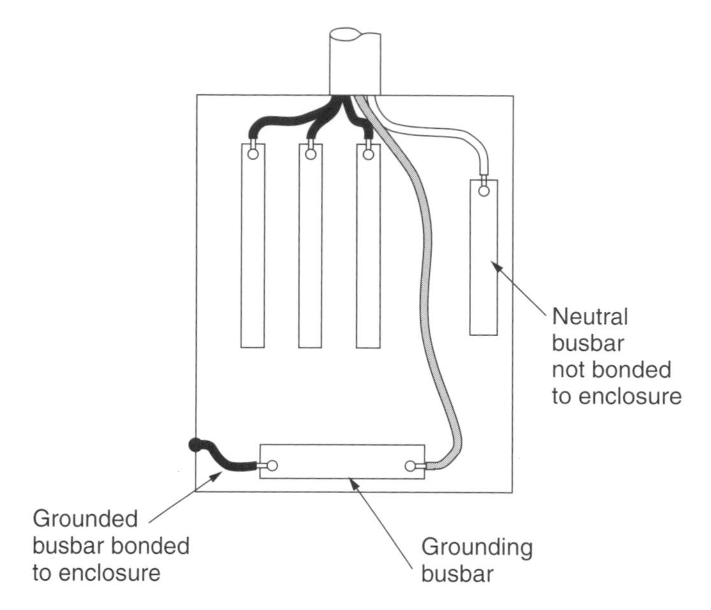

Figure 11-11. At the distribution panel, the grounding busbar is bonded to the panel enclosure and the neutral busbar is separated from the enclosure by insulation. The neutral and grounding busbars can be connected *only* at the service disconnect.

NEC NOTE 250.24(C)

A grounding electrode conductor shall be used to connect the equipment grounding conductors, the service-equipment enclosures, and, where the system is grounded, the grounded service conductor to the grounding electrode(s).

Grounding Cord-and-Plug–Connected Devices

There are three common methods used to ground the enclosures of cord-and-plug–connected devices:

- If the cord has an outer metal covering, one end is connected to the equipment and the other end is connected to the grounding prong of the plug.
- A grounding conductor within the cord can be connected to the equipment at one end and the grounding prong of the plug at the other end.
- A separate, external, flexible conductor can be run from the equipment case to a grounding electrode.

Grounding Permanently Wired Equipment

If armored cable, BX cable, metal conduit, or EMT is the wiring method, it can also serve as the equipment grounding conductor. If a nonmetallic wiring method is used, such as multiconductor cable or PVC, a separate conductor must be used as the equipment grounding conductor. This conductor is bare, has green insulation, or has green insulation with a yellow stripe.

Grounding Ranges, Ovens, and Clothes Dryers

If the existing circuit supplying ranges, ovens, and clothes dryers originates at the service panel, the grounded conductor can function as both the neutral and the equipment grounding conductor. The grounded conductor must be no smaller than 10 AWG (copper) or 8 AWG (aluminum). It must be insulated if it is part of NM cable or in nonmetallic conduit. If part of service-entrance cable, it can be bare. See *Section 250.140* for more information.

Multiple Buildings Supplied by a Single Service

When several buildings are supplied from a single main service, each service must be separately grounded. Each of the additional buildings must have a unique grounding electrode and grounding electrode conductor.

Grounding Computer Equipment

Special methods are used to ground computer equipment to minimize unwanted "electronic noise." The exposed noncurrent-carrying parts of the computer must be connected to the grounding electrode of the service or the transformer feeding the computer. The equipment grounding conductor can be the conduit wiring method or an equipment grounding conductor of the cable feeding the equipment. All the grounds and neutrals should be run with the power conductors.

Review Questions

Answer the following questions. Do not write in this book.

1. Explain briefly the difference between system grounding and equipment grounding.
2. What is the primary purpose for grounding an electrical system?
3. What is a *bonding jumper*?
4. Which type of grounding electrode conductor system *must* be supplemented by another system?
5. Why would it be dangerous to have an ungrounded piece of equipment located next to a grounded piece of equipment?
6. What is the name of the conductor that connects the grounding busbar with the grounding electrode?
7. What does a ground-fault circuit-interrupter measure in the circuit?
8. What is the minimum burial depth and exposed area for a metal plate electrode?
9. Compare the grounded conductor to the grounding conductor.
10. What two sets of conductors are connected by the main bonding jumper and where is this jumper located?

USING THE NEC

Refer to the National Electrical Code to answer the following questions. Do not write in this book.

1. The service-entrance conductors supplying a small commercial structure are 3/0 AWG THW copper conductors. What size copper grounding electrode conductor is required?
2. Define *separately derived system.*
3. What is the minimum allowable size for the copper equipment grounding conductor in a 30-amp circuit?
4. What is the minimum thickness for a steel plate electrode?
5. *Article 200—Uses and Identification of Grounded Conductors* requires grounded conductors to be marked for easy identification. How are conductors 6 AWG and smaller marked, and in what additional manner can sizes larger than 6 AWG be marked?
6. *Part III* of *Article 250* contains requirements for grounding electrode systems. All grounding electrodes must be bonded together. The bonding jumpers must be installed, sized, and connected in accordance with which *Code* sections?
7. In general, the connection between the grounding electrode conductor and the grounding electrode must be accessible. In what situations does the *Code* allow this connection to be inaccessible?
8. *Part I* of *Article 250* discusses general grounding requirements. List the three requirements of an effective fault current path.
9. The service-entrance feeder conductors for a system are 300 kcmil THW copper. What is the minimum size of the copper grounding electrode conductor?
10. According to *Section 250.56*, when is a supplemental electrode required?

Chapter 12

Branch Circuits and Feeders

Technical Terms

Branch circuit
Continuous load
Feeders
General lighting load
Show-window lighting load
Track lighting

Objectives

After completing this chapter, you will be able to:

- Identify the feeder and branch circuit portions of a distribution system.
- Describe the various types of branch circuits.
- Define the functions of a feeder and the functions of branch-circuit conductors.
- Calculate lighting and receptacle loads using *Code* requirements.
- Size branch circuits in accordance with the *Code.*
- Determine branch circuit overcurrent protection required by the *Code.*
- Use the *Code* to size feeder conductors.

In an electrical system, power must be transferred from the service equipment to the lights, machines, and outlets. Regardless of the wiring methods used, the conductors carrying the power fall into one of two categories: feeders or branch-circuit conductors. This chapter will explore the characteristics of these two conductor types.

Definitions

Several definitions are essential to understanding branch circuits and feeders. The following items are illustrated in **Figure 12-1:**

- **Service conductors**—These conductors extend from the power company terminals to the main service disconnect.
- **Feeder**—A conductor that originates at the main distribution or main disconnect device and terminates at another distribution center, panelboard, or load center.
- **Subfeeders**—These conductors originate at distribution centers other than the main distribution center and extend to panelboards, load centers, and disconnect switches that supply branch circuits.
- **Panelboard**—This can be a single panel or multiple panels containing switches, fuses, and circuit breakers for switching, controlling, and protecting circuits.
- **Branch circuits**—The portion of the wiring system extending past the final overcurrent device. These circuits usually originate at a panel and transfer power to load devices.

Branch Circuits

Any circuit that extends beyond the final overcurrent protective device is called a ***branch circuit.*** This includes circuits servicing single motors (individual) and circuits serving many lights and receptacles (multiwire). Branch circuits are usually low current (30 amps or less), but can also supply high currents.

A basic branch circuit is made up of conductors extending from the final overcurrent protective device to the load. Some branch circuits originate at safety switches (disconnects), but most originate at a panelboard. The following are several branch circuit classifications (**Figure 12-2**):

- **Individual branch circuit**—A branch circuit that supplies a single load.
- **Multioutlet branch circuit**—A branch circuit with multiple loads.
- **General purpose branch circuit**—A multioutlet branch circuit that supplies multiple outlets for appliances and lighting.

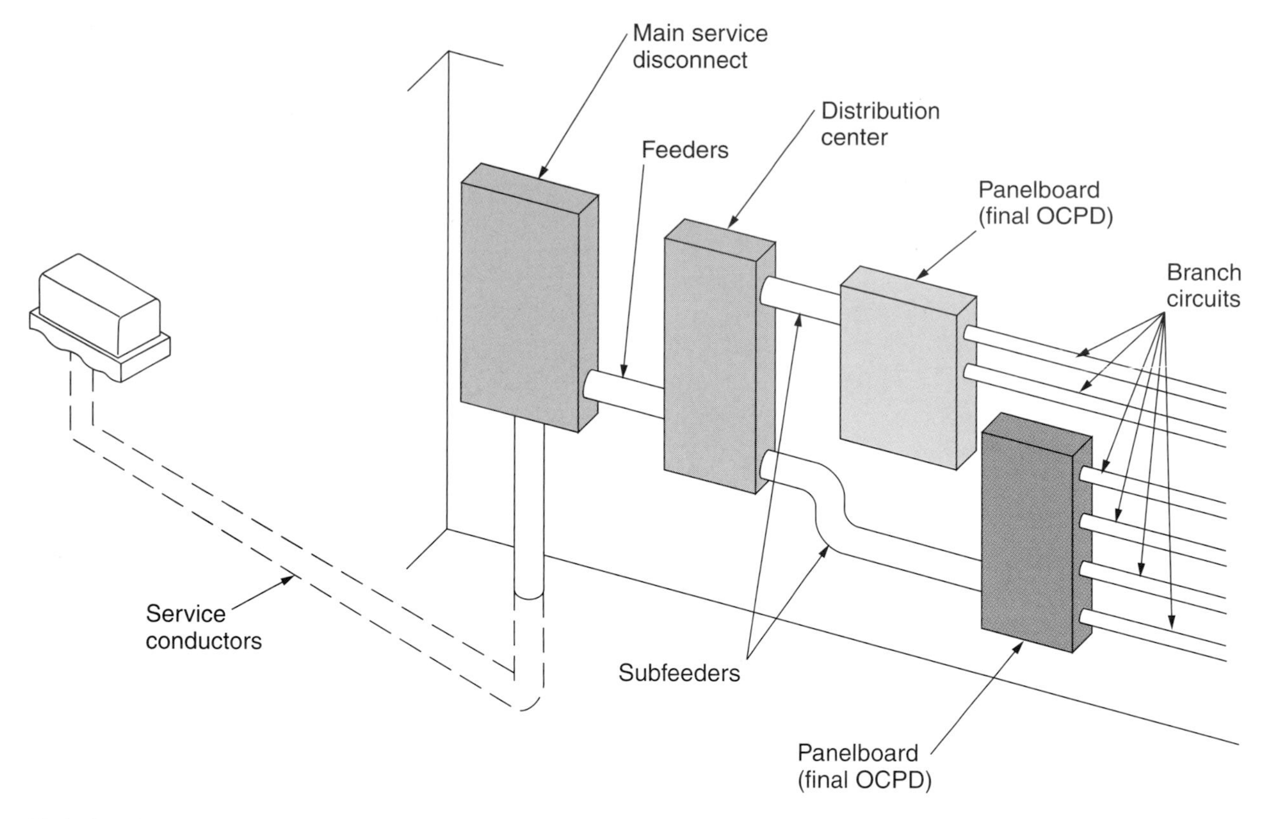

Figure 12-1. Conductors are classified based on their location in the electrical supply system.

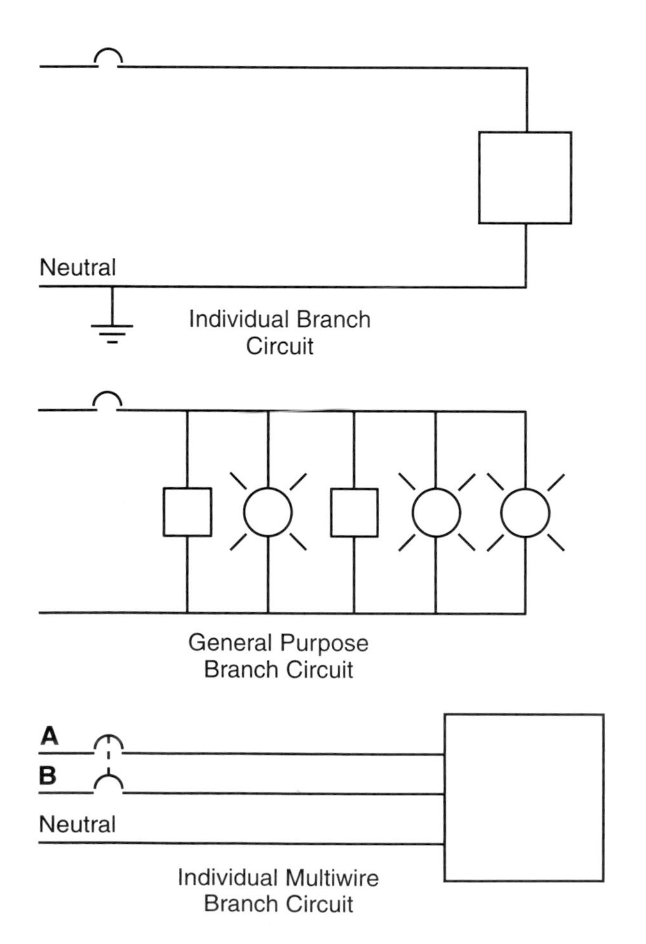

Figure 12-2. Branch circuits are classified as individual or multioutlet, appliance or general.

- **Appliance branch circuit**—A branch circuit that supplies a single appliance load.
- **Multiwire branch circuit**—A branch circuit with two or more ungrounded conductors and one grounded conductor.

Branch-Circuit Rating

A branch circuit is sized for the load it will supply. Sizing the circuit for additional future loads is good practice. The rating of a branch circuit depends on the rating of the overcurrent device protecting the circuit.

Branch circuits serving only one device can have any rating, while a circuit supplying more that one load is limited to ratings of 15, 20, 30, 40, or 50 amps.

Branch-Circuit Voltage

Branch-circuit voltage limits are contained in *Section 210.6* of the *Code*. These limits are based on the equipment being supplied by the circuit:

- In residences and hotel rooms, circuits supplying lighting fixtures and small receptacle loads cannot exceed 120 volts.
- Circuits that are 120 volts and less may be used to supply lampholders, auxiliary equipment of electric-discharge lamps, receptacles, and permanently wired equipment.

- Branch circuits exceeding 120 volts but not exceeding 277 volts may supply mogul-base screw-shell lampholders, ballasts for fluorescent lighting, ballasts for electric-discharge lighting, plug-connected appliances, and hard-wired appliances. Incandescent lighting operating over 150 volts is permitted in commercial construction.
- Circuits exceeding 277 volts and up to 600 volts can supply mercury-vapor and fluorescent lighting, provided the lighting units are installed at heights not less that 22′ above grade and in tunnels at heights no less than 18′.

Conductor Size and Ampacity

The amperage rating of branch-circuit conductors must be greater than the maximum load the circuit will provide. For multiple-load branch circuits, the conductor ampacity must correspond to the rating of the overcurrent protective device. However, for circuits supplying hard-wired devices (such as electric heaters, air-conditioning units, and water heaters), the fuse or circuit breaker can be rated at the next higher rating. The conductor is acceptable if its rating is at least that of the load current, even if the overcurrent protective device rating is higher.

The smallest general-purpose conductor for branch circuits is 14 AWG. Tap conductors can be smaller. See *Section 210.19* for more information.

Multiwire Branch Circuits

A branch circuit can be either a two-wire or multiwire branch circuit. A multiwire branch circuit consists of a grounded conductor and two or more ungrounded conductors. A multiwire circuit can be an individual circuit or a multioutlet circuit.

Conductor Color Code

Grounded conductors of branch circuits are identified by color. If the grounded conductor is 6 AWG or smaller, it is white, gray, white with a color stripe, or has three continuous white stripes on other than green insulation. If wires from different systems are contained in the same raceway, the neutrals of different systems are distinguished from one another. For example, the neutral of one system would be white, the neutral of the another system would be gray, and the neutral of a third system would be white with a colored stripe. The equipment grounding conductor must be green, green with yellow stripes, or bare (without any insulation).

Hot conductors can be any color except white, gray, green, and white with a color stripe. Normally, hot conductors are black, blue, and red. In a three-phase, four-wire delta system with a neutral connected at the midpoint of a winding, the "high leg" phase conductor should be identified with orange markings.

Branch-Circuit Loads

The *Code* places load limitations on branch circuits with ***continuous loads*** (loads with a duration longer than three hours, such as lighting). The continuous load must not exceed 80% of the circuit rating allotted for it. If the overcurrent protective device is listed for continuous operation at 100% of its rating, the 80% factor is not used.

Branch-circuit loads are classified into five categories:

- Lighting loads.
- Receptacle loads.
- Equipment loads.
- Heating and cooling loads.
- Motor loads.

NOTE

Motor loads are discussed in Chapter 13 of this text.

Lighting Loads

In the broad sense, lighting loads may be categorized as follows:

- General lighting.
- Show-window lighting.
- Track lighting.
- Sign and outline lighting.
- Other lighting.

Each lighting load is computed separately and then combined to determine the total lighting load.

General lighting

General lighting is the overhead lighting within a building. Its intensity should be adequate for any type of work performed in the area. Determining the general lighting load can be based on either the load per area method or the actual full-load current of the fixtures used, whichever is greater. Within a structure, there are normally several different types of areas—storage, office, hallways, and cafeterias—and these must be considered separately.

Most commercial structures have continuous lighting loads and the branch circuits must be adequate for carrying 125% of the calculated load. *Code* requirements for general lighting loads are found in several sections:

- ***Section 220.14***—General lighting load requirements.
- ***Table 220.12***—Minimum volt-amperes per area requirements.
- ***Sections 220.12, 220.14, and 220.16***—Demand factors for general lighting.

Table 220.12 of the *Code* contains minimum general lighting loads (in VA/ft²) for various types of buildings. A condensed version of this table is shown in **Figure 12-3.** The general lighting load is calculated by multiplying the floor area (in ft²) by the unit load (in VA/ft²). If the load is continuous, the calculated load is multiplied by 1.25 (the inverse of 80%) to determine the circuit requirements.

General Lighting Loads

Type Building	Unit Load (VA/ft²)
Auditoriums	1
Banks	3 1/2
Barber shops	3
Churches	1
Dwelling units	3
Hospitals	2
Hotels	2
Office buildings	3 1/2
Restaurants	2
Schools	3
Stores	3
Warehouses	1/4

Figure 12-3. Minimum general lighting loads are dependent on the type of area being lit.

Sample Problem 12-1

Problem: A 25,000 ft² office building is being designed. What is the general lighting load and what load does the circuit need to supply?

Solution: From *Table 220.12*, the unit load for an office building is 3 1/2 VA/ft². The general lighting load is determined by multiplying this value by the square footage of the building:

$$3\ 1/2\ \text{VA/ft}^2 \times 25{,}000\ \text{ft}^2 = 87{,}500\ \text{VA}$$

The general lighting load is 87,500 volt-amperes. However, the load is continuous and can only be 80% of the load supplied by the circuit. This value must be multiplied by 1.25 to determine the circuit requirements:

$$87{,}500\ \text{VA} \times 1.25 = 109{,}375\ \text{VA}$$

The circuit is designed to supply 109.375 kilovolt-amperes.

The general lighting load isn't required if the load for each lamp is determined separately. If the individual load is continuous, it must be multiplied by 1.25. When determining the current draw of fluorescent fixtures, use the current rating of the ballast, not the tube wattage.

Sample Problem 12-2

Problem: A 4′ long, two-lamp fluorescent fixture ballast draws 0.7 amps at 120 volts. How many of these fixtures can be connected on a 20-amp circuit?

Solution: This is a continuous load, so the current used by the lights can only be 80% of the circuit current rating:

$$\begin{aligned}\text{Allowable current} &= 20\ \text{A} \times 0.80\\ &= 16\ \text{A}\end{aligned}$$

By dividing the allowable load by the load of each lamp, the total number of lamps is determined:

$$\frac{16\ \text{A}}{0.7\ \text{A}} = 22.8\ \text{fixtures}$$

The maximum number of fixtures on the circuit is 22.

Show-window lighting

The ***show-window lighting load*** is not considered as part of the general lighting load. *Section 220.43(A)* of the *Code* requires that show-window lighting be computed as 200 volt-amperes per linear foot or as the maximum volt-ampere rating of the equipment and lights, whichever is greater.

NEC NOTE 100

Show Window: Any window used or designed to be used for the display of goods or advertising material, whether it is fully or partly enclosed or entirely open at the rear and whether or not it has a platform raised higher than the street floor level.

The *Code* includes several sections applicable to show-window lighting:

- ***Section 220.43(A)***—Show-window lighting.
- ***Section 220.14(G)***—Load computation.

The *Code* also requires at least one receptacle outlet for every 12′ of show-window space measured horizontally with the load computed at 180 volt-amperes per outlet. This receptacle load is in addition to the show-window lighting load.

NEC NOTE 410.7

Chain-supported fixtures used in a show window shall be permitted to be externally wired. No other externally wired fixtures shall be used.

Sample Problem 12-3

Problem: A department store has two lighted show windows, one 25′ long and the other 20′ long. What are the branch-circuit requirements for the show-window load?

Solution: Compute the load based on linear feet of show window:

$$\text{Total length} = 25' + 20' = 45'$$

$$\text{Show-window load} = 45' \times 200 \text{ VA/ft} = 9000 \text{ VA}$$

The lighting is a continuous load, so the show-window load is multiplied by 1.25 to determine the circuit load requirements:

$$\text{Circuit requirements} = 9000 \text{ VA} \times 1.25 = 11{,}250 \text{ VA}$$

The circuits supplying power for the show-window lighting must have a minimum capacity of 11,250 volt-amperes.

In addition, receptacles are required for every 12′ of show window. A total of five receptacles (two for the 20′ window and three for the 25′ window) are needed. The receptacle load can then be computed:

$$5 \times 180 \text{ VA} = 900 \text{ VA}$$

$$\text{Total load} = 11{,}250 \text{ VA} + 900 \text{ VA} = 12{,}150 \text{ VA}$$

Track lighting

Track lighting is often used in commercial buildings for accent lighting. It is discussed in *Part XV* of *Article 410* of the *Code*. The minimum load for track lighting is 150 volt-amperes for every 2′ of track. To compute the track lighting load requirements, simply determine the total length of track lighting, divide by two, and multiply by 150 volt-amperes.

NEC NOTE 410.100

Lighting track is a manufactured assembly designed to support and energize luminaires (lighting fixtures) that are capable of being readily repositioned on the track. Its length can be altered by addition or subtraction of sections of track.

Sample Problem 12-4

Problem: Determine the track lighting load for a 22′ long section of track.

Solution: Every 2′ requires 150 volt-amperes, so the length (in feet) is divided by 2′, and then multiplied by 150 volt-amperes:

$$\text{Track lighting load} = \frac{\text{length}}{2'} \times 150 \text{ VA} = \frac{22'}{2'} \times 150 \text{ VA} = 1650 \text{ VA}$$

If this is a continuous load, the circuit requirements would also include a 1.25 factor.

Sign and outline lighting

Sign and outline lighting is discussed in *Article 600* of the *Code*. A structure must have at least one circuit exclusively used to supply sign or outline lighting. The circuit must be designed for a minimum load of 1200 volt-amperes.

Sign and outline lighting loads are considered continuous loads. Therefore, if the rating of the sign and outline fixtures is greater than 960 volt-amperes ($1200 \times 0.8 = 960$), the circuit will be greater than 1200 volt-amperes.

Sample Problem 12-5

Problem: A hardware store measures 80′ × 120′. A portion of the building (80′ × 40′) is used for storage. The remainder of the building is used as a showroom. There is a total of 45′ of show windows, and there is one outdoor sign. What is the total lighting load for this building?

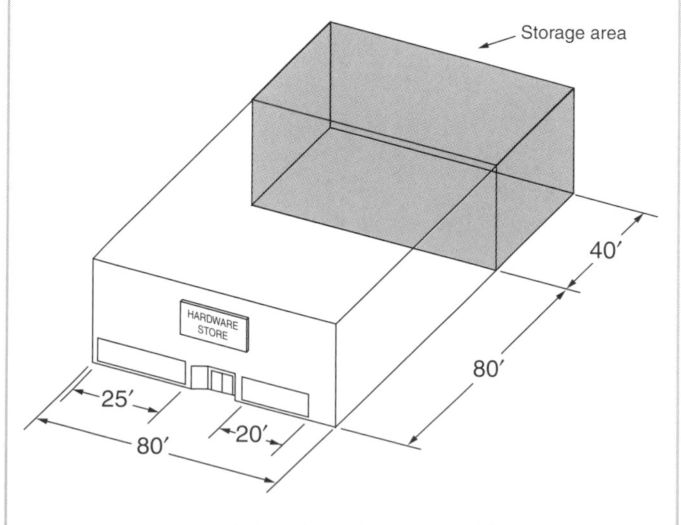

(Continued on the following page.)

Sample Problem 12-5 *Continued*

Solution: Each type of lighting load is computed separately and then combined to determine the total lighting load. First, the general lighting loads for the two areas of the store are calculated:

$$\text{Storage area} = 80' \times 40' = 3200 \text{ ft}^2$$

$$\text{Showroom area} = 80' \times 80' = 6400 \text{ ft}^2$$

Table 220.3(A) lists unit loads for storage and showroom as 1/4 VA/ft^2 and 3 VA/ft^2, respectively:

$$\text{General lighting load (storage)} = 3200 \text{ ft}^2 \times 1/4 \text{ VA/ft}^2 = 800 \text{ VA}$$

$$\text{General lighting load (showroom)} = 6400 \text{ ft}^2 \times 3 \text{ VA/ft}^2 = 19{,}200 \text{ VA}$$

$$\text{General lighting load} = 800 \text{ VA} + 19{,}200 \text{ VA} = 20{,}000 \text{ VA}$$

The show-window lighting load is based on 200 VA per linear foot:

$$\text{Show-window lighting load} = 45' \times 200 \text{ VA/ft} = 9000 \text{ VA}$$

The minimum load for the sign lighting is used:

$$\text{Sign lighting} = 1200 \text{ VA}$$

Now the total lighting load can be calculated by adding the parts together:

$$\text{Total lighting load} = 20{,}000 \text{ VA} + 9000 \text{ VA} + 1200 \text{ VA} = 30{,}200 \text{ VA}$$

These loads are all continuous, so the total load is multiplied by a factor of 1.25 to determine the circuit requirements.

$$30{,}200 \text{ VA} \times 1.25 = 37{,}750 \text{ VA}$$

Additional lighting loads

Additional lighting loads should be computed separately from the general lighting load and then added to the general lighting load. Loads for additional lighting such as security lighting, parking area lighting, sidewalk lighting, roadway lighting, and stadium lighting are calculated using the actual load. These are considered continuous loads where appropriate. The additional lighting load must be treated separately from the general lighting load for computation purposes.

Demand factors for feeder loads

Due to the great diversity of lighting loads on commercial premises, *Section 220.42* and *Table 220.42* allow the general lighting load to be *derated* (reduced) for feeder, panel, or service computations. See **Figure 12-4.** For example, it is highly unlikely that every light in a hospital would be operating at the same time. Of course, there are areas within a hospital where the derating factors should not be applied as these areas (such as operating rooms, emergency rooms, intensive care units, nurses stations, stairways, and cardiac units) are likely to have lighting units on at all times.

NOTE

Derating factors do not apply to branch-circuit conductor or branch-circuit overcurrent protective device calculations.

Demand Factors for Lighting Loads

Type of Occupancy	Porton of Load (VA)	Demand Factor (%)
Dwelling Unit		
	0–3000	100
	3001–120,000	35
	Over 120,000	25
Hospital		
	0–50,000	40
	Over 50,000	20
Hotels and Motels		
	0–20,000	50
	20,000–100,000	40
	Over 100,000	30
Warehouses		
	0–12,500	100
	Over 12,500	50
All Others	**Total VA**	**100**

Figure 12-4. Lighting loads can be derated for structures where all lights are not in use continuously.

Sample Problem 12-6

Problem: A hotel has 250 rooms, each with an area of 400 ft². Determine the general lighting load and then calculate the derated load to be used for feeder calculations.

Solution: First, determine the total area of the hotel rooms:

$$250 \text{ rooms} \times 400 \text{ ft}^2\text{/room} = 100{,}000 \text{ ft}^2$$

The general lighting load for hotel rooms (from *Table 220.12*) is 2 VA/ft². Determine the general lighting load:

$$100{,}000 \text{ ft}^2 \times 2 \text{ VA/ft}^2 = 200{,}000 \text{ VA}$$

The branch circuits are required to supply the full general load. However, the feeder can be derated using the factors in *Table 220.42.*

Table 220.42 Lighting Load Demand Factors

Type of Occupancy	Portion of Lighting Load to Which Demand Factor Applies (Volt-Amperes)	Demand Factor (Percent)
Dwelling units	First 3000 or less at From 3001 to 120,000 at Remainder over 120,000 at	100 35 25
Hospitals*	First 50,000 or less at Remainder over 50,000 at	40 20
Hotels and motels, including apartment houses without provision for cooking by tenants*	First 20,000 or less From 20,001 to 100,000 Remainder over 100,000	50 40 30
Warehouses (storage)	First 12,500 or less at Remainder over 12,500 at	100 50
All others	Total volt-amperes	100

*The demand factors of this table shall not apply to the computed load of feeders or services supplying areas in hospitals, hotels, and motels where the entire lighting is likely to be used at one time, as in operating rooms, ballrooms, or dining rooms.

The general lighting load must be divided into three parts to correspond to the three different demand factors. The first 20,000 volt-amperes have a demand factor of 50%. The next 80,000 volt-amperes (20,000 to 100,000) have a demand factor of 40%. The final 100,000 volt-amperes (100,000 to 200,000) have a demand factor of 30%. Each section is calculated individually:

$$20{,}000 \text{ VA} \times 0.50 = 10{,}000 \text{ VA}$$

$$80{,}000 \text{ VA} \times 0.40 = 32{,}000 \text{ VA}$$

$$100{,}000 \text{ VA} \times 0.30 = 30{,}000 \text{ VA}$$

To determine the derated feeder load, add the three totals together:

$$10{,}000 \text{ VA} + 32{,}000 \text{ VA} + 30{,}000 \text{ VA} = 72{,}000 \text{ VA}$$

The derated load used for sizing the feeder is 72,000 volt-amperes.

Receptacle Loads

The majority of receptacles installed in commercial structures do not supply continuous loads. It is difficult to predict what size load will be supplied at a receptacle, unless the receptacle is dedicated (assigned a specific purpose). The *Code* does not require a minimum number of outlets for commercial buildings. Normally, many receptacles are required.

When a receptacle is the load supplied by an individual branch circuit, the receptacle ampere rating must be equal to or greater than that of the branch circuit. When there are multiple receptacles on a branch circuit, the receptacle rating varies with the current rating. See **Figure 12-5,** which reflects *Table 210.21(B)(3).*

Receptacles connected to a 15-amp or 20-amp circuit are grounded. Grounded receptacles should not be used if the circuit is not actually grounded. A GFCI receptacle

Receptacle Rating

Circuit Rating (A)	Receptacle Rating (A)
15	Not over 15
20	15 or 20
30	30
40	40 or 50
50	50

Figure 12-5. Receptacle ratings are determined by the circuit rating.

can be used as a replacement for an ungrounded receptacles outlet.

A load of 180 volt-amperes is assigned to each receptacle, whether it is single, duplex, or triplex. If a receptacle is dedicated for a specific device, then the actual load is used. If the dedicated load is continuous, then the 125% overrate is appropriate.

To calculate the allowable number of receptacles on a branch circuit, multiply the circuit voltage and amperage, then divide by 180 volt-amperes. The receptacle load can be included with the general lighting load by adding a value of 1 VA/ft² to the general lighting unit loads found in *Table 220.12*. However, this method should only be used when the number of receptacles is unknown.

Sample Problem 12-7

Problem: How many receptacles can be placed on a 120-volt, 20-amp circuit? How many can be placed on a 120-volt, 15-amp circuit?

Solution: Determine the maximum circuit power:

$$P = E \times I$$
$$= 120 \text{ V} \times 20 \text{ A}$$
$$= 2400 \text{ VA (for 20-amp circuit)}$$

$$P = 120 \text{ V} \times 15 \text{ A}$$
$$= 1800 \text{ VA (for 15-amp circuit)}$$

Then divide the power by the load per receptacle (180 volt-amperes):

20-amp circuit:

$$\frac{2400 \text{ VA}}{180 \text{ VA}} = 13.3$$

15-amp circuit:

$$\frac{1800 \text{ VA}}{180 \text{ VA}} = 10$$

A 120-volt, 20-amp circuit can supply 13 receptacles. A 120-volt, 15-amp circuit can supply 10 receptacles.

Multioutlet assemblies are frequently installed in repair shops, lighting display areas, electronics departments, and other locations where many outlets are needed. These multioutlet assemblies require 180 volt-amperes for each 5′ of length. In stores, repair shops, and laboratories, the *Code* allows the overall load to be derated in accordance with *Table 220.44* (if the load exceeds 10,000 volt-amperes).

The diversity and inconsistent loading of general purpose receptacles allows the total receptacle load to be derated (see *Section 220.44*). If the load exceeds 10,000 volt-amperes, the first 10 kilowatts are counted at 100%, but additional load is counted at 50%. This may not be used if the *Code* dictates that the specific appliances cannot be derated. Refer to *Sections 220.12* and *220.44* and *Table 220.44*.

Sample Problem 12-8

Problem: Determine (a) the receptacle load for an 80′ × 120′ hardware store and (b) the number of 15-amp circuits needed to supply the load. The number of receptacles is unknown.

Solution: (a) The number of receptacles is unknown, so a receptacle load of 1 VA/ft² can be calculated:

$$\text{Area} = 80' \times 120'$$
$$= 9600 \text{ ft}^2$$

$$\text{Receptacle load} = 1 \text{ VA/ft}^2 \times 9600 \text{ ft}^2$$
$$= 9600 \text{ VA}$$

(b) To determine the number of circuits required, first calculate the allowable load for a single circuit:

$$\text{Max load} = 120 \text{ V} \times 15 \text{ A}$$
$$= 1800 \text{ VA}$$

Divide the total receptacle load by the maximum load per circuit to determine the minimum number of circuits:

$$\text{Circuits} = \frac{9600 \text{ VA}}{1800 \text{ VA}}$$
$$= 5.33$$

This is the minimum number, so round up to six circuits.

Sample Problem 12-9

Problem: The hardware store introduced in Sample Problem 12-5 will have one duplex receptacle for every 12′ of wall around the storage area and showroom. A total of twelve floor receptacles will be used. Determine the receptacle load, including show-window receptacles for 20′ and 25′ show windows.

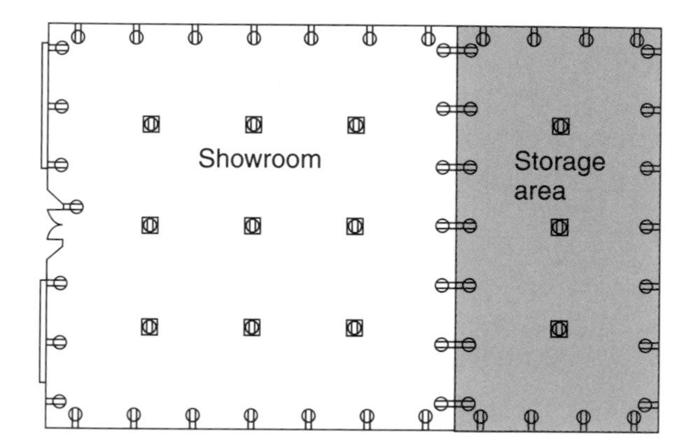

Solution: Receptacles will be placed along six 80′-long walls and two 40′-long walls. Determine the number of receptacles needed for each 80′-long wall:

$$\text{Receptacles per } 80' \text{ wall} = \frac{80'}{12'} = 6.67$$

Seven receptacles are needed for each 80′ wall. Next, determine the number of receptacles needed for each 40′ wall:

$$\text{Receptacles per } 40' \text{ wall} = \frac{40'}{12'} = 3.33$$

Four receptacles are needed for each 40′ wall. The total number of wall receptacles can now be calculated:

$$\text{Wall receptacles} = (6 \times 7) + (2 \times 4) = 42 + 8 = 50$$

One receptacle is needed for every 12′ of show window length, so the 20′ show window requires two receptacles and the 25′ show window requires 3 receptacles. Therefore, the show windows require a total of five receptacles. The total number of receptacles can be found by adding together the wall receptacles, floor receptacles, and show-window receptacles:

$$\text{Total receptacles} = 50 + 12 + 5 = 67 \text{ receptacles}$$

The total receptacle load can then be determined:

$$\text{Receptacle load} = 67 \times 180 \text{ VA} = 12{,}060 \text{ VA}$$

The total receptacle load is 12,060 VA. For feeder sizing, the first 10,000 VA must be counted at 100%, but only 50% of the additional load needs to be considered. Determine 50% of 2060 VA:

$$2060 \text{ VA} \times 0.50 = 1030 \text{ VA}$$

$$\text{Feeder load} = 10{,}000 \text{ VA} + 1030 \text{ VA} = 11{,}030 \text{ VA}$$

The receptacle load on the feeder is 11,030 VA.

Equipment Loads

Within this category of branch-circuit loads is a vast array of equipment, such as appliances, water heaters, washers, dryers, and cooking equipment. Most of these items are used for brief periods and are considered noncontinuous, so the required load supplied by the circuit is identical to the equipment requirement.

Equipment is often hard-wired, but can also be cord-and-plug connected to a receptacle. Branch circuits for appliance loads must have conductors with an ampacity equal to or exceeding the ampacity of the appliance. The ampacity of the appliance is marked on the unit by the manufacturer. If the appliance has a motor, the ampacity of the branch-circuit conductors must be 125% of the current rating of the motor.

Commercial kitchen equipment loads

Loads for commercial cooking equipment are discussed in *Section 220.56* of the *Code*. The total feeder load is simply the sum of the nameplate ratings of the appliances. If there are three or more pieces of cooking equipment, the feeder load can be derated in accordance with *Table 220.56*. The table in **Figure 12-6** lists the demand factors. The branch-circuit loads cannot be derated using these factors.

Ovens, grills, fryers, food warmers, large vat blending machines, booster heaters, conveyors, and tray assemblies are considered kitchen equipment and may be derated in accordance with *Table 220.56*. Auxiliary equipment such as exhaust fans, space heaters, and air-conditioning units are not counted as kitchen equipment and cannot be derated.

Feeder Demand Factors for Kitchen Equipment

Units of Equipment	Demand Factors (%)
1–2	100
3	90
4	80
5	70
6+	65

Figure 12-6. The load for multiple pieces of commercial cooking equipment can be derated in accordance with the demand factors listed in *Table 220.56.*

NEC NOTE 220.56

Demand factors for kitchen equipment shall be applied to all equipment that has either thermostatic control or intermittent use as kitchen equipment. They shall not apply to space-heating, ventilating, or air-conditioning equipment.

Heating and Cooling Loads

Regardless of the type of structure—residential, commercial, or industrial—the heating load must be computed at 100% of the nameplate rating of the unit. Depending on the type of heating unit, the branch circuit may require other considerations.

Fixed electric space heating is covered in detail in *Article 424—Fixed Electric Space-Heating Equipment.* This article includes fixed equipment, such as central heating systems, boilers, heating cable, and unit heaters (baseboard, panel, and duct heaters). For information regarding the installation, control, and specifics about each type of heater, refer to *Article 424.*

The *Code* also requires a disconnect for the heater and motor controller, as well as supplementary overcurrent protection for any fixed electric space-heating units. The disconnect is for safety during maintenance. The disconnecting means must be within sight of the unit and must disconnect all components of the heating unit, including any overcurrent protective devices, contactors, elements, and motor controllers.

The rules for sizing the branch-circuit wiring and overcurrent protection are very specific. The rating on the equipment nameplate is used to determine the load. If the equipment operates continuously for at least three hours, its rating must be increased by a factor of 1.25.

Fixed electric space heating shall be considered continuous load. Several general rules must be followed when sizing the overcurrent protective devices for branch circuits supplying heating equipment (see *Section 424.22*):

- Heating equipment is protected by the branch-circuit overcurrent protective device. This can be a set of fuses or circuit breakers.
- Motors used in conjunction with the heating equipment must also have overcurrent protection.
- Heating units having resistance elements exceeding 48 amps must have their load subdivided. Each of the loads must have overcurrent protection provided by the manufacturer. Conductors from the overcurrent devices to the heating unit must be sized at 125% if the load is 50 kilowatts or less.
- If the unit load for the heating equipment is greater than 50 kilowatts, the conductors can be sized at 100% provided the heating unit has a controller (thermostat). If this is not the case, then the conductors must be sized at 125% of the load.
- For conditions other than those described above, size the conductors at 125% of the load.

Sample Problem 12-10

Problem: A 30-kilowatt, 240-volt heating unit with a 6-amp fan motor is being installed for a shoe store. Determine the size of the THW conductors and overcurrent protective device for the circuit supplying the equipment.

Solution: First, determine the current required for the heating unit:

$$I = \frac{P}{E} = \frac{30{,}000 \text{ VA}}{240 \text{ V}} = 125 \text{ A}$$

Combining the current for the heating unit and the motor:

$$\text{Required current} = 125 \text{ A} + 6 \text{ A} = 131 \text{ A}$$

The conductors must be designed for 125% of this current (*Section 424.3(B)*):

$$\text{Required conductor capacity} = 131 \text{ A} \times 1.25 = 164 \text{ A}$$

The overcurrent protective device must be the next highest standard size (*Section 240.6*), which is 175 amps. Using *Table 310.16*, 2/0 AWG copper conductors are needed. The equipment grounding conductor can be a 6 AWG copper conductor (*Section 250.122*).

An air-conditioning branch-circuit load is determined from the data provided on the nameplate affixed to the unit. The nameplate will list the phase, voltage, frequency, full-load current, and other pertinent information of the hermetic motor compressor. The full-load current rating shown on the nameplate serves as the basis for determining the branch-circuit conductor size, overcurrent protection requirement, controller rating, and disconnect size.

Sometimes the branch-circuit current rating is also included on the nameplate. In such instances, use the larger of the branch-circuit current rating or full-load current rating for sizing the circuit. If a unit has two or more motors, then the circuit ampacity must be computed at 125% of the largest motor plus the sum of the other motors. For a single-motor unit disconnect, use 115% of the full-load current.

NOTE

A disconnecting means must be within sight of the unit and it must be correctly sized at 115% of the total current rating.

The overcurrent protective device cannot exceed 175% of the motor full-load current or branch-circuit rating, whichever is larger. This can be increased to 225% if the motor will not start or come up to full speed without tripping an overcurrent protective device rated for 175%.

The label may include the rating of the motor in horsepower. This rating must be converted prior to determining the branch-circuit load rating. Use *Tables 430.247* through *430.250* for converting horsepower to amperes. These tables are contained in the Reference Section at the end of the text.

Sample Problem 12-11

Problem: For an air-conditioning unit with a three-phase, 230-volt, 20-hp motor, what is the required size of the branch-circuit overcurrent protective device?

Solution: Use *Table 430.250* to convert horsepower to amperes. 20 horsepower with a 230-volt supply is equal to 54 amps. This load must be increased by 125% for the conductor ampacity:

$$\text{Conductor ampacity} = 54 \text{ A} \times 1.25$$
$$= 67.5 \text{ A}$$

The next highest standard overcurrent protective device rating is 70 amps.

Feeders

The conductors between the service equipment and the branch-circuit overcurrent devices are called ***feeders***. *Article 215—Feeders* provides information regarding the safe and adequate sizing and installation of these conductors. This article also applies to subfeeders, which provide power to branch-circuit panels but originate at power distribution centers rather than the service equipment.

Feeder loading is dependent on the total power requirement of the system. If it is possible for all connected loads to operate simultaneously, then the feeder must be of sufficient ampacity to meet that demand. If only 75% of the connected loads will ever be operating at the same time, then the feeder would be sized *larger* than the service conductors.

Prior to installation, certain factors must be considered to ensure the feeder size, type, and overcurrent protection is correct for the application:

- **Material**—The feeder can be copper or aluminum.
- **Location**—The environment around the feeder (damp, hot, corrosive) must be considered.
- **Wiring method**—The feeder can be run in conduit, cable trays, or other systems.
- **Cable**—Single or multiconductor cable can be used.
- **Type**—Feeders can be paralleled or individual.
- **Length**—Voltage drop becomes a consideration in long feeders.
- **Derating factor**—Conductor sizing includes several factors: conduit fill, ambient temperature, and connected load demand.
- **Neutral**—A neutral wire may not be necessary with the feeder.
- **Demand factor**—Continuous loads will affect feeder size.
- **Protection**—Various overcurrent protective devices can be used with feeders.

Summary of Commercial Service Load Computation Procedure

1. Determine the general lighting load based on the total square footage multiplied by the volt-ampere load indicated in *Table 220.12*.
2. If the lighting is a continuous load, as is true of most commercial lighting, increase the load by 125%. That is, multiply by 1.25. (*Section 210.20(A)*)
3. For feeder and service loads, use *Table 220.42* and apply the lighting demand factor required for the type of building.
4. After computing both air-conditioning and heating loads, omit the smaller of the two. (*Section 220.60*)

5. Compute receptacle outlets as follows:
   - Receptacles as 180VA each. (*Section 220.14(I)*)
   - Multioutlet assemblies at 180 VA per each 5′ portion. (*Section 220.14(H)(1)*)
   - Show windows at either of the following (*Section 220.14(G)*):
     (1) The unit load per outlet as required in other provisions of *Section 220.14*
     (2) At 200 VA per 300 mm (1 foot) of show window
   - Heavy-duty lamp holders at 600 VA each. (*Section 220.14(E)*)
   - Others as outlined in *Section 220.14*
6. Apply any of the demand factors as shown in *Table 220.44* for receptacle loads and *Table 222.56* for kitchen equipment loads.
7. Add the sign lighting load (1200 VA minimum). (*Section 220.14(F)*)
8. Compute the motor loads using the appropriate tables in *Article 430.*
9. Increase largest motor load based on full-load current by 25%. (*Section 430.24*)
10. Size the service and service conductors. Compute by dividing the total load by the line voltage. Conductor size is selected from *Table 310.16.*
11. Using the guideline given in *Section 250.24(C)*, size the grounded service conductor. Be sure the grounded service conductor is not smaller than the grounding electrode conductor as given in *Table 250.66.*
12. Provided there is no discharge lighting, the neutral load—if over 200 amperes—can be derated by 70%. Refer to *Section 220.61.*

Sample Problems

The following examples represent a small sampling of situations requiring computations to determine correct sizing of equipment, conductors, overcurrent protection, and other load demands encountered in commercial wiring. The main purpose is to introduce the concept and give the reader a general feel for the overall procedure. Numerous other factors such as wiring methods, routing, distances, and voltage drop (to name a few) have been purposefully ignored here for the sake of simplicity. Still, by reviewing and understanding these examples and the step-by-step methodology, the commercial electrician, designer, engineer, and student will be better equipped to move on to the "real" calculations required in practical design situations.

The following problems illustrate load calculation and feeder sizing for commercial structures:

- **Small retail store**—Sample Problem 12-12.
- **Office building**—Sample Problem 12-13.
- **Restaurant**—Sample Problem 12-14.
- **Hotel**—Sample Problem 12-15.

Sample Problem 12-12

Problem: A small retail store is being constructed. Its power supply is single-phase, 120/240-volt. The store is 80′ × 60′ and has several loads:

50-kVA heating equipment
25-kVA air-conditioning unit
1/2-hp, 240-volt ventilating unit
60 duplex receptacles
40 linear feet of show-window lighting
1.2-kVA outdoor sign lighting

Copper THW conductors are used as feeders. What size should the current-carrying feeders be?

Solution: First, all the loads must be computed. The heating unit and air-conditioning unit will not be used at the same time, so only the larger load is needed. The 1/2-hp, 240-volt motor draws a current of 4.9 amps, as shown in *Table 430.248.*

$$\text{General lighting load} = 80' \times 60' \times 3\ \text{VA/ft}^2 = 14{,}400\ \text{VA}$$

$$\text{Show-window lighting load} = 40' \times 200\ \text{VA} = 8000\ \text{VA}$$

$$\text{Sign lighting load} = 1200\ \text{VA}$$

$$\text{Receptacle load} = 60 \times 180\ \text{VA} = 10{,}800\ \text{VA}$$

$$\text{Heating/air-conditioning load} = 50{,}000\ \text{VA}$$

$$\text{Motor load} = 4.9\ \text{A} \times 240\ \text{V} = 1176\ \text{VA}$$

The total lighting load is the combination of the three separate lighting loads.

$$\text{Total lighting load} = 14{,}400\ \text{VA} + 8000\ \text{VA} + 1200\ \text{VA} = 23{,}600\ \text{VA}$$

The motor load and lighting loads are continuous, so the conductors must supply 125% of the full-load current:

$$\text{Total lighting load} = 23{,}600\ \text{VA} \times 1.25 = 29{,}500\ \text{VA}$$

$$\text{Motor load} = 1176\ \text{VA} \times 1.25 = 1470\ \text{VA}$$

(Continued on the following page.)

Sample Problem 12-12 *Continued*

All loads are added to determine the total load:

$$\text{Total load} = 29{,}500 \text{ VA} + 10{,}800 \text{ VA} + 50{,}000 \text{ VA} + 1470 \text{ VA} = 91{,}770 \text{ VA}$$

The total current can be calculated:

$$I = \frac{P}{E} = \frac{91{,}770 \text{ VA}}{240 \text{ V}} = 382 \text{ A}$$

Using *Table 310.16*, 500 kcmil conductors will be sufficient.
The neutral feeder conductor is sized for the 120-volt loads (lighting and receptacles) only.

$$\text{Neutral feeder load} = \text{Lighting load} + \text{Receptacle load} = 29{,}500 \text{ VA} + 10{,}800 \text{ VA} = 40{,}300 \text{ VA}$$

The current can be calculated:

$$I = \frac{P}{E} = \frac{40{,}300 \text{ VA}}{240 \text{ V}} = 168 \text{ A}$$

A 2/0 AWG copper conductor should be used for the neutral feeder conductor.

Sample Problem 12-13

Problem: A two-story office building (80′ × 80′) is supplied with 120/208-volt service. Several loads are supplied:

- 75-kVA air-conditioning unit
- 85-kVA heating unit
- 200 duplex receptacles
- 50 linear feet of show window
- 30 exterior light fixtures (175 VA each)
- 3 blower motors (3/4-hp, 208-volt)

Copper THW conductors are used as feeders. What size should the current-carrying conductors be?

Solution: First, all loads must be computed. The heating unit and air-conditioning unit will not be used at the same time, so only the larger load is

(Continued)

Sample Problem 12-13 *Continued*

needed. The 3/4-hp, 208-volt motors draw 3.5 amps, as shown in *Table 430.250*. The actual motor voltage is 208 V × 1.732 (or 360 volts) because the service is three-phase.

$$\text{General lighting load} = 2 \times 80' \times 80' \times 3\ 1/2 \text{ VA/ft}^2 = 44{,}800 \text{ VA}$$

$$\text{Show-window lighting load} = 50' \times 200 \text{ VA} = 10{,}000 \text{ VA}$$

$$\text{Exterior lighting load} = 30 \times 175 \text{ VA} = 5250 \text{ VA}$$

$$\text{Receptacle load} = 200 \times 180 \text{ VA} = 36{,}000 \text{ VA}$$

$$\text{Heating/air-conditioning load} = 85{,}000 \text{ VA}$$

$$\text{Motor load} = 3 \times 3.5 \text{ A} \times 360 \text{ V} = 3780 \text{ VA}$$

The total lighting load is the combination of the three separate lighting loads.

$$\text{Total lighting load} = 44{,}800 \text{ VA} + 10{,}000 \text{ VA} + 5250 \text{ VA} = 60{,}050 \text{ VA}$$

The motor load and lighting loads are continuous, so the conductors must supply 125% of the full-load current:

$$\text{Total lighting load} = 60{,}050 \text{ VA} \times 1.25 = 75{,}063 \text{ VA}$$

$$\text{Motor load} = 3780 \text{ VA} \times 1.25 = 4725 \text{ VA}$$

All loads are added to determine the total load:

$$\text{Total load} = 75{,}063 \text{ VA} + 36{,}000 \text{ VA} + 85{,}000 \text{ VA} + 4725 \text{ VA} = 200{,}788 \text{ VA}$$

The total current can be calculated:

$$I = \frac{P}{E} = \frac{200{,}788 \text{ VA}}{360 \text{ V}} = 557 \text{ A}$$

Using *Table 310.16*, 1250 kcmil conductors will be sufficient.

Sample Problem 12-14

Problem: A restaurant is supplied with three-wire, 120/240-volt service. The restaurant is 50′ × 80′ and has several loads:

(1) 12-kW electric range (120-volt)
(1) 10-kW water heater (240-volt)
(2) 8-kW fryers (240-volt)
(2) 1.5-kW coffeemakers (120-volt)
(1) 2.5-kW steam table (120-volt)
(2) 3.0-kW toasters (120-volt)
(1) 2.5-kW disposal unit (120-volt)
(1) 1.5-kW outdoor sign (120-volt)
(1) 40-kW heating unit
(26) duplex receptacles

Copper THW conductors are used as feeders. What size should the current-carrying feeders be?

Solution: First, the loads must be computed.

General lighting load = 50′ × 80′ × 2 VA/ft²
= 8000 VA

Sign lighting load = 1500 VA

Receptacle load = 26 × 180 VA
= 4680 VA

Heating load = 40,000 VA

The cooking equipment will be combined as the cooking equipment load:

Electric range: 12,000 VA
Water heater: 10,000 VA
Fryers: 16,000 VA
Coffeemakers: 3000 VA
Steam table: 2500 VA
Toasters: 6000 VA
Disposal unit: 2500 VA

Cooking equipment load = 52,000 VA

Table 220.56 contains load-reduction factors for commercial kitchen equipment. For six or more pieces of equipment, the feeder conductors can be designed for 65% of the load.

Reduced cooking equipment load = 52,000 VA × 0.65
= 33,800 VA

The total lighting load is the combination of the two separate lighting loads.

Total lighting load = 8000 VA + 1500 VA
= 9500 VA

The lighting loads are continuous, so the conductors must supply 125% of the full-load current:

Total lighting load = 9500 VA × 1.25
= 11,875 VA

All loads are added to determine the total load:

Load = 11,875 VA + 4680 VA + 40,000 VA + 33,800 VA
= 90,355 VA

The total current can be calculated:

$$I = \frac{P}{E} = \frac{90{,}355\text{ VA}}{240\text{ V}} = 377\text{ A}$$

Using *Table 310.16*, 500 kcmil conductors will be sufficient.

Sample Problem 12-15

Problem: A 200-unit hotel is supplied with three-phase, 120/208-volt service. Each unit is 300 ft², has a 30-amp air conditioner (208-volt) and has a heating load of 8 kilowatts. The hotel also has a 30-kilowatt continuous load for general equipment. What is the total current that must be supplied to the hotel?

Solution: First, all loads must be computed. The heating unit and air-conditioning unit will not be used at the same time, so only the larger load is needed. *Section 220.14* explains that the receptacle loads do not need to be included.

General lighting load = 200 × 300 ft² × 2 VA/ft²
= 120,000 VA

This load can be reduced in accordance with *Table 220.42*:

50% of 0–20 kW = 10,000 VA
40% of 20–100 kW = 32,000 VA
30% of 100–120 kW = 8000 VA

(Continued on the following page.)

Sample Problem 12-15 *Continued*

The total reduced lighting load is determined:

$$\text{Lighting load} = 10{,}000\text{ VA} + 32{,}000\text{ VA} + 8000\text{ VA}$$
$$= 50{,}000\text{ VA}$$

The air-conditioning load must be calculated. The voltage is 208 V × 1.732 (360 volts) due to the three-phase supply:

$$\text{Air-conditioning load} = 360\text{ V} \times 30\text{ A}$$
$$= 10{,}800\text{ VA}$$

This is larger than the heating load, so the air-conditioning load is used in the total load. The load per room is 10,800 VA, so it must be multiplied by 200 to determine the total load:

$$\text{Total air-conditioning load} = 10{,}800\text{ VA} \times 200$$
$$= 2{,}160{,}000\text{ VA}$$

$$\text{General equipment load} = 30{,}000\text{ VA} \times 1.25$$
$$= 37{,}500\text{ VA}$$

All loads are added to determine the total load:

$$\text{Total load} = 50{,}000\text{ VA} + 2{,}160{,}000\text{ VA} + 37{,}500\text{ VA}$$
$$= 2{,}247{,}500\text{ VA}$$

The total current can be calculated:

$$I = \frac{P}{E}$$
$$= \frac{2{,}247{,}500\text{ VA}}{360\text{ V}}$$
$$= 6243\text{ A}$$

The total current is 6243 A.

Review Questions

Answer the following questions. Do not write in this book.

1. What type of conductors extend beyond the final overcurrent device?
2. What type of conductors originate at the main disconnect device and terminate at panelboards and load centers?
3. How is the rating of a branch circuit determined?
4. What is the smallest conductor permitted for use in a branch circuit?
5. What color insulation is used on grounding conductors?
6. Define *continuous* load.
7. What are the five general types of branch-circuit loads?
8. If the exact number of receptacles is unknown, how can the receptacle load be calculated?
9. How is the show-window lighting load determined?
10. How many receptacles must be installed for the nondwelling receptacle loads derating to reduce the total receptacle load?
11. A large hotel has 2500 ft^2 of hallway space. Assuming continuous load, how many volt-amperes should be allowed for this load for service computation?
12. Calculate the general lighting load for a commercial warehouse having an area of 90,000 ft^2. Assume continuous duty.
13. What is the demand factor used for four kitchen equipment loads in a commercial cafeteria?
14. Compute the receptacle load for a department store having 165 duplex receptacles.
15. A furniture store has 60 linear feet of show window. Calculate this continuous load.
16. What is the minimum outside sign lighting load required by the *Code*?
17. What is the general load for a 50-room motel where each room is 12′ × 20′?
18. A 24-hour service restaurant has 75 duplex receptacles. Calculate the total load for feeder calculations.
19. What is the service demand for a small bank measuring 2500 ft^2 and supplied with 208/120-volt, three-phase service? Assume lighting as continuous. The bank has the following loads:
    - 15 kW heating (208-volt)
    - 7.5 hp air conditioning
    - 50 linear feet of show window
    - 25 duplex receptacles
20. A new high school is to be erected, supplied with 208/120-volt, single-phase power. The school has the following dimensions and loads. Calculate the total feeder conductor and feeder neutral loads. Assume all lighting loads are continuous.
    - 30,000 ft^2 classroom space
    - 6000 ft^2 auditorium
    - 5000 ft^2 cafeteria
    - 10 kW outside lighting (120-volt)
    - 225 duplex receptacles (120-volt)

Kitchen equipment:

(2) 15 kW ovens (208-volt)
(2) 10 kW ranges (208-volt)
(4) 4 kW fryers (208-volt)
(1) 10 kW water heater (208-volt)
(1) 3 kW dishwasher (208-volt)
(2) 2.5 kW toasters (208-volt)
(1) 1/2 hp vent fan (120-volt)
(1) 3/4 hp blower motor (120-volt)

Heating:

40 kW electric heat (208-volt)

USING THE NEC

Refer to the National Electrical Code to answer the following questions. Do not write in this book.

1. In addition to *Article 210,* which other section contains information on branch circuits supplying pipe organs?
2. What is the minimum branch-circuit current rating for cooking ranges with ratings of 8 3/4 kW or higher?
3. How near to the appliance does a dedicated appliance receptacle need to be installed?
4. How is the size for a feeder overcurrent protective device determined?
5. *Article 220* addresses branch-circuit, feeder, and service calculations. How are fractions of amperes handled in calculations?
6. Show-window loads can be computed in two ways. Name the ways.

Chapter 13

Motors

Technical Terms

Apparent power
Capacitive reactance
Capacitor motors
Capacitor-start-run motor
Exciter
Induction motor
Motor
Motor service factor
Phase angle
Power factor
Repulsion motors
Rotor
Running winding
Shading coils
Starting torque
Starting winding
Stator
Synchronous motors
Torque
True power
Unity
Universal motors
Windings

Objectives

After completing this chapter, you will be able to:

- Explain the basic components of motors.
- List various classes of motors.
- Calculate motor starting currents.
- Use the *Code* to design motor branch circuits, including overcurrent protection.
- Define and compute true power, apparent power, and power factor.

Motors are found in nearly every industry, ranging in size from fractional horsepower to thousands of horsepower. Electricians, designers, and engineers must have a firm knowledge of motor types and circuit requirements to properly install and maintain motors.

This chapter will briefly explore some of the basic varieties of motors and the basics of motor installation and motor circuits. *Article 430—Motors, Motor Circuits, and Controllers* of the *Code* covers the design and installation of motors.

Motor Theory

A ***motor*** is a device that converts electrical energy to mechanical energy. The two principal components of a motor are the stator and the rotor. The ***stator*** consists of one or more ***windings*** (coils of wire) and surrounds the rotor. The windings in the stator are sometimes called field windings or field coils.

The ***rotor*** is a cylindrical component housed within the stator. Both ends of a rod extending through the rotor are set in bearings, allowing the assembly to turn. There are two types of rotors: squirrel-cage and wound. A squirrel-cage rotor houses a ring of copper or aluminum bars parallel to the rotor axis. A wound rotor has coils (called armature coils) connected to commutator segments. The commutator receives current from brushes, and is used to vary the current through the windings.

Motor Types

Motors differ in a variety of ways. The current supplying the motor can be single-phase ac, polyphase ac, or dc. There are different ways of starting the motor and different ways to keep the motor running. The internal mechanisms vary, depending on these factors and the size of the motor.

Alternating-Current Motors

There are two classifications of ac motors: single-phase and polyphase. Single-phase motors are used in residential and light commercial applications where the service is single-phase voltage. The most common polyphase motors are used with three-phase current. The three-phase motors are more efficient and can be used for larger loads than single-phase motors.

Single-phase motors

Single-phase motors can be found throughout buildings supplied with single-phase power. Power tools, vacuum cleaners, refrigerators, and air-conditioning units all use single-phase motors. Single-phase motors can be divided into two broad categories: induction motors and repulsion motors.

An ac ***induction motor*** has a primary winding (stator) connected to the power supply. The secondary winding in the rotor carries the induced current.

Unlike other motors, induction motors have no physical connection between the energy source and the rotor. The current used by the rotor is derived from electromagnetic induction from the stator winding. **Figure 13-1** illustrates an induction motor.

There are several types of single-phase induction motors:

- **Split-phase motor**—These motors have dual windings on the stator: the ***starting winding*** and the ***running winding***. These windings are spread or separated by about 30°. The two windings are connected in parallel, with a centrifugal switch in series with the starting winding. The starting switch opens when the motor reaches its full running speed. This removes the starting winding from the circuit. See **Figure 13-2A.**
- **Shaded-pole motor**—The shaded-pole motor has auxiliary coils (***shading coils***) embedded in each field pole. Each coil forms a loop within the running winding. This causes the motor to have a very low starting torque.

Capacitor motors are another type of single-phase induction motors:

- **Capacitor motor**—A capacitor motor has an auxiliary winding connected in series with a capacitor. These motors have low starting torque, but high maximum torque.
- **Capacitor-start motor**—This type of single-phase ac motor is identical to the split-phase motor with a capacitor added to create a higher starting torque. The capacitor is connected in series with the starting winding and a centrifugal switch, **Figure 13-2B.** When the motor reaches a set running speed, the switch opens and the capacitor and auxiliary winding are removed from the circuit.
- **Permanent-split capacitor motor**—Also called a ***capacitor-start-run motor,*** these motors have no switch, so the capacitor remains in the circuit while the motor is running. This type of motor has high running torque and high starting torque. See **Figure 13-2C.**

Repulsion motors operate using the principle of magnetic repulsion. Current flows through the field windings on the stator and through the coils in the wound rotor. The interaction of the magnetic fields produced by the field and armature coils cause the rotor to turn.

Repulsion motors are sometimes called series motors because the field and armature windings are connected in series. See **Figure 13-3.** The following are the three common types of repulsion motors:

- **Repulsion motor**—This motor uses repulsion for starting and running. Repulsion motors have high starting torque.
- **Repulsion-induction motor**—In addition to the repulsion motor winding, this motor also has a squirrel-cage winding in the rotor.
- **Repulsion-start induction motor**—This motor is similar to a repulsion-induction motor, but at a set speed the commutator bars are short-circuited and the motor operates on induction alone.

Figure 13-1. Cutaway view of an induction motor. (Reliance Electric)

Three-phase motors

Three-phase induction motors do not require a commutator to change the polarity of the current. The varying alternating phases produce a magnetic field that drives the rotor.

Synchronous motors have rotor poles that are energized by a dc generator (called an ***exciter***). These motors operate at a constant speed, have a controllable power factor, and can be adjusted by the motor operator. They are rarely smaller than 25 horsepower.

Direct-Current Motors

There are three types of dc motors: shunt-wound, series-wound, and compound-wound. These motors differ in the connection between the field and armature windings. See **Figure 13-4.**

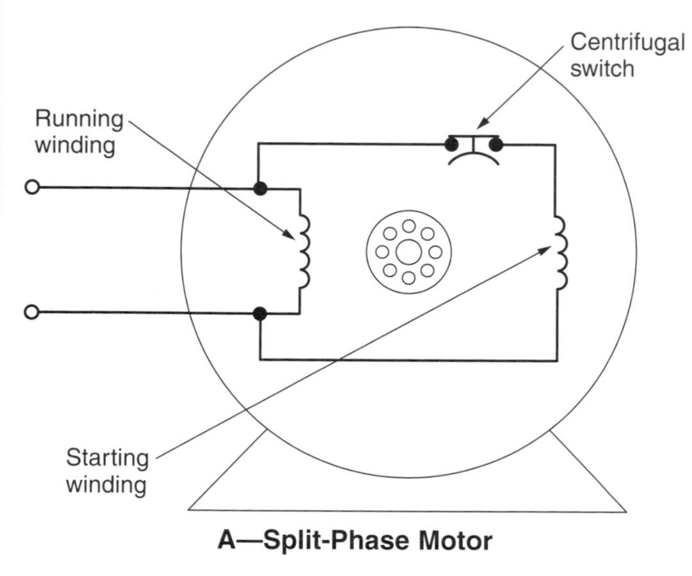

A—Split-Phase Motor

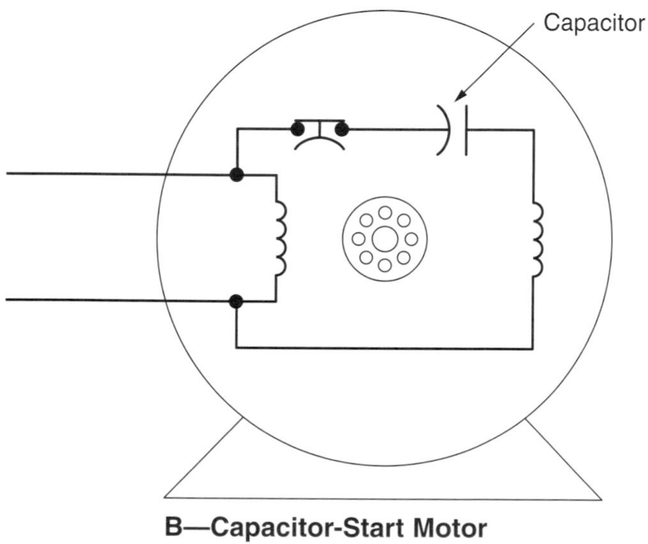

B—Capacitor-Start Motor

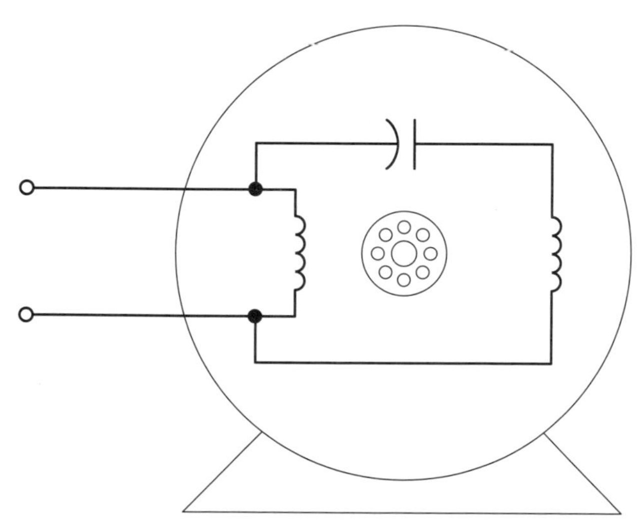

C—Permanent-Split Capacitor Motor

Figure 13-2. AC induction motors with starting and running windings. A—The centrifugal switch in a split-phase motor opens as the motor approaches normal speed. B—A capacitor is added to the design in a capacitor-start motor to increase starting torque. C—There is no centrifugal switch in a permanent-split capacitor motor, so the starting winding remains energized.

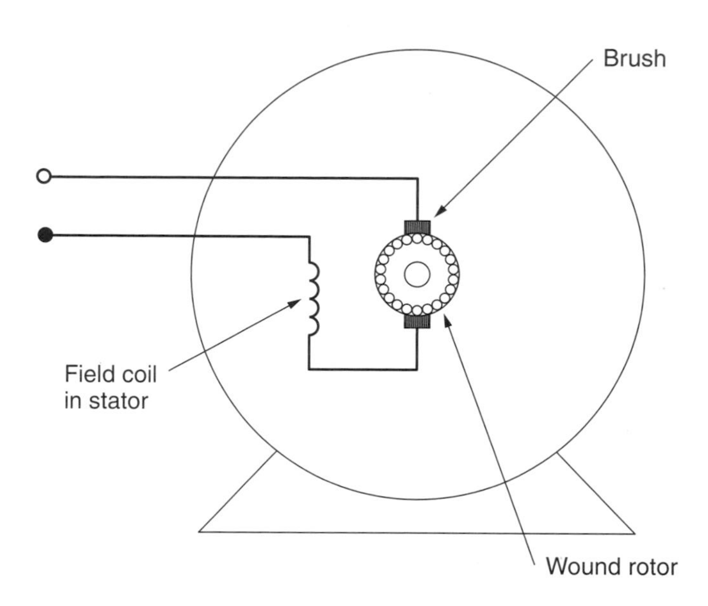

Figure 13-3. A single-phase ac repulsion motor has the windings connected in series.

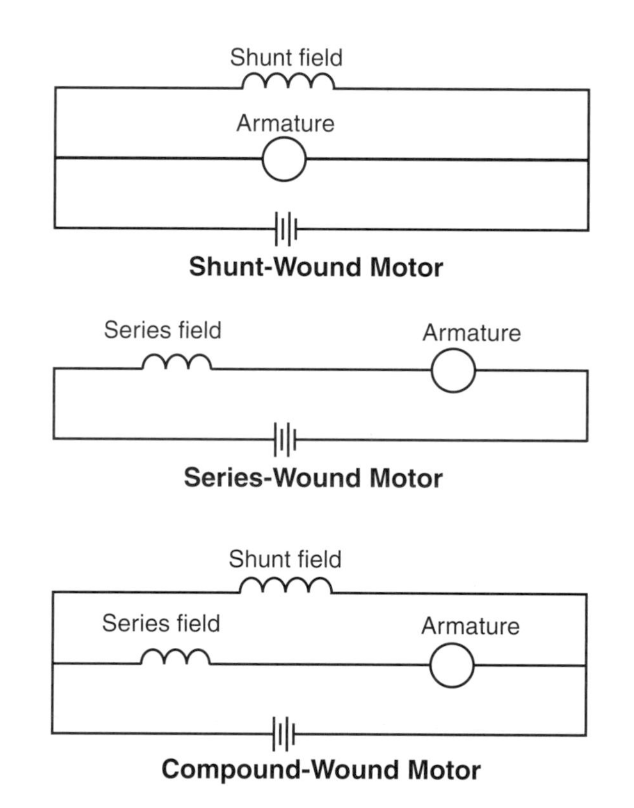

Figure 13-4. Wiring diagrams for dc motor varieties.

- **Shunt-wound motor**—The field and armature windings are connected in parallel.
- **Series-wound motor**—The field and armature windings are connected in series.
- **Compound-wound motor**—This motor has two field windings: one connected in parallel and one connected in series.

Universal motors are usually small, series-wound motors that operate on dc voltage or single-phase ac voltage.

The motor's performance is independent on the type of voltage. These motors are available in sizes up to 3/4 hp.

Universal motors have an armature, brushes, field windings, and a commutator. See **Figure 13-5.** The field windings are connected in series with the armature and brushes. The speed of universal motors can be modified by adding resistance to one of the field windings. The higher the resistance added, the slower the motor speed. See **Figure 13-6.**

Motor Torque

Torque is a force that produces rotation. There are several kinds of motor torque, all of which are related to the speed at which a motor will run:

- **Locked-rotor torque**—Also referred to as ***starting torque***, this is the torque produced by a motor when it is initially started. It is also the force required to get the motor running.
- **Pull-up torque**—Torque required to bring the motor to full speed after it has been started.
- **Full-load torque**—Torque necessary to keep the motor running at the rated speed.
- **Breakdown torque**—The maximum torque a motor can achieve before stalling.

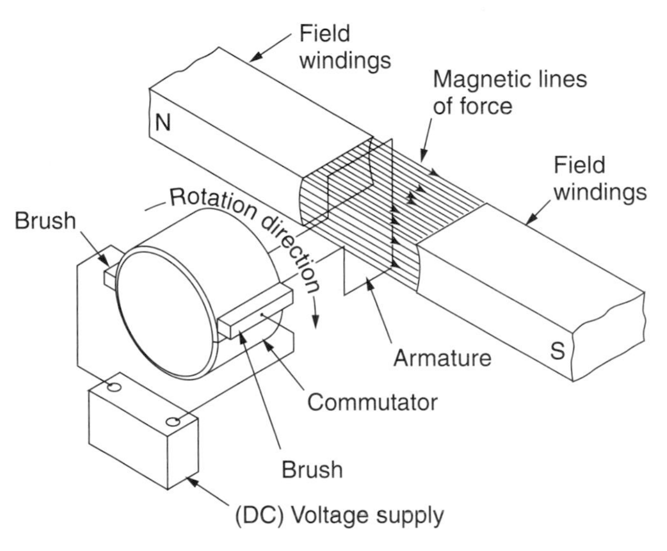

Figure 13-5. Components of a universal, series-wound motor.

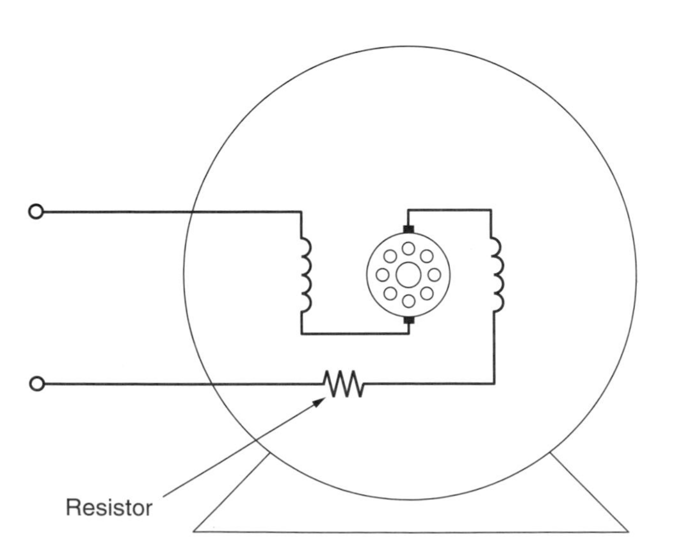

Figure 13-6. Varying the speed of a universal motor by adding resistance.

Torque calculations

Torque is calculated using a simple formula:

$$T = \frac{hp \times 5252}{rpm}$$

where

T = Torque (in lb-ft)

hp = Horsepower of the motor

rpm = Revolutions per minute

Sample Problem 13-1

Problem: Determine the full-load torque developed by a 50-hp motor having a 240-volt supply and turning at 1800 rpm.

Solution: Using the torque equation:

$$T = \frac{hp \times 5252}{rpm} = \frac{50 \times 5252}{1800} = 146 \text{ lb-ft}$$

The full-load torque is 146 lb-ft.

NEMA torque classification

The National Electrical Manufacturers Association (NEMA) classifies motors based on their torque characteristics and minimum starting torque. Motors are classified as Class A through Class F, each class having a different range of starting torque. For example, Class B (the most common class of motors) has a starting torque that is 150% of the full-load torque. Some examples of common starting torque for motors are shown in the table in **Figure 13-7.**

Motor Selection Factors

Determining which type of motor to use in a particular installation is an interesting challenge. To make the proper motor selection, several general factors are considered:

- **Power source**—Most motors use either a single-phase alternating-current (ac), three-phase ac, or direct-current (dc) source. In some instances, either

Motor Starting Torque

Motor Type	Percent of Full-Load
Single-phase ac motors	
Shaded-pole	50–100
Split-pole	80–200
Capacitor-start	200–350
Capacitor-run	350–400
Three-phase ac motors	
Induction	100–250
Wound-rotor	200–300
Synchronous	50–150
Universal motors	300–400
Direct-current motors	
Series-wound	400–500
Shunt-wound	125–250
Compound-wound	300–400

Figure 13-7. Motor starting torque expressed as percent of full-load torque for various motor types.

ac or dc can be used. Normally the available power source dictates the motor type. For each type of service, there are a variety of motors available.

- **Load**—A motor, like any machine, accomplishes a task. For a motor to perform the task efficiently and dependably, it must be properly matched to the load. If the motor is too large or too small, it will not perform economically.
- **Environment**—The location can affect motor selection. Some motors are more durable than others. Moisture, dust, explosive gases, corrosive chemicals, radiation, temperature extremes, and vibrations can be harmful to motors.
- **Temperature**—Overheating is the largest cause of motor failure. High-temperature areas do not allow the heat generated within the motor to dissipate. The most common ambient temperature limit for motors is 104°F (40°C). The ambient temperature limit is shown on the motor nameplate.

Motor Enclosures

Motors can be classified as either open or enclosed. An open motor allows air to flow freely through the enclosure. In an enclosed motor, the enclosure seals the motor from the surrounding air. Within the general classifications of open and enclosed, there are a variety of enclosure types available. See **Figure 13-8.**

Many factors affect motor enclosure selection. The following are some of the most important factors:

- **Environment**—The area where the motor is installed will most likely dictate the type of enclosure needed. Hazardous locations require explosion-proof enclosures. Areas where liquid may be present could require drip-proof or splash-proof enclosures.
- **Supplied equipment**—A protective enclosure may be specified for motors supplying power to critical equipment, even if the enclosure is not required based on the environment.
- **Availability**—A given size and type of motor may be available with only certain enclosure types.

Motor Enclosures

Enclosure	Description
Open	
Drip-proof	Designed so water or fluids falling at angle up to 15° from vertical cannot enter the motor.
Splash-proof	Designed so water or fluids falling at angle up to 100° from vertical cannot enter the motor.
Guarded	Openings that allow access to rotating or energized parts are protected to prevent the passage of any object as large as a 0.75″ diameter rod.
Semiguarded	Some openings are guarded while others are left open. Normally the openings on the upper part of the motor are guarded.
Externally ventilated	Motor has a separate blower used for ventilation. The blower is mounted to the enclosure.
Enclosed	
Water-cooled	A motor having a water jacket surrounding its casing to cool the motor surface.
Pipe-ventilated	Pipe-ventilated motor with an exhaust fan forcing clean air through the system.
Fan-cooled	Motor cooled by an exhaust fan outside the enclosure.
Nonventilated	Motor with no ventilation.
Explosion-proof	Designed so an internal explosion will be contained and will not produce an external explosion or spark that could cause explosive vapors or gases in the surrounding area to ignite.
Dust-ignition-proof	Designed so dust particles cannot enter the motor and cause an explosion.

Figure 13-8. Select the proper motor enclosure for the environment where the motor will be located.

Motor Code Letters

Section 430.7(B) identifies code letters used to indicate the motor input with locked rotor. This is an indication of the current drawn by the motor when it starts. This starting current is much larger than the current needed once the motor is running. See **Figure 13-9,** which reflects *Table 430.7(B)*.

Locked-Rotor Indicating Codes

Code Letter	kVA per HP with Locked Rotor	Code Letter	kVA per HP with Locked Rotor
A	0–3.14	L	9.0–9.99
B	3.15–3.54	M	10.0–11.19
C	3.55–3.99	N	11.2–12.49
D	4.0–4.49	P	12.5–13.99
E	4.5–4.99	R	14.0–15.99
F	5.0–5.59	S	16.0–17.99
G	5.6–6.29	T	18.0–19.99
H	6.3–7.09	U	20.0–22.39
J	7.1–7.99	V	above 22.4
K	8.0–8.99		

Figure 13-9. A code letter on the motor nameplate represents the motor input current at starting.

Service Factor

Most motors are built to withstand minor overloads. This extra margin of safety is indicated by the ***motor service factor*** (SF). A motor with a service factor of 1.25 can safely operate at 125% of its full-load current rating without any harmful effects to the insulation. The motor service factor is shown on the motor nameplate.

Sample Problem 13-2

Problem: Find the minimum and maximum starting current for a three-phase, 240-volt, 25-horsepower motor having a code letter F on the nameplate.

Solution: Using *Table 430.7(B)*, the range of kilovolt-amperes per horsepower is 5.0–5.59. This range is multiplied by the horsepower to determine the range for this particular motor:

$$5.0 \text{ kVA/hp} \times 25 \text{ hp} = 125 \text{ kVA}$$

$$5.59 \text{ kVA/hp} \times 25 \text{ hp} = 139.75 \text{ kVA}$$

Convert kilovolt-amperes to volt-amperes:

$$125 \text{ kVA} = 125{,}000 \text{ VA}$$
$$139.75 \text{ kVA} = 139{,}750 \text{ VA}$$

The voltage is three-phase, so the nominal value must be multiplied by 1.732 to determine the actual voltage:

$$240 \text{ V} \times 1.732 = 416 \text{ V}$$

The current is calculated by dividing the power by the voltage. For the minimum current:

$$I = \frac{P}{E} = \frac{125{,}000 \text{ VA}}{416 \text{ V}} = 300 \text{ A}$$

For the maximum current:

$$I = \frac{P}{E} = \frac{139{,}750 \text{ VA}}{416 \text{ V}} = 336 \text{ A}$$

The starting current range for this motor is 300–336 amps.

Sample Problem 13-3

Problem: What is the maximum output of a motor rated at 30 hp with a service factor of 1.25?

Solution: The maximum horsepower is the rating multiplied by the service factor:

$$\text{Max hp} = 30 \text{ hp} \times 1.25 = 37.5 \text{ hp}$$

The maximum safe output is 37.5 horsepower.

Motor Circuit Conductors

The requirements for branch-circuit conductors supplying power to a motor are defined in *Part II* of *Article 430*. The requirements vary, depending on the loads supplied by the circuit. Circuits are sized differently for the following situations:

- Circuit supplying a single motor.
- Circuit supplying several motors or a motor and other loads.
- Multimotor and combination-load equipment.
- Ratings and size of motor load.

Conductors Supplying a Single Motor

Conductors for an individual branch circuit supplying a single motor are sized in accordance with *Section 430.22(E)*. Motor circuit conductors are sized for 125% of the full-load current rating of the motor. Conductors that supply power to motors that are not in continuous use may be sized for a lesser current in accordance with *Table 430.22(E)*. See **Figure 13-10.** The full-load currents for various motors are found in several tables:

- ***Table 430.247***—Full-load current for dc motors.
- ***Table 430.248***—Full-load current for single-phase ac motors.
- ***Table 430.249***—Full-load current for two-phase ac motors (4-wire).
- ***Table 430.250***—Full-load current for three-phase ac motors.

These tables list the full-load current based on the motor's horsepower and voltage ratings. All four tables are included in the Reference Section at the end of this text.

Conductors Supplying a Motor and Other Loads

If conductors supply more than one motor, or if there are other loads on the circuit with a motor, the conductor sizing is more complex than when there is only a single motor. The conductors must have enough capacity for 125% of the full-load current of the largest motor, plus the full-load current of other motors, plus the ampere rating of any other loads.

NEC NOTE 430.24

Conductors supplying several motors, or a motor(s) and other load(s), shall have an ampacity not less than 125 percent of the full-load current rating of the highest rated motor plus the sum of the full-load current ratings of all the other motors in the group, as determined by *Section 430.6(A)*, plus the ampacity required for the other loads.

Table 430.22(E) Duty-Cycle Service

	Nameplate Current Rating Percentages			
Classification of Service	**5-Minute Rated Motor**	**15-Minute Rated Motor**	**30- & 60-Minute Rated Motor**	**Continuous Rated Motor**
Short-time duty operating valves, raising or lowering rolls, etc.	110	120	150	—
Intermittent duty freight and passenger elevators, tool heads, pumps, drawbridges, turntables, etc. (for arc welders, see Section 630.11)	85	85	90	140
Periodic duty rolls, ore- and coal-handling machines, etc.	85	90	95	140
Varying duty	110	120	150	200

Note: Any motor application shall be considered as continuous duty unless the nature of the apparatus it drives is such that the motor will not operate continuously with load under any condition of use.

Figure 13-10. Conductors supplying power to motors that are not used continuously are sized as a percentage of the nameplate current rating.

Sample Problem 13-4

Problem: Determine the minimum ampere rating for conductors supplying power to a 3/4-horsepower, 115-volt, single-phase motor.

Solution: The conductors must be sized for 125% of the full-load current. The full-load current is found as 13.8 amps in *Table 430.248.*

Horsepower	115 Volts	200 Volts	208 Volts	230 Volts
1/6	4.4	2.5	2.4	2.2
1/4	5.8	3.3	3.2	2.9
1/3	7.2	4.1	4.0	3.6
1/2	9.8	5.6	5.4	4.9
3/4	13.8	7.9	7.6	6.9
1	16	9.2	8.8	8.0
1 1/2	20	11.5	11.0	10
2	24	13.8	13.2	12
3	34	19.6	18.7	17
5	56	32.2	30.8	28
7 1/2	80	46.0	44.0	40
10	100	57.5	55.0	50

The required conductor amperage can now be calculated:

$$13.8\ \text{A} \times 1.25 = 17.25\ \text{A}$$

The conductors must be sized to carry 17.25 amps.

Sample Problem 13-5

Problem: A branch circuit supplies three motors with ratings of 22 amps, 42 amps, and 68 amps. What size copper THWN conductors are needed?

Solution: *Section 430.24* requires that an additional 25% be added to the largest motor:

$$68\ A \times 1.25 = 85\ A$$

Adding the ratings of the other motors:

$$\text{Total load} = 85\ A + 42\ A + 22\ A = 149\ A$$

Using *Table 310.16*, 1/0 AWG conductors are needed.

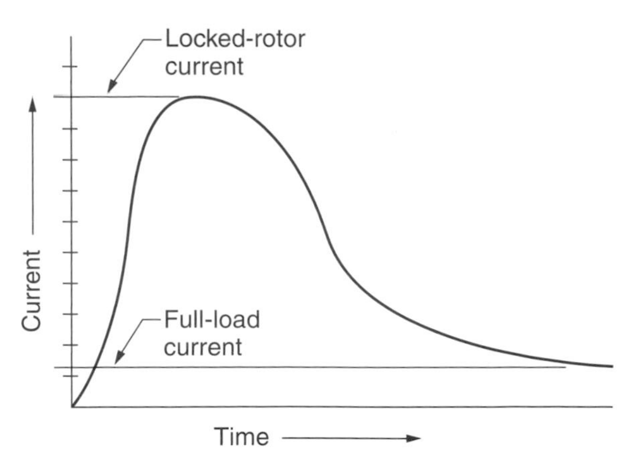

Figure 13-11. The current drawn by a motor is initially large but stabilizes to the full-load current.

Conductors Supplying Multimotor and Combination-Load Equipment

Conductors supplying multimotor equipment must have enough capacity to supply the minimum ampacity listed on the equipment nameplate. If each motor has an individual nameplate (rather than one for the entire assembly), the conductors are sized using the rules for multiple motors on a circuit.

Overcurrent Protection

In the normal course of motor operation, occasional overloads are not uncommon. Motor starting, additional motor loading, and other conditions that create high torque draw high currents but pose no threat to the motor as long as they occur for a brief duration. Longer, sustained overloads create abnormal heating of the circuit and can damage the conductor insulation and result in motor failure. Motors must be protected from these overcurrents with circuit breakers, fuses, and overload relays.

The current drawn by the motor is not constant. The locked-rotor current is relatively high (four to six times the running current). This current is needed to start the motor running. Once the motor is started, the current draw decreases until the full-load current is reached. See **Figure 13-11.**

The overcurrent protective device is located at the motor control center, safety switch, combination starter, or panel serving the motor circuit. The maximum ratings for motor branch-circuit overcurrent protective devices are listed in *Table 430.52*. See **Figure 13-12.** These ratings are based on a percentage of the full-load current of the motor. The type of circuit breaker or fuse used affects the setting in the following manner:

- **Nontime-delay fuse**—Fuses of this type hold 500% of their rating for one-quarter second. Therefore, a motor must attain its full-load speed within one-quarter second or the fuse will open the circuit.
- **Time-delay fuse**—A time-delay fuse carries 500% of its rating for up to ten seconds. This allows more time for the current drawn by the motor to stabilize. Therefore, a relatively low fuse rating is acceptable.
- **Instantaneous-trip circuit breaker**—These circuit breakers trip the instant the set rating (ranging from about three to ten times the base rating) is reached. This type of breaker will only trip due to short circuit—*not* overload or overheating. When using an instantaneous-trip circuit breaker, it is important to provide overload protection for the motor. These breakers should be installed before the disconnect and starter, and set above the locked-rotor current of the motor.
- **Inverse-time circuit breakers**—Breakers of this type are often used in commercial and industrial applications. The inverse-time breaker will hold a 300% overload for up to a minute, depending on the circuit breaker rating and the system voltage.

NEC NOTE 430.52(C)(1)

Exception No. 1: Where the values for branch-circuit, short-circuit, and ground-fault protective devices determined by Table 430.52 do not correspond to the standard size or ratings of fuses, nonadjustable circuit breakers, thermal protective devices, or possible settings for adjustable circuit breakers, the next higher standard size, rating, or possible setting shall be permitted.

When there are unusually high starting currents, the overcurrent protective device may still open the circuit during motor start-up, even when it is set at its maximum rating. For these situations, the exceptions in *Section 430.52*

Table 430.52 Maximum Rating or Setting of Motor Branch-Circuit Short-Circuit and Ground-Fault Protective Devices

	Percentage of Full-Load Current			
Type of Motor	Nontime Delay Fuse[1]	Dual Element (Time-Delay) Fuse[1]	Instantaneous Trip Breaker	Inverse Time Breaker[2]
Single-phase motors	300	175	800	250
AC polyphase motors other than wound-rotor				
Squirrel cage—				
Other than Design B energy-efficient	300	175	800	250
Design B energy-efficient	300	175	1100	250
Synchronous[3]	300	175	800	250
Wound rotor	150	150	800	150
Direct current (constant voltage)	150	150	250	150

Note: For certain exceptions to the values specified, see 430.54.

[1]The values in the Nontime Delay Fuse column apply to Time-Delay Class CC fuses.

[2]The values given in the last column also cover the ratings of nonadjustable inverse time types of circuit breakers that may be modified as in 430.52(C), Exception No. 1 and No. 2.

[3]Synchronous motors of the low-torque, low-speed type (usually 450 rpm or lower), such as are used to drive reciprocating compressors, pumps, etc., that start unloaded, do not require a fuse rating or circuit-breaker setting in excess of 200 percent of full-load current.

Figure 13-12. The required rating for a motor branch-circuit overcurrent protective device depends on the type of device used.

allow larger percentages of the full-load current to be used to size the fuse or breaker. These exceptions are only valid for a circuit that supplies a single motor.

Branch Circuits with More Than One Motor

Requirements for protecting circuits that supply more than a single motor load are listed in *Section 430.53*. In all cases, each motor must have overcurrent protection. The *Code* lists three sets of criteria for allowing other loads besides a motor on a branch circuit. If any of the three are satisfied, the circuit is allowed:

- ***Section 430.53(A)***—Motors less than 1 hp with individual full-load ratings less than 6 amps can be placed on the same branch circuit if the branch-circuit protection does not exceed that of the controllers. The *Code* also limits the voltage and current in these circuits.
- ***Section 430.53(B)***—If the overcurrent protection on the circuit is less than the smallest rating of all the motors, multiple motors are allowed. However, the branch-circuit overcurrent protective device must allow enough current for the motors to start.
- ***Section 430.53(C)***—This section contains the requirements for group installations on a single branch circuit. Special requirements for the motor controllers and overload devices are included.

Feeder for Several Motors

Overcurrent protective devices are installed for each motor (usually at the motor starter controller equipment). In addition, group overload protection for the common feeder must be provided. Feeder overload protection is addressed in *Section 430.62*.

If a feeder supplies only a single branch circuit that supplies a single motor, the feeder requires the same overcurrent protection as the branch circuit. If the feeder supplies anything in addition to a single motor, the overcurrent protection should be no greater than the largest rating for any motor plus the full-load current of the other motors.

Sample Problem 13-6

Problem: A three-phase, 460-volt feeder supplies three motors. The sizes of the squirrel-cage motors are 10 hp, 20 hp, and 40 hp, and each is on an individual branch circuit. What size inverse-time circuit breaker is needed for the feeder?

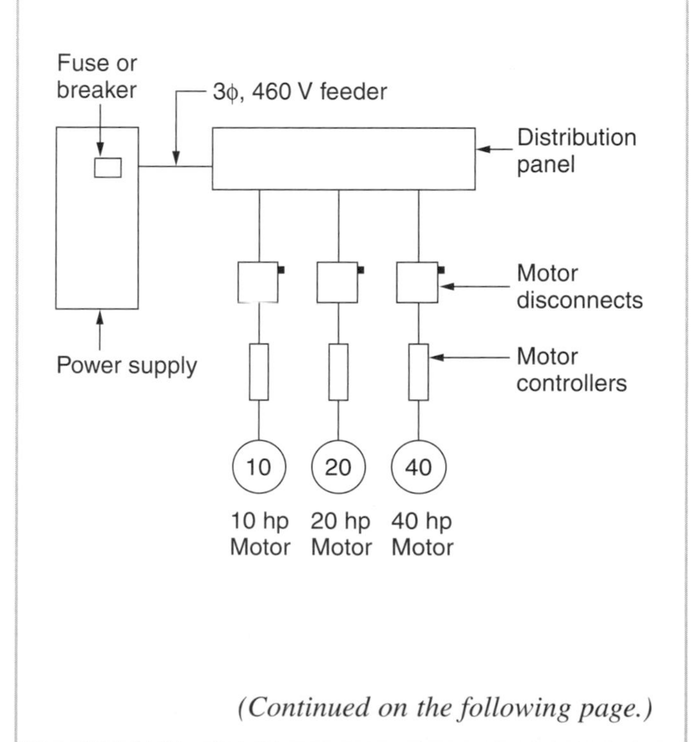

(Continued on the following page.)

Sample Problem 13-6 *Continued*

Solution: Using *Table 430.250*, determine the full-load current for each motor:

10 hp: Full-load current = 14 A
20 hp: Full-load current = 27 A
40 hp: Full-load current = 52 A

From *Table 430.52*, the maximum breaker setting is 250% of the full-load current. Therefore, the breaker on the 40 hp motor can be determined:

$$52 \text{ A} \times 250\% = 130 \text{ A}$$

The next highest standard device size is 150 amps. Therefore, the feeder breaker must be less than the breaker size of the largest motor plus the full-load currents of the other motors:

$$\text{Max. Breaker} = 150 \text{ A} + 27 \text{ A} + 14 \text{ A} = 191 \text{ A}$$

This is the maximum rating, so a 175-amp circuit breaker (the next lower standard size) is used.

Sample Problem 13-7

Problem: A three-phase, 40-hp motor is connected to a 230-volt supply. What size time-delay fuse is needed to protect the motor?

Solution: Using *Table 430.250*, the full-load current is 104 amps. *Table 430.52* states that the maximum fuse rating is 175% of the full-load current.

$$\text{Breaker rating} = 104 \text{ A} \times 175\% = 182 \text{ A}$$

The next larger standard fuse size is 200 amps.

Power Factor

Motor control circuits create a significant amount of inductive reactance. This results in the current lagging behind the voltage. That is, the current and voltage are out of phase. The current does not peak at the same time the voltage peaks. The amount that the current lags behind the voltage is measured in degrees and is called the ***phase angle.***

When calculating the power consumed by an out-of-phase, three-phase motor circuit, another factor must be added to the power formula:

$$\text{Power} = \text{Current} \times \text{Volts} \times \text{Cosine of Phase Angle} \times 1.732$$

The cosine of the phase angle is called the ***power factor.*** The power calculated by this formula is called the ***true power,*** and represents the actual power used by the circuit. This true power is read by the power companies wattmeters for billing purposes.

Sample Problem 13-8

Problem: A 20-hp induction motor has a full-load current of 54 amps on a 240-volt circuit with a phase angle of 30°. What is the true power used by the circuit?

Solution: Using the formula for true power:

$$\begin{aligned} P &= I \times E \times \cos 30^\circ \times 1.732 \\ &= 54 \text{ A} \times 240 \text{ V} \times 0.866 \times 1.732 \\ &= 19{,}440 \text{ VA} \end{aligned}$$

The true power is 19,440 volt-amperes, or 19.44 kilovolt-amperes.

The actual power supplied to a motor is normally greater than the true power. The power supplied is calculated without including the power factor and is called the ***apparent power.***

Sample Problem 13-9

Problem: Determine the apparent power of the motor in Sample Problem 13-8.

Solution: Using the basic power formula without the power factor:

$$\begin{aligned} P &= I \times E \times 1.732 \\ &= 54 \text{ A} \times 240 \text{ V} \times 1.732 \\ &= 22{,}450 \text{ VA} \end{aligned}$$

The apparent power is 22,450 volt-amperes, or 22.45 kilovolt-amperes.

The power factor is also the ratio of true power to apparent power and is often expressed as a percentage. Therefore, the power factor (PF) can be computed from the true power and the apparent power:

$$PF = \frac{\text{True power}}{\text{Apparent power}}$$

If the true power and apparent power are equal (under perfectly ideal conditions) the power factor would be 1, or 100%. This is called ***unity*** and is, for all practical purposes, impossible to attain.

Sample Problem 13-10

Problem: Calculate the power factor from the true power and apparent power calculated in Sample Problems 13-8 and 13-9.

Solution: Divide the true power by the apparent power:

$$PF = \frac{19.44 \text{ kVA}}{22.45 \text{ kVA}}$$
$$= 0.866$$
$$= 87\%$$

The power factor is 0.866, or 87%. This is identical to the cosine of the phase angle.

Power companies only charge for the actual power being used by the motor. However, they impose penalties or higher rates for customers whose power factor is below a set limit, normally 90%.

Power Factor Improvement

Ideally a motor circuit would have a power factor of one. However, this is realistically unattainable. Low power factors can be improved by adding capacitors to the circuit, or by reducing the lag between the current and the voltage. This introduces ***capacitive reactance,*** which counteracts the inductive reactance and improves the power factor. Capacitors are rated for their reactance power, normally in units of reactive kilovolt-amperes (kvar). A capacitor bank is shown in **Figure 13-13.**

Motor manufacturers provide charts that can be used to determine the capacitor size needed to change the power factor of a motor. See **Figure 13-14.** To use this chart, first locate the existing power factor in the far left column. Then find the desired power factor along the top row. Multiply the number found at the intersection of these numbers by the true power of the motor (in kVA) to determine the capacitor rating (in kvar).

Sample Problem 13-11

Problem: A 50-hp, three-phase induction motor draws a full-load current of 65 amps at 460 volts. The power factor is 65%. What size capacitor is needed to raise the power factor to 90%?

Solution: First, find the true power usage of the motor circuit:

$$\text{True power} = I \times E \times 1.732 \times PF$$
$$= 65 \text{ A} \times 460 \text{ V} \times 1.732 \times 0.65$$
$$= 33{,}661 \text{ VA}$$
$$= 33.7 \text{ kVA}$$

Determine the multiplier from the chart in Figure 13-14. The multiplier at the intersection of an existing 0.65 power factor and a desired 0.90 power factor is 0.685. To determine the required capacitor rating, multiply the true power by the multiplier:

$$\text{Capacitor rating} = 33.7 \text{ kVA} \times 0.685$$
$$= 23.1 \text{ kvar}$$

A capacitor rated 23.1 kvar can be used to correct the power factor of this motor circuit to 90%.

Figure 13-13. A capacitor bank improves motor performance. (Cooper Industries, Cooper Power Systems)

Power Factor Multiplier Chart

Existing Power Factor (percent)	Desired Power Factor (percent)												
	80	85	90	91	92	93	94	95	96	97	98	99	100
50	0.982	1.112	1.248	1.276	1.306	1.337	1.369	1.403	1.442	1.481	1.529	1.590	1.732
55	.769	.899	1.035	1.063	1.090	1.124	1.156	1.190	1.228	1.268	1.316	1.377	1.519
60	.584	.714	.850	.878	.905	.939	.971	1.005	1.043	1.083	1.131	1.192	1.334
65	.419	.549	.685	.713	.740	.774	.806	.840	.878	.918	.966	1.027	1.169
70	.270	.400	.536	.564	.591	.625	.657	.691	.729	.769	.811	.878	1.020
75	.132	.262	.398	.426	.453	.487	.519	.553	.591	.631	.673	.740	.882
76	.105	.235	.371	.399	.426	.460	.492	.526	.564	.604	.652	.713	.855
77	.079	.209	.345	.373	.400	.434	.466	.500	.538	.578	.620	.687	.829
78	.053	.183	.319	.347	.374	.408	.440	.474	.512	.552	.594	.661	.803
79	.026	.156	.292	.320	.347	.381	.413	.447	.485	.525	.567	.634	.776
80	—	.130	.266	.294	.321	.355	.387	.421	.459	.499	.541	.608	.750
81	—	.104	.240	.268	.295	.329	.361	.395	.433	.473	.515	.582	.724
82	—	.078	.214	.242	.269	.303	.335	.369	.407	.447	.489	.556	.698
83	—	.052	.188	.216	.243	.277	.309	.343	.381	.421	.463	.530	.672
84	—	.026	.162	.190	.217	.251	.283	.317	.355	.395	.437	.504	.645
85	—	—	.136	.164	.191	.225	.257	.291	.329	.369	.417	.478	.620
86	—	—	.109	.137	.167	.198	.230	.265	.301	.343	.390	.451	.593
87	—	—	.082	.111	.141	.172	.204	.238	.275	.317	.364	.425	.567
88	—	—	.056	.084	.114	.145	.177	.211	.248	.290	.337	.398	.540
89	—	—	.028	.056	.086	.117	.149	.183	.220	.262	.309	.370	.512
90	—	—	—	.028	.058	.089	.121	.155	.192	.234	.281	.342	.484
91	—	—	—	—	.030	.061	.093	.127	.164	.206	.253	.314	.456
92	—	—	—	—	—	.031	.063	.097	.134	.176	.223	.284	.426
93	—	—	—	—	—	—	.032	.066	.103	.145	.192	.253	.395
94	—	—	—	—	—	—	—	.034	.071	.113	.160	.221	.363
95	—	—	—	—	—	—	—	—	.037	.079	.126	.187	.328
96	—	—	—	—	—	—	—	—	—	.042	.089	.150	.292
97	—	—	—	—	—	—	—	—	—	—	.047	.108	.251
98	—	—	—	—	—	—	—	—	—	—	—	.061	.203
99	—	—	—	—	—	—	—	—	—	—	—	—	.142

Figure 13-14. The column on the left is the existing power factor. To attain the power factor in the top row, multiply the value at the row-column intersection by the true power of the motor to determine the required capacitor rating. (General Electric Co.)

Review Questions

Answer the following questions. Do not write in this book.

1. What is the difference between a capacitor-start motor and a permanent-split capacitor motor?
2. List the three types of direct-current motors.
3. What type of motor can be run with either direct current or alternating current?
4. Define *locked-rotor torque.*
5. Determine the full-load torque developed by a 35-hp motor turning at 1250 rpm.
6. Calculate the minimum and maximum starting torque for a three-phase, 480-volt, 60-hp motor with an H for its locked-rotor input code.
7. What does a motor service factor of 1.2 mean?
8. Determine the minimum ampere rating for conductors supplying power to a 3-hp, 230-volt, single-phase motor.
9. Copper THHN conductors supply power to three single-phase, 208-volt motors. The horsepower ratings of the motors are 1/4 hp, 1/2 hp, and 1 hp. What size conductor is required?
10. Why are the maximum ratings for instantaneous-trip circuit breakers significantly higher than the maximum ratings for inverse-time circuit breakers when protecting motor branch circuits?
11. What size inverse-time circuit breaker is required to protect a three-phase, 5-hp, 208-volt squirrel-cage motor?
12. What causes the true power of a motor to differ from the apparent power?
13. How can the power factor of a motor be improved?

USING THE NEC

Refer to the National Electrical Code to answer the following questions. Do not write in this book. Questions 1–4 refer to Part I (General) of Article 430.

1. In addition to *Article 430,* what other article addresses motors and controllers used for cranes and hoists?
2. Motors are normally marked with a time rating. What are the possible time ratings for a motor?
3. What does the code letter on a motor nameplate represent?
4. Which two types of multispeed motors are *not* required to have the full-load amperes for each speed marked on the motor?

Questions 5–7 refer to Part II (Motor Circuit Conductors) of Article 430.

5. What percent of the motor full-load current should conductors supplying a 15-minute rated motor classified for short-term duty be sized as?
6. Explain *Exception No. 3* of *Section 430.24.*
7. When capacitors are installed in the motor circuit, with which other *Code* sections must the conductors comply?

Questions 8–10 refer to Part III (Motor and Branch-Circuit Overload Protection) of Article 430.

8. How is *overload* defined in *Part III* of *Article 430*?
9. Under what conditions can the motor overload sensing device set off an alarm when an overload occurs, rather than stopping current to the motor?
10. For a motor larger than 1 horsepower with a full-load current greater than 20 amperes, what percentage of the full-load current should a thermal protector integral with the motor be sized?

Questions 11–13 refer to Part IV (Motor Branch-Circuit Short-Circuit and Ground-Fault Protection) of Article 430.

11. Which table in the *Code* is used to determine the required rating for motor ground-fault and short-circuit protection?
12. To what percent of full-load current can a time-delay fuse be increased if the maximum rating specified in the table mentioned in Question 11 is not sufficient for the starting current of the motor?
13. Motor branch-circuit short-circuit and ground-fault protection and motor overload protection can be combined in a single protective device if the setting of the device provides the protection specified in which *Code* section?

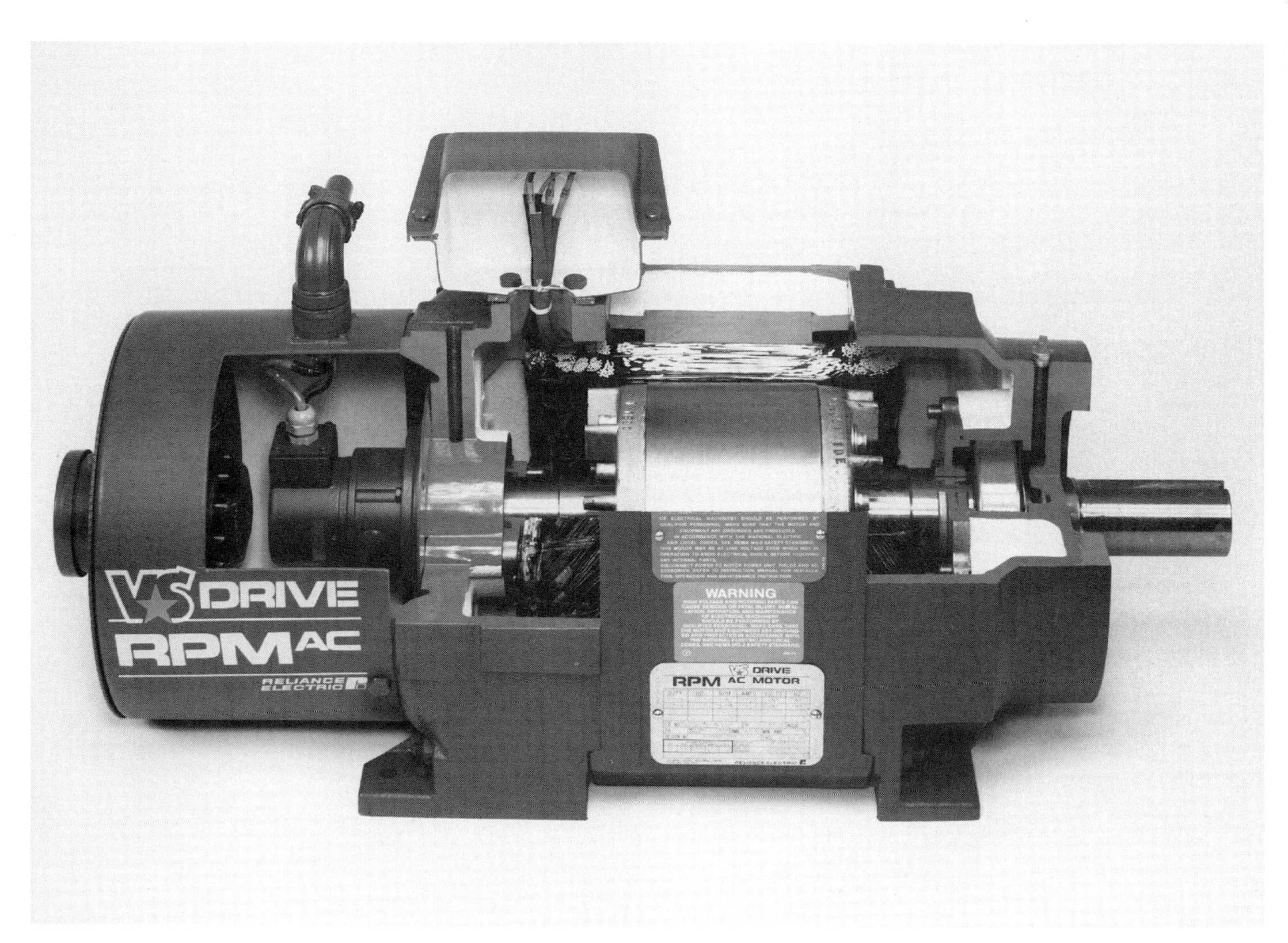

Cutaway of a variable speed ac motor. The windings, stator, rotor, and bearings are visible, along with the motor nameplate. (Reliance Electric)

Chapter 14

Motor Controls

Technical Terms

Control transformer
Drawout unit
Full-voltage reversing starter (FVR)
Manual starters
Magnetic starters
Momentary contacts
Motor control center (MCC)
Motor control circuit
Motor controller
Motor starters
Normally closed (NC)
Normally open (NO)
Pilot device
Remote-control circuit
Seal-in contact

Objectives

After completing this chapter, you will be able to:

- ❍ Describe manual and magnetic starters.
- ❍ List *Code* requirements for motor controls and motor control circuits.
- ❍ Determine the proper overcurrent protection and conductor sizes for a control circuit.
- ❍ Identify several types of pilot devices.
- ❍ Read control circuit diagrams.
- ❍ Reverse motor direction by switching connections.
- ❍ Describe basic motor control center setup.

To make practical use of motors, they must be controlled. Control can be accomplished by many methods. A snap switch can control small motors; large motors use more complex controllers requiring programmable sequencing. In addition, some motors require reduced-voltage control or reversing ability.

Motor Controllers

A ***motor controller*** regulates the power supply to a motor. Contacts within the controller open or close the motor circuit, thereby starting or stopping the motor. Motor controllers are often referred to as ***motor starters.***

Motor starters are rated for the motor horsepower and full-load current. The National Electrical Manufacturers Association (NEMA) has a rating system for starters, with numbers ranging from 00 to 8.

NEC NOTE 430.81(A)

Motor Controller: Any switch or device normally used to start and stop a motor by making and breaking the motor circuit current.

The *Code* addresses motor controllers in *Part VII* of *Article 430*. There are several noteworthy sections:

- ***Section 430.82(A)***—A starter must be able to interrupt the locked-rotor current of the motor, which may be eight times as large as the full-load current. It must also be able to start and stop the motor during normal operation.
- ***Section 430.83(E)***—If the conductor has a straight voltage rating (480 volts, for example), the nominal voltage between any two conductors cannot exceed the rating. If the controller has a slash rating (120/240 volts), the voltage between two conductors cannot exceed the upper value and the voltage between a conductor and ground cannot exceed the lower value.
- ***Section 430.84***—The controller does not need to open all motor conductors, unless it also serves as the disconnecting means.
- ***Section 430.87***—Each motor must be provided with an individual controller.

There are two common classes of starters:

- Manual starters
- Magnetic starters

Manual Starters

Mechanically, ***manual starters*** are the simplest type of motor control. These hand-operated On/Off switches are used for motors rated up to 10 hp. See **Figure 14-1.** The contacts remain closed if the power supply is interrupted, so the motor automatically starts when power is resumed. If a motor is rated at 2 hp or less, a general use ac snap switch can be used as the control, as long as the full-load current does not exceed 80% of the snap switch design rating.

Magnetic Starters

Magnetic starters operate using electromagnetism. An iron core within the controller is wrapped in a coil of wire. When current runs through the coil, the core becomes magnetized and an armature is attracted. As the armature moves toward the core, it closes the contacts, allowing current to flow to the motor.

As long as the coil is energized, the armature is held, the contacts remain closed, and the motor continues to run. When the power to the coil is stopped, the core loses its magnetism, the armature returns to its resting position, the contacts are opened, and the motor stops running. See **Figure 14-2.**

The contacts are copper or silver-plated metal. Maintenance of the contacts is fairly simple. They require periodic cleaning and realignment, and must be replaced when they become worn or burned.

A magnetic starter also contains an integral overload relay. The relay can be thermal, magnetic, or solid-state electronics, **Figure 14-3.** A thermal relay can be either bimetallic or consist of a melting metal alloy. Magnetic relays are also of two common varieties: inverse time and instantaneous trip. Solid-state relays are sealed units with current sensors through which the motor circuit conductors pass. These devices have a high-speed trip.

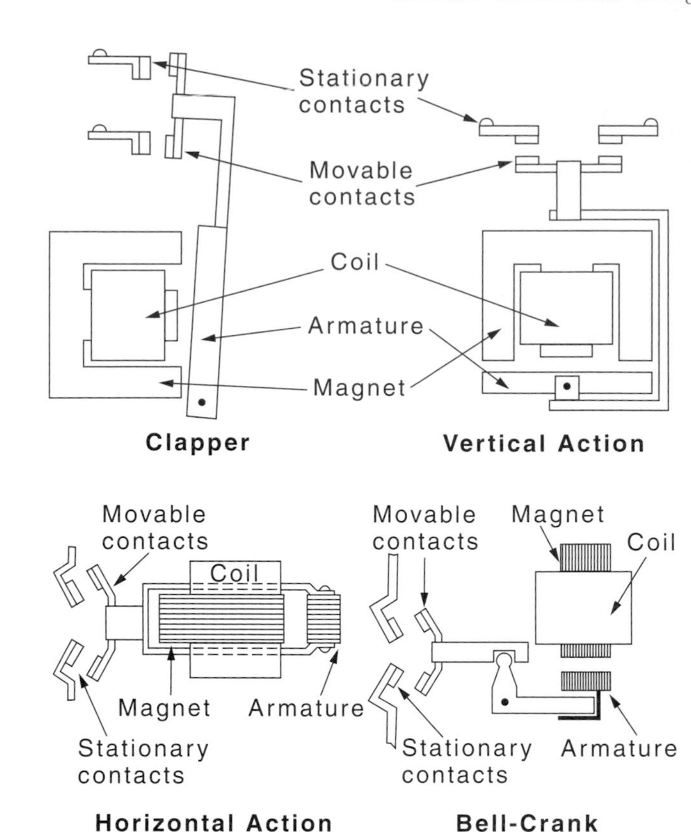

Figure 14-2. Four types of magnetic starters. All work using the same principles and perform the same function, but the internal mechanics differ slightly. (Square D Co.)

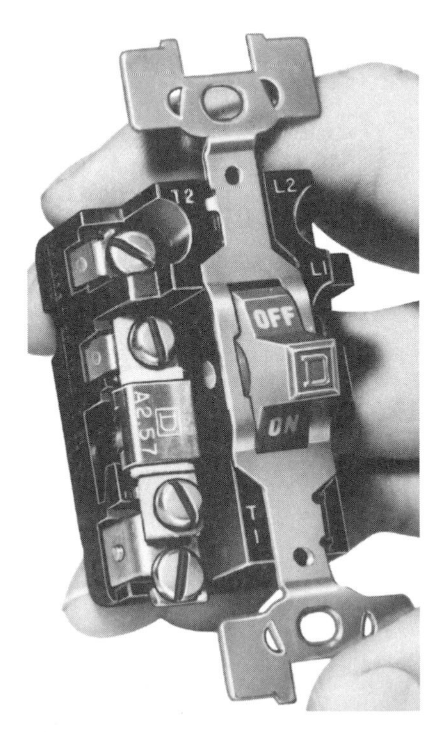

Figure 14-1. A manual starter is a simple, mechanical On/Off switch. (Square D Co.)

Motor Control Circuits

Power for a magnetic starter is supplied by a ***motor control circuit.*** This circuit is separate from the circuit supplying power to the motor. The conductors used for the motor control circuit normally originate at the service equipment or tap into the line side of the motor power supply.

Motor control circuits do not carry motor current, only signal current to direct the functioning of the controller. In most cases, the circuit is rated for 120 volts or less. A ***control transformer*** is used to reduce high-voltage supplies.

NEC NOTE 430.71

Motor Control Circuit: The circuit of a control apparatus or system that carries the electric signals directing the performance of the controller, but does not carry the main power current.

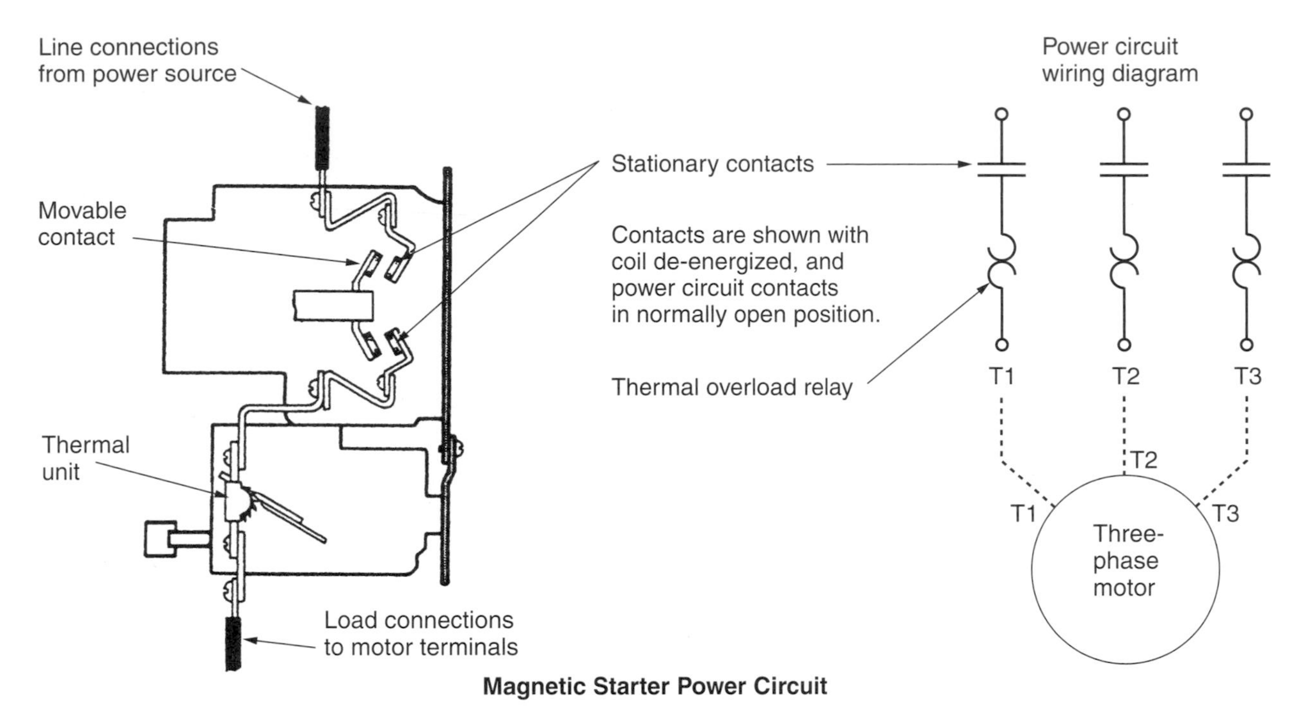

Figure 14-3. A magnetic starter with thermal overload relay. (Square D Co.)

Control Circuit Protection

Control circuit conductors must have overcurrent protection. When the control circuit becomes overloaded and opens, the motor circuit is also opened.

If the conductors are tapped from the motor power conductors on the load side of the protection device, the control circuit conductors must have supplemental protection. The maximum rating for the control circuit overcurrent protective device is based on the conductor size. These values are listed in *Table 430.72(B)* in the *Code*.

Control circuit conductors can also be supplied from a separate source, such as service equipment or a power panel. In these cases, the circuit is considered a ***remote-control circuit*** and the overcurrent protection is based on the conductor ampacity. *Section 725.23* specifies that the conductor ampacity derating factors of *Section 310.15* are *not* considered when sizing the overcurrent protection. The *Code* also limits the size of the overcurrent protection on small conductors:

- **18 AWG conductor**—7 amperes maximum.
- **16 AWG conductor**—10 amperes maximum.

Separate disconnecting devices can be used for the motor power supply and the control circuit. However, the devices must be adjacent to one another.

NEC NOTE 100

Remote-control circuit: Any electric circuit that controls any other circuit through a relay or an equivalent device.

Motor Control Connections

A wiring diagram is used to show the connections and components of a motor control circuit, **Figure 14-4.** The wiring diagram shows the control circuit as thin lines and the motor power circuit as thicker lines. These diagrams are used to determine where connections are made and for solving troubleshooting problems.

However, a wiring diagram does not readily lend itself to understanding how the control circuit functions.

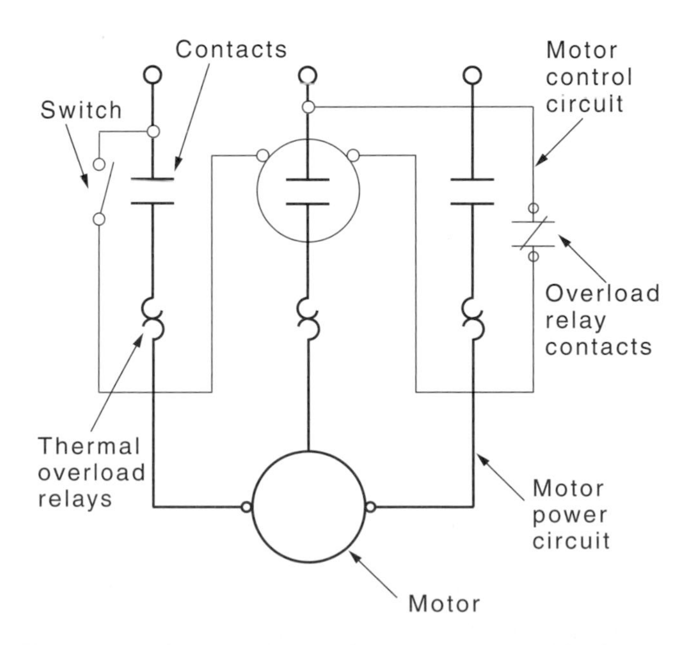

Figure 14-4. A wiring diagram for a motor control circuit.

To illustrate the sequence of operations, an elementary, or ladder, diagram is used. See **Figure 14-5.** Both figures represent the same circuit in different ways.

Control circuit diagrams always depict the circuit in the Off condition. Contacts shown open, referred to as ***normally open (NO)***, will close when energized. Those contacts shown closed in the diagram will open when energized and are called ***normally closed (NC).*** Overload relays can be open or closed, depending on whether other associated relays are open or closed.

The control circuit in its most basic form (without the power circuit) is illustrated in **Figure 14-6.** The three components of the starter are connected in series.

Pilot Devices

The control circuit is operated by a switch, or ***pilot device.*** See **Figure 14-7.** Two varieties of switches are available: single-pole switches and double-pole switches.

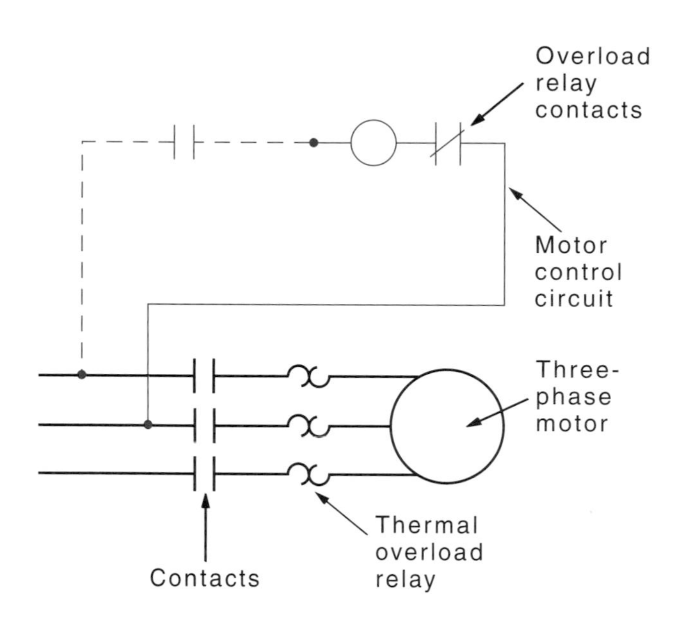

Figure 14-5. An elementary diagram of the circuit shown in Figure 14-4. This is a two-wire control circuit.

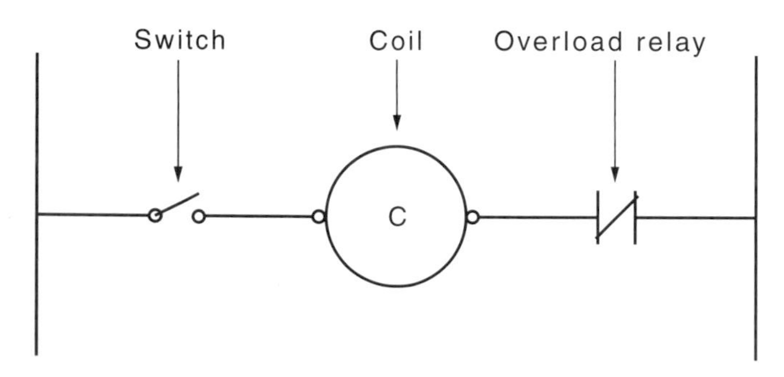

Figure 14-6. A schematic of the components of a control circuit.

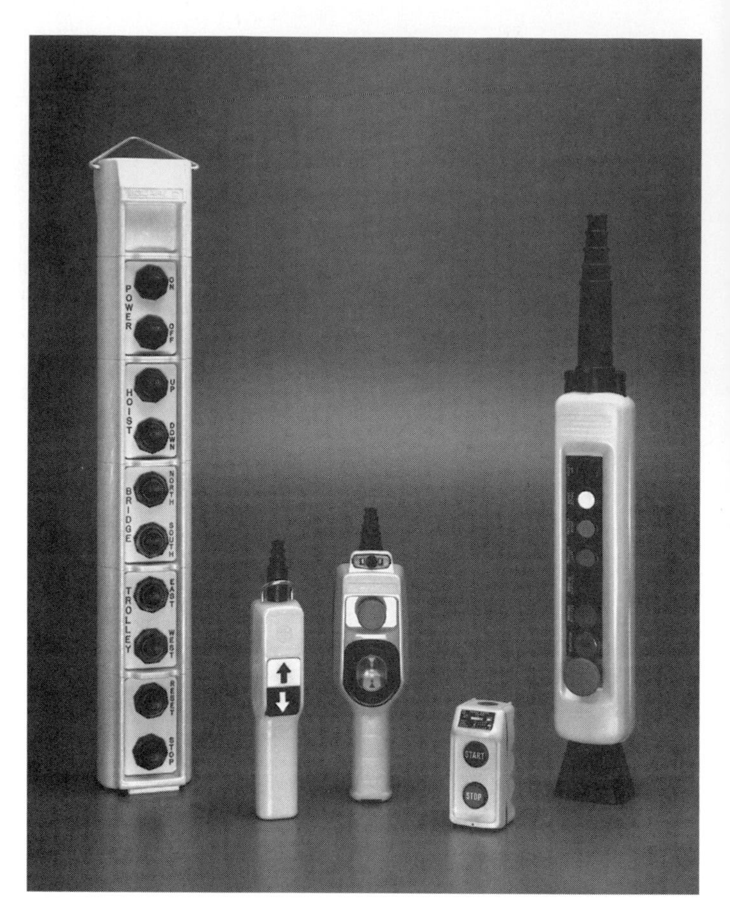

Figure 14-7. Pilot devices, or switches, are available in various styles and complexities. (Square D Co.)

Closing the switch completes the circuit and energizes the main coil. When the switch is open, the circuit is broken and the main coil de-energizes, stopping the motor. The following types of pilot devices are used:

- Float switch
- Limit switch
- Snap switch
- Foot switch
- Pressure switch

Two-Wire Control Circuits

A two-wire circuit provides low-voltage release but does not provide low-voltage protection. When power is lost to the control circuit, the contacts will release, stopping the motor. When power returns, the coil is reenergized, starting the motor. Low-voltage protection is missing—there is no way to anticipate when the motor will start. Therefore, extreme caution must be exercised when servicing two-wire circuits.

Figure 14-8 illustrates a two-wire control circuit. These circuits are used to control equipment operated by a remote device or equipment with no operator, such as water pump motors, air conditioner motors, and sump pump motors.

Three-Wire Control Circuits

A three-wire control circuit allows additional switches to be added to the circuit. An example of this type of circuit is shown in **Figure 14-9.**

The control circuit is tapped from the line-side terminals of the motor starter. The branch circuit that is supplying power to the motor also powers the control circuit. If power to the motor is interrupted, the power to the control circuit is also interrupted.

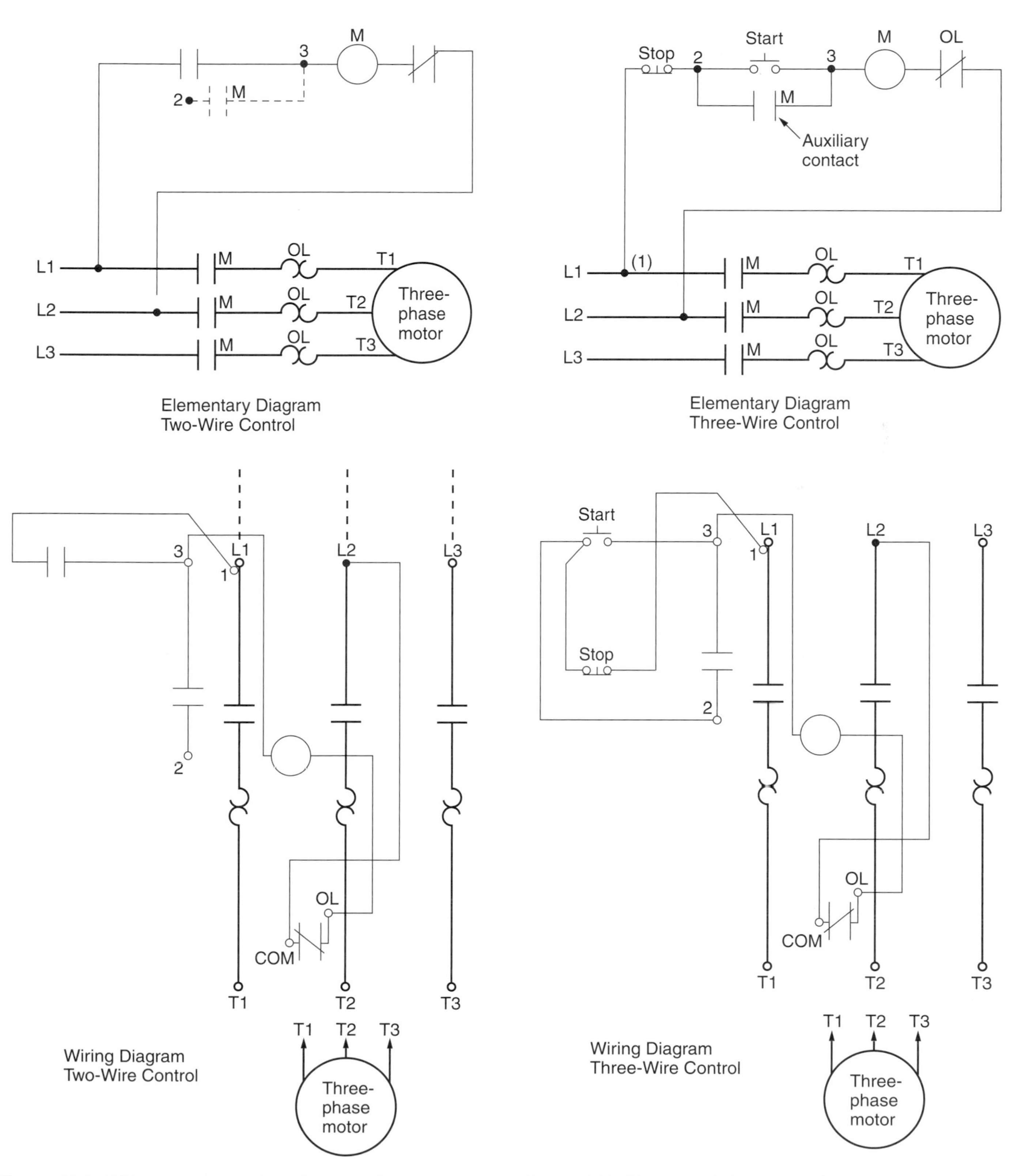

Figure 14-8. Wiring and elementary diagrams for a two-wire control circuit. (Square D Co.)

Figure 14-9. Wiring and elementary diagrams for a three-wire control circuit. (Square D Co.)

The push-button switches have ***momentary contacts.*** They are spring-loaded and return to their original position when released. When the Start button is pushed, the circuit to the coil is closed, the coil is energized, and the starter contacts close to start the motor. At the same moment, the auxiliary contact closes and completes the circuit parallel to the Start button. This remains closed when the button is released, so the coil remains energized. Momentary contacts are useful if a motor needs to be started and stopped from several locations, **Figure 14-10.**

The auxiliary contact is referred to as the ***seal-in contact.*** The control circuit is "sealed in" and remains energized until the Stop button is pushed and the circuit is opened. The momentary pressing of the Stop button opens the control circuit and releases the auxiliary contacts, removing power from the motor. The motor can be restarted by pushing the Start button.

If power is interrupted, a three-wire circuit will not automatically restart a motor when power is restored. A three-wire circuit also provides low-voltage protection. A few cycles of low voltage will cause the starter contacts to open and stop the motor. This prevents a motor from trying to run on low voltage, a major cause of motor failure.

Control Circuit Voltage Source

A motor control circuit normally operates at 120 volts or less. If a higher voltage is used to power the motor, the controls cannot use this feed directly. There are two options for supplying the control circuit:

- Reduce the voltage between the motor circuit and the control circuit using a control transformer, **Figure 14-11.**
- Operate the control circuit from a separate source, **Figure 14-12.**

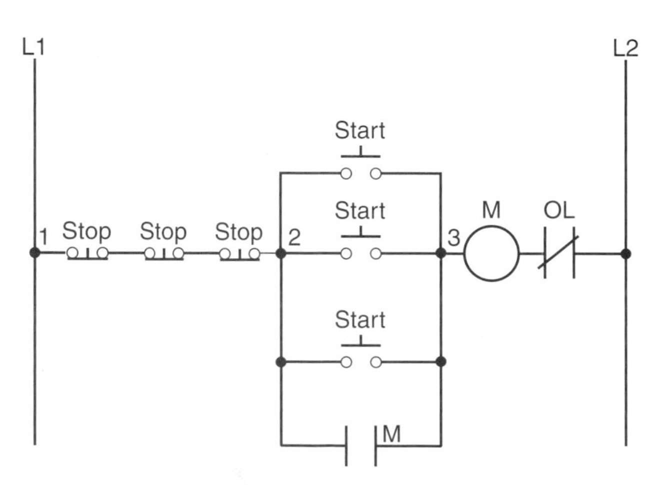

Figure 14-10. Elementary diagram of a motor that can be controlled from several locations. This configuration requires momentary contacts. (Square D Co.)

Normally, a control transformer is used. Standard control transformers are rated from 25 volt-amperes to 5000 volt-amperes. They most commonly have a 240/480-volt primary and an isolated 120-volt secondary.

NOTE

When a control transformer is used, the transformer must be protected from overcurrent in accordance with the requirements of *Article 450.*

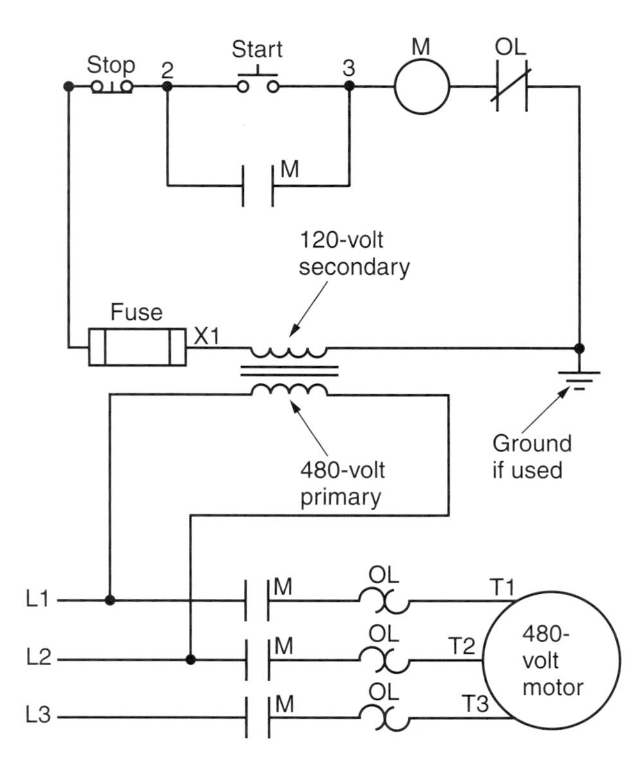

Figure 14-11. Elementary diagram for a control circuit supplied through a control transformer. (Square D Co.)

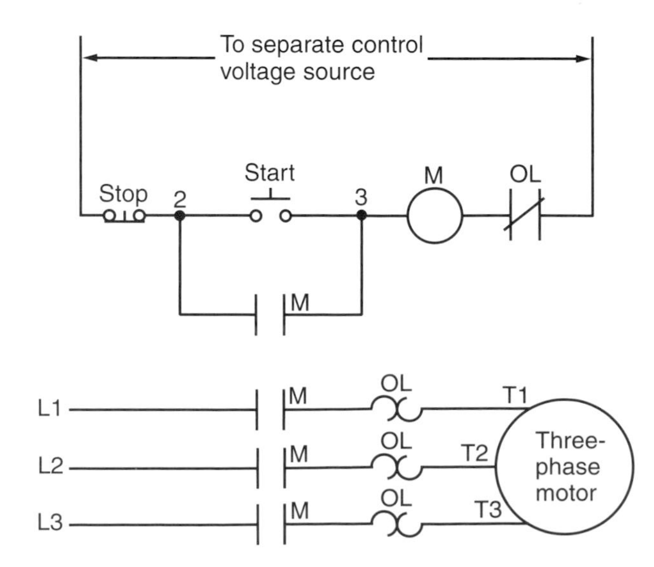

Figure 14-12. Elementary diagram for a motor control circuit supplied by a separate power source. (Square D Co.)

Reversing Motor Rotation

It is often necessary to be able to run a motor in forward and reverse. Motor direction can be manually reversed by switching connections. Some motor controls can reverse motor direction automatically.

Reversing ac motors

Single-phase, three-wire motors can be reversed by switching the line connections. For three-phase motors, any two of the line leads can be switched (normally L1 and L3 are switched). This also works for synchronous motors. For split-phase motors, the connections on the main winding or the auxiliary winding can be switched. Single-phase capacitor motors are reversed the same way. A single-phase repulsion motor is reversed by shifting the brushes to the opposite side of the neutral axis.

A ***full-voltage reversing starter (FVR)*** has two magnetic contacts within the starter enclosure. One contact makes the motor run forward, the other causes the motor to run reversed. Interlocks are present so that both contacts cannot operate at the same instant—this would cause a dead short. An FVR control circuit is illustrated in **Figure 14-13.**

Reversing dc motors

To reverse a shunt-wound or series-wound dc motor, the connections of the field or armature winding are reversed. If there are no commutating poles, the brushes can be shifted to the opposite side of the neutral axis.

For compound-wound dc motors, there are several ways to reverse rotation:

- Switch the connections of the two armature lead wires.
- Switch the connections of the two shunt-field lead wires and the connections of the two series-field lead wires.
- If the motor does not have commutating poles, shift the brushes to the opposite side of the neutral axis.

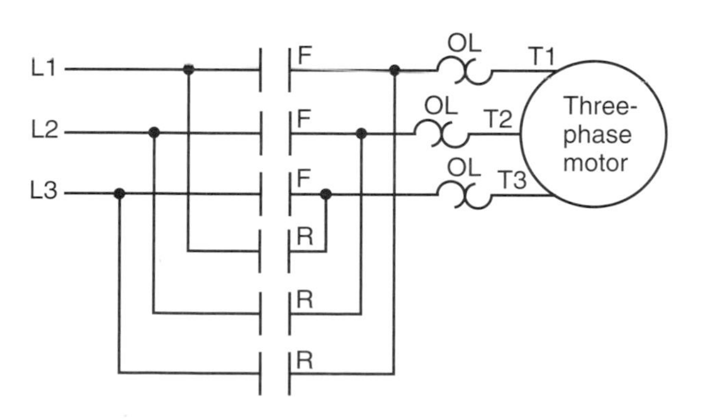

Figure 14-13. A full-voltage reversing starter. The upper three contacts close to run forward. The lower three contacts close to run the motor in reverse. Interlocks prevent both sets of contacts from closing at the same time. (Square D Co.)

Motor Control Centers

Often, several motors are controlled from a central location, called a ***motor control center (MCC).*** An MCC contains a group of motor starters.

The MCC structure is made of vertical sections (usually 20″ × 20″ × 90″ tall) bolted together. A single main horizontal bus runs across its entire length. Each vertical section has a vertical bus tapped from the main bus to supply power to the controllers within the section. Both vertical and horizontal wireways provide space for the conductors serving the starters, remote devices, and motors.

Motor control centers have no live parts exposed on the operating side of the equipment (this is referred to as "dead-front"). The MCC is normally freestanding and mounted to the floor. It is located in the room where the power supply enters or in a central location relative to the major operating equipment.

Combination starters are normally used in motor control centers. The MCC can also house feeder taps for remote equipment other than motors, such as heaters and lighting panels. Each control unit, called a ***drawout unit,*** is modular and independent of the other modules. The drawout units can be removed for repair or replacement without affecting any of the other units.

Each unit typically has a disconnect device, magnetic starter, and lockable handle. Other devices that may be present include Start/Stop buttons, pilot lights, metering devices, and selector switches. See **Figure 14-14.** Control transformers can be mounted within drawout units.

Figure 14-14. A typical MCC module includes motor starter, overcurrent protection, disconnect switch, and pilot devices. (Furnas Electric Co.)

NEC NOTE 100

Motor Control Center: An assembly of one or more enclosed sections having a common power bus and principally containing motor control units.

MCC Design Considerations

When designing an MCC, space is left open for future components. The available fault current is used to determine the correct size for the bus and bus bracing. Minimum clearances specified in *Section 110.26* of the *Code* must be maintained.

Horizontal wireway
20 HP 2X | 30 HP 4X | Future Space 3X | 7½ HP 2X | Feeders
20 HP 2X | 40 HP 4X | 7½ HP 2X | Future Space 3X
10 HP 2X | 100 HP 10X
10 HP 2X | Future Space 6X | 10 HP 2X | 10 HP 2X
Horizontal wireway
7′-6″
6′-0″
8′-4″

Figure 14-15. Diagram of a motor control center.

Figure 14-15 illustrates a diagram for a motor control center for a commercial building. Each vertical section is divided into 6″ spaces. The sections are 6′ tall, so there are twelve spaces per section. The horizontal wireways at the top and the bottom are 9″ high. The MCC in the figure contains several starters:

- **7 1/2 hp motor starter**—Two, each requiring two spaces.
- **10 hp motor starters**—Four, each requiring two spaces.
- **20 hp motor starters**—Two, each requiring two spaces.
- **30 hp motor starters**—One requiring four spaces.
- **40 hp motor starter**—One, requiring four spaces.
- **100 hp motor starter**— One, requiring ten spaces.

There are three sections open for future expansion. Three spaces are open in each of the first two columns, and six spaces are open in the far right section. A typical motor control unit wiring diagram is shown in **Figure 14-16.** The wiring diagram shows the physical relationships of the devices within a controller.

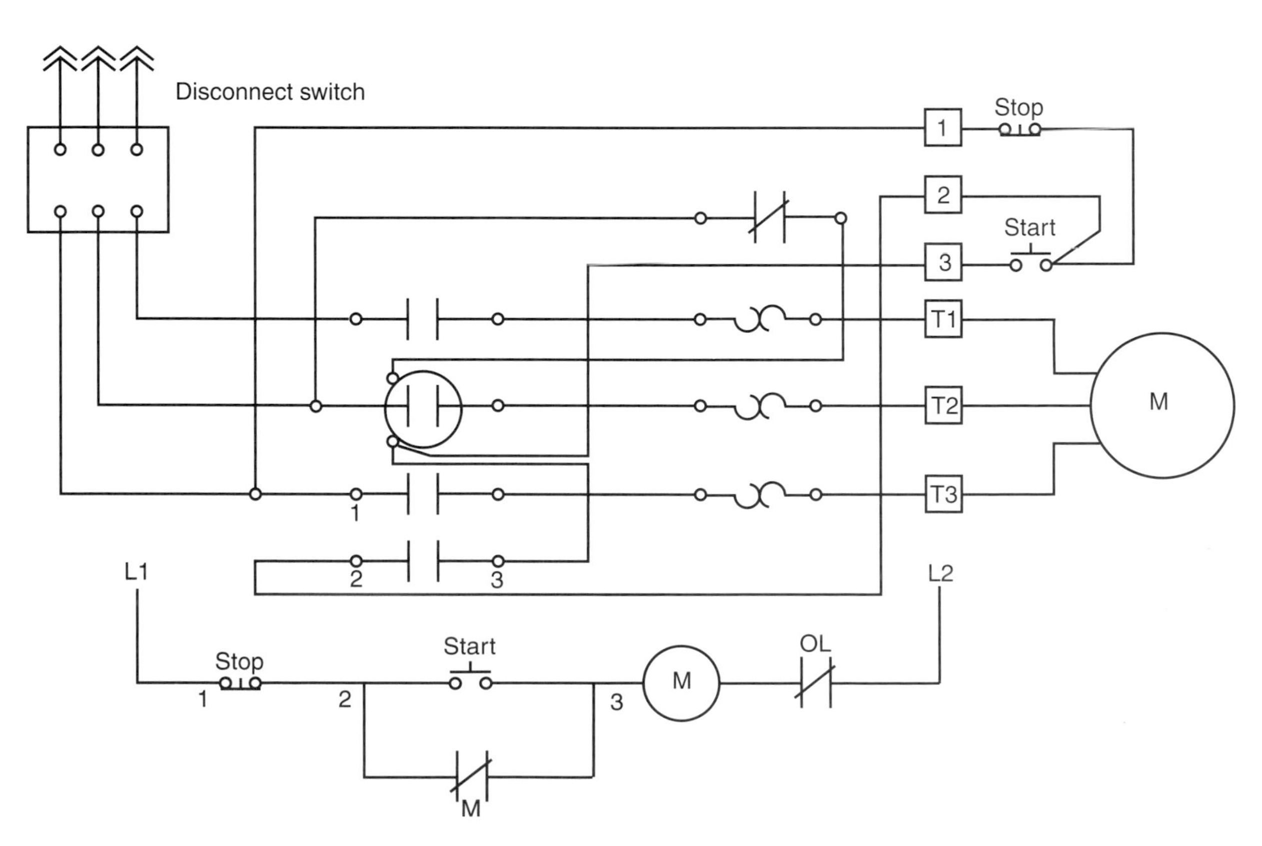

Figure 14-16. A motor control unit wiring diagram.

Review Questions

Answer the following questions. Do not write in this book.

1. Explain the purpose of a momentary contact.
2. What is the difference between a control circuit and a branch circuit?
3. In what voltage range do most control circuits operate?
4. How does a magnetic starter operate?
5. Motor control circuits receive power from one of two sources. What are these two sources?
6. Do motor control diagrams illustrate a circuit that is On or Off?
7. What is a pilot device?
8. What are the restrictions for using a general snap switch as a manual starter?
9. A motor with a three-wire control circuit suddenly loses power. Explain why the motor will not automatically start when power is resumed.
10. How can the direction of a three-phase motor be reversed?

USING THE NEC

Refer to the National Electrical Code to answer the following questions. Do not write in this book. Questions 1–5 refer to Part VI (Motor Control Circuits) of Article 430.

1. Under what circumstances should motor control circuit conductors be contained in raceway?
2. The overcurrent protection requirements in *Section 430.72* are applicable for tapped control circuits. If the circuit is separately connected to a power panel, which section must be referenced to determine the required overcurrent protection?
3. What are the basic overcurrent protection requirements for control circuits separately connected to a power panel?
4. The Fine Print Note following *Section 430.71* references another section for equipment device terminal requirements. What does the referenced section require?
5. If two devices are used as a means of disconnecting, what specifically do each of the devices disconnect and how must the devices be located?

Questions 6–10 refer to Part VII (Motor Controllers) of Article 430.

6. What types of machines must be provided with speed-limiting devices?
7. In general, should the horsepower rating of a motor controller be higher or lower than the motor's horsepower rating?
8. How is "controller" defined by the *Code* for this article?
9. What type of motor controller enclosure should *not* be used for a controller installed in an environment with windblown dust?
10. Under what condition must the motor controller open all ungrounded conductors?

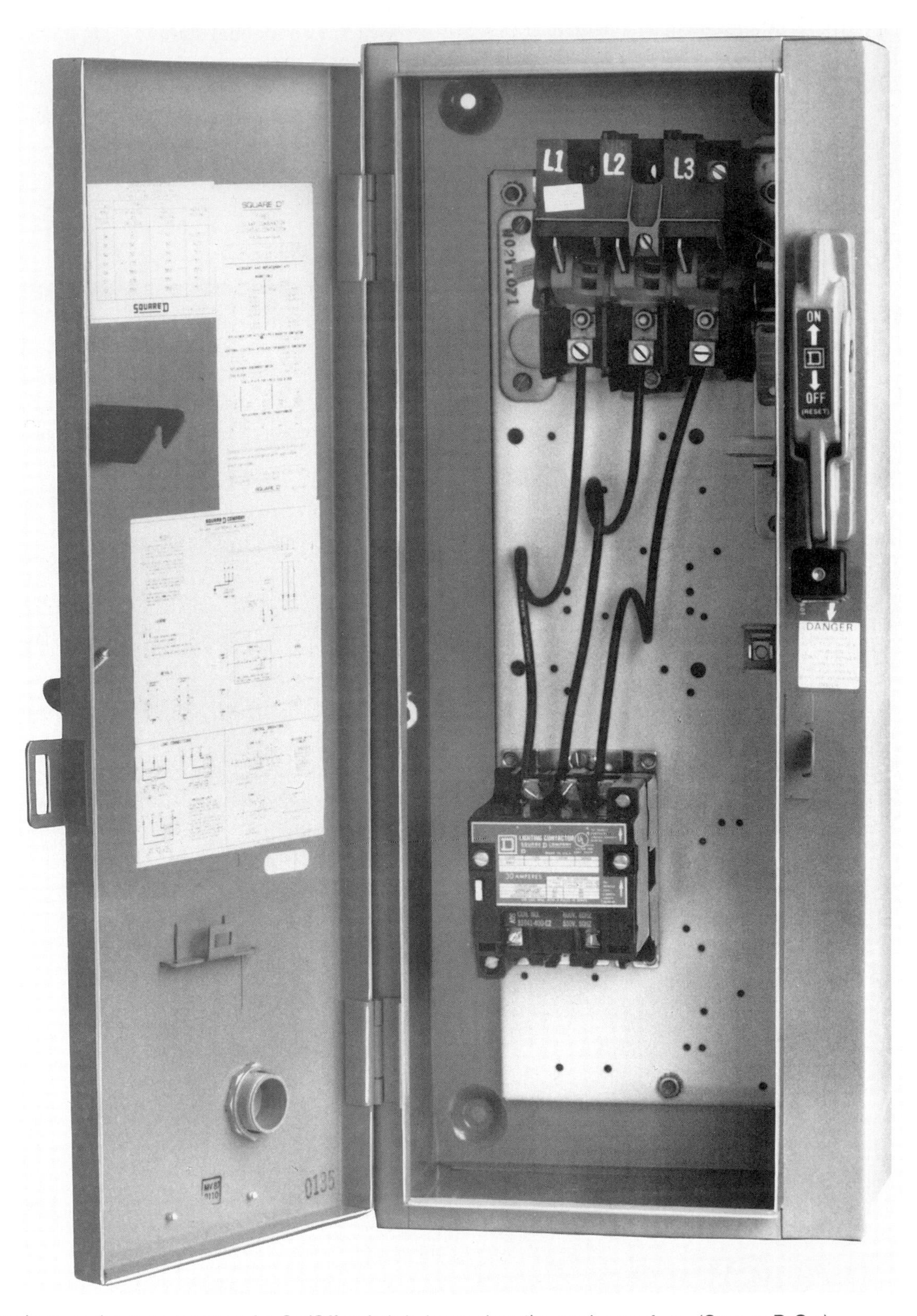

This enclosure houses the motor starter. An On/Off switch is located on the enclosure face. (Square D Co.)

Chapter 15

Emergency Power

Technical Terms

Automatic transfer switch
Emergency system
Legally required standby system
Optional standby system
Standby systems

Objectives

After completing this chapter, you will be able to:

- Identify equipment to be supplied by emergency power and standby power.
- Explain the operation of an automatic transfer switch.
- Describe various sources of emergency power.
- Compare optional standby systems to legally required standby systems.

Emergency power is needed in applications where a loss of power could result in hazards to life, property, or production. This chapter will address the requirements provided in the *Code*, as well as methods of installing emergency power supply components. Standby power systems are also presented. These systems are important for use with computers, complex communications systems, and data processing networks. The following are brief descriptions of the different types of backup systems defined by the *Code*:

- **Emergency systems (*Article 700*)**—Systems where an interruption of power would create health or life hazards. These systems also include lighting to reduce panic and aid in exiting the building in an emergency. Transfer of loads to the emergency power source is automatic.
- **Legally required standby systems (*Article 701*)**—Systems that could create hazards or hamper rescue or fire-fighting operations in the event of loss of normal power. Transfer of loads to the standby power source is automatic.
- **Optional standby systems (*Article 702*)**—Systems that could cause discomfort, interrupt processes, or cause damage in the event of loss of normal power. Transfer of loads may be automatic or manual.

Emergency Power Systems

In *Article 700—Emergency Systems*, the *Code* contains requirements for the emergency power system. These rules apply only when such systems are required by law. However, the building codes governing most commercial construction will require emergency power.

An emergency power system normally provides power for the following loads:

- **Emergency lighting**—The emergency power supplies lighting in areas of assembly to reduce panic in the event of an emergency. Also, emergency lighting is located to aid persons exiting the structure.
- **Fire pumps**—Any electrical system involved with fire extinguishing is normally supplied by emergency power. The fire alarm system is often included as an emergency system load.
- **Ventilation**—When a nonoperative ventilating system could create a life-threatening situation, the ventilating system is supplied by emergency power.
- **Other**—Any system that would create a health or life hazard if power is interrupted is supplied by emergency power.

Automatic Transfer Equipment

When the standard service fails, emergency loads must be automatically transferred to the emergency power system within ten seconds. This is accomplished by an ***automatic transfer switch,*** **Figure 15-1.**

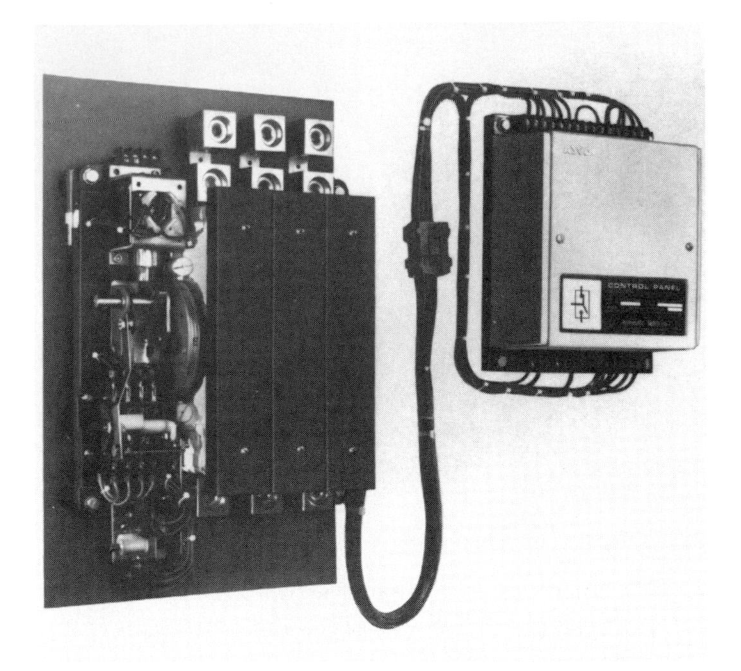

Figure 15-1. This automatic transfer switch monitors the standard service conductors. When normal power fails, the switch activates the emergency power source. (Automatic Switch Co.)

The automatic transfer switch contains a voltage sensor that monitors the standard service conductors. When the standard power source is supplying little or no voltage, the automatic transfer switch signals the generator supplying the emergency power and the generator starts. The transfer switch monitors the generator output. Once the output reaches the acceptable voltage and frequency, the transfer switch disconnects the emergency loads from the normal power supply and connects them to the emergency power supply.

NOTE

An audible and visual signal device indicates when loads are being supplied by the emergency power source.

Standard ratings for automatic transfer switches are listed in **Figure 15-2.** Some transfer switches have integral overcurrent protection, but most have the overcurrent protective devices located outside of the switch.

Standard Ratings for Automatic Transfer Switches

Voltage			Amperage		
120	208	240	30	70	100
480	600		150	260	400
			600	800	1000
			1200	1600	2000
			3000	4000	

Figure 15-2. Standard ratings for automatic transfer switches.

While emergency power is being supplied, the transfer switch continues to monitor the standard power source. When this source is restored, the transfer switch automatically switches the loads from the emergency power source back to the standard power source.

Emergency Power Sources

Sources of emergency power are discussed in *Part III* of *Article 700.* There are several ways in which emergency power can be supplied:

- Storage batteries
- Generators
- Separate service
- Supply tap ahead of main
- Individual unit equipment

NEC NOTE 700.8(A)

A sign shall be placed at the service entrance equipment indicating type and location of on-site emergency power sources.

Storage batteries

Storage batteries are often used to supply emergency power. Batteries must be capable of supplying at least 87 1/2% of the normal system voltage for at least 1 1/2 hours. **Figure 15-3** illustrates the general layout for a battery-backed emergency power system.

NOTE

Article 480—Storage Batteries contains *Code* requirements for the installation of battery-supplied power systems. Refer to this article when working with storage batteries.

Generators

Emergency power can be supplied by an engine-driven generator, **Figure 15-4.** Normally, the generator is powered by a four-cycle engine. A battery may be used as an ignition source. The two most important factors when selecting a generator are the engine type and generator capacity.

Generator engines are normally powered by one of the following fuels: LP gas, gasoline, natural gas, or diesel fuel. The type of generator used in a given application often depends on the cost, availability, and storage of its fuel.

Generators may be water-cooled or air-cooled. If the engine is water-cooled, the water supply must be

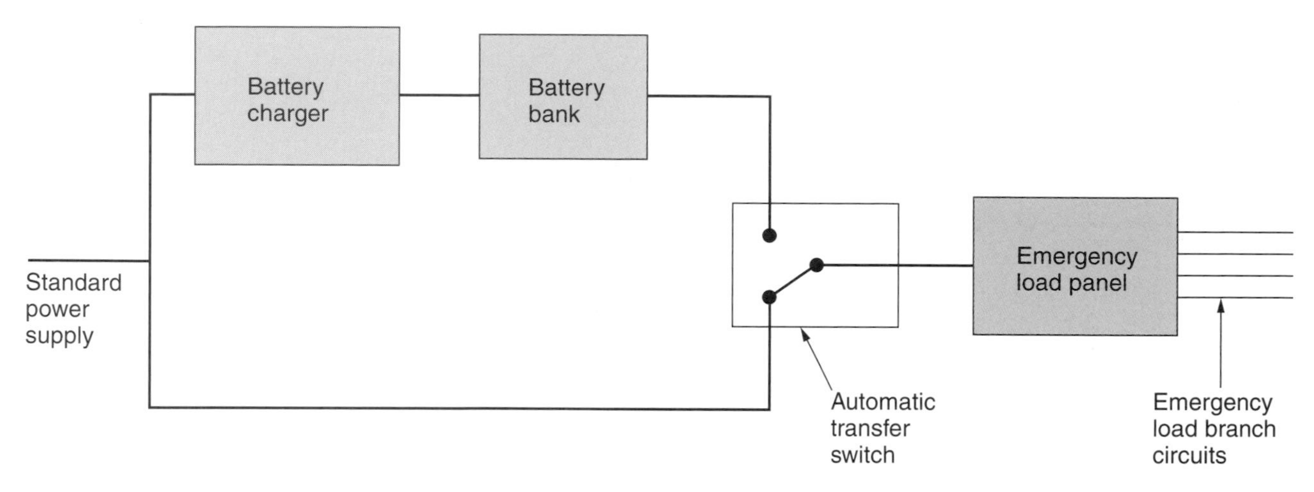

Figure 15-3. Batteries can be used to supply an emergency power system.

Figure 15-4. This 30-kW gasoline-fueled generator supplies emergency power. (Kohler Co.)

independent of the public utility water supply. The water pump must be supplied by emergency power as well. If the engine is air-cooled, the area where the generator is located must have proper ventilation. Exhausted air must be removed to help keep the generator cool and to prevent air recirculation.

NOTE

Code requirements for generator installation are contained in *Article 445—Generators.*

The emergency power generator must produce voltage identical to the normal building supply voltage. The generator engine must have a fuel supply sufficient to sustain at least two hours of operation at full demand.

Generators should be sized larger than required. A generator should be sized for the amount of power needed to start all loads, rather than the power needed to run all loads. If the generator is based on the running kVA, then each should be multiplied by 1.25 (125%) to obtain a built-in margin of spare capacity.

An emergency power generator should be located near the service equipment, in an area with moderate temperatures. If the temperature is too high, the engine will not cool properly. If the temperature is too low, the engine may have trouble starting. Refer to the generator specifications for specific temperature limits.

NOTE

A minimum clearance of 2′ should be kept around the generator to allow for maintenance.

Separate service

Emergency power can be supplied from a separate service. The emergency service and main service should be supplied from separate source transformers. Ventilation blower motors, elevator motors, lighting circuits, and other critical items should be alternated on both systems to ensure partial service, regardless of which supply fails.

Supply tap ahead of main

An emergency power supply can be tapped ahead of the main service disconnect, **Figure 15-5.** However, this method only supplies emergency power if there is a power failure at the main disconnect or main service distribution equipment. If the power failure originates at the utility, the tapped system is useless.

Individual equipment power

Some items, such as emergency lights, are equipped with an individual battery for use in emergency situations.

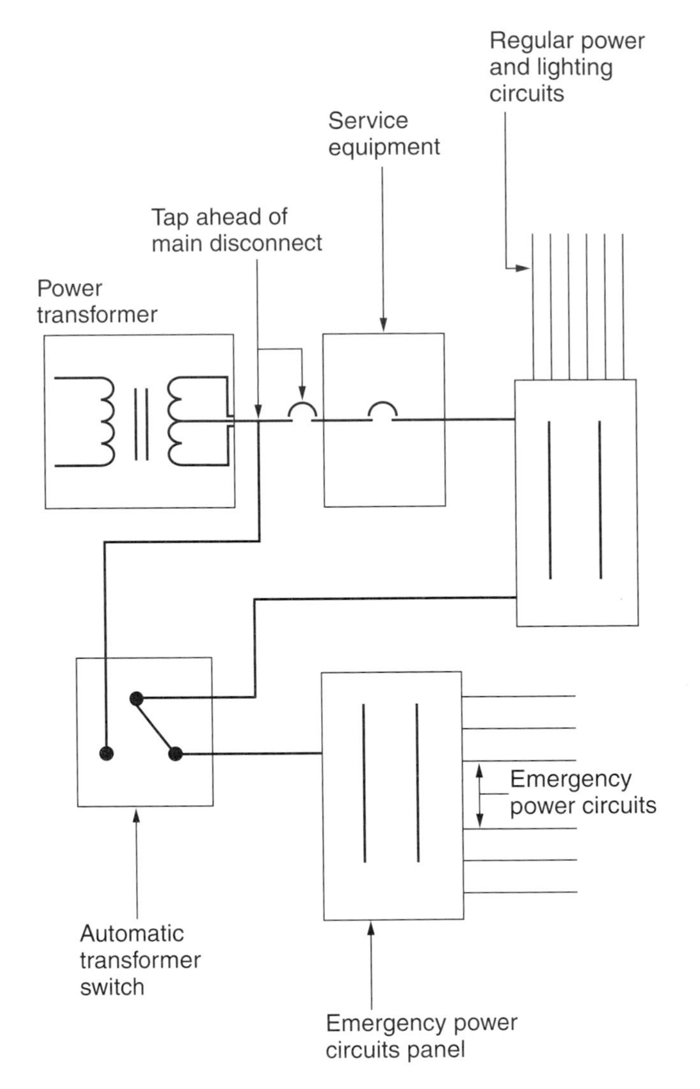

Figure 15-5. Emergency power tapped ahead of the main service disconnect switch.

Figure 15-6. Battery pack emergency lighting may be plugged into any regular lighting branch circuit provided it is not a wall-switched circuit. (Chloride Systems)

Figure 15-7. Exit signs are normally supplied by the emergency power source. (Chloride Systems)

These units should provide 87 1/2% of the battery rating for at least 1 1/2 hours of operation. The units can be plugged into a regular lighting branch circuit, but cannot be controlled by a wall switch. The units must be permanently attached and regularly maintained. See **Figure 15-6.**

Exit Lights

Both the Occupational Safety and Health Administration (OSHA) and the National Fire Protection Agency (NFPA) have requirements for exit lights. These requirements extend to all places of employment in both new and existing facilities.

Neither OSHA nor NFPA require exit lights to be supplied by emergency power circuits. However, the *Code* considers exit lights as emergency lighting equipment, and requires that they be supplied by emergency power circuits. See **Figure 15-7.**

NEC NOTE 700.16

Emergency illumination shall include all required means of egress lighting, illuminated exit signs, and all other lights specified as necessary to provide required illumination.

Approval and Identification

Section 700.3 requires that equipment used as emergency power equipment be listed for that purpose by a recognized testing laboratory. Boxes, enclosures, and circuits of the emergency power system must be identified with labels. Emergency circuit conductors must be located in raceways separate from the nonemergency system.

NEC NOTE **700.4(A)**

The authority having jurisdiction shall conduct or witness a test of the complete system upon installation and periodically afterward.

Standby Power Systems

Standby systems serve loads that are not life threatening in the event of power failure. The *Code* recognizes two types of standby systems: legally required and optional. The transfer of power from the normal source to the emergency source must be automatic for legally required standby systems. For optional standby systems, the transfer may be automatic or manual.

Legally Required Standby Systems

A legally required standby system, as the name suggests, is an emergency system supplying loads required by law in the event of normal power loss. The loads supplied by a legally required standby system could create hazards or hamper rescue and fire-fighting operations if normal power is interrupted. Some loads that may be supplied by this system include the following:

- Heating systems
- Ventilation and cooling systems
- Sewerage systems
- Lighting

Equipment used in standby systems shall be used in its intended capacity and shall be approved by and acceptable to the inspection authority. A sign clearly stating the type and location of the standby power source must be placed at the service entrance.

Article 701—Legally Required Standby Systems contains the *Code* requirements for these systems. Included are the following requirements:

- The *Code* specifies testing and maintenance procedures.
- There must be a signal when the system is not functioning.
- The system power source must meet the requirements of *Part III* of *Article 701*.
- The overcurrent protective devices must be accessible to only qualified personnel.
- Transfer equipment must be automatic and marked as part of the standby system.

Optional Standby Systems

Optional standby systems are not intended to prevent dangerous or hazardous situations. Their intent is to protect property and business operations. These systems provide power to selected loads by either manual or automatic transfer. Typically, the following loads may be included in an optional standby system:

- Refrigeration units
- Heater units
- Computers
- Data processing systems
- Communication systems
- Manufacturing processes
- Alarm systems

Manual transfer switches can be used for an optional standby system. The manual switches should only be used in situations where operating personnel are present and the supplied load is not critical.

The need for continuous service is becoming more and more important, especially with the expansion of computer use and more sophisticated communications systems. As time goes on, more intricate and sophisticated back-up systems will be designed and installed to ensure that provision for power will be there regardless of utility failure.

NOTE

Optional standby systems are addressed in *Article 702—Optional Standby Systems* in the *Code*.

Review Questions

Answer the following questions. Do not write in this book.

1. What types of loads are normally supplied by the emergency power system?
2. What types of loads are normally supplied by the legally required standby system?
3. What types of loads are normally supplied by the optional standby system?
4. What types of fuels can be used to power a generator?
5. Briefly describe what an automatic transfer switch does.
6. List five methods of supplying emergency power.
7. For how long must a battery-powered emergency system supply the emergency loads?
8. What is the disadvantage of using a separate tap before the main service disconnect as the emergency power supply?
9. In what situations can a manual switch be used for an optional standby system?
10. In generator-supplied emergency power systems, how long must the fuel supply be able to run the generator?

USING THE NEC

Refer to the National Electrical Code to answer the following questions. Do not write in this book.

1. What is the maximum amount of time allowed by the *Code* before power is available from the emergency power source?
2. Under what conditions can a generator take longer than the time specified in Question 1 to begin supplying power?
3. Who determines which loads are included in the legally required standby system?
4. Compare the testing requirements for an emergency system with those for a legally required standby system.
5. What is the maximum amount of time allowed by the *Code* before power is available from the legally required standby power source?
6. In general, can normal branch circuit conductors occupy the same raceway with conductors for the emergency system? With legally required standby system conductors? With optional standby system conductors?
7. Which *Code* section contains the requirements for sizing conductors connected to a generator?

2500-kVA/2000-kW, 480-volt S&C PureWave UPS System installed at a large vaccine-manufacturing facility. (S&C Electric Company)

Chapter 16
Hazardous Locations

Technical Terms

Dust-ignitionproof
Dusttight
Hazardous area
Seals

Objectives

After completing this chapter, you will be able to:

- Distinguish between various classes of hazardous areas.
- Summarize ignition and combustion principles.
- Cite criteria defining hazardous areas.
- Identify hazardous areas based on *Code* specifications found in *Article 500.*

Safety is an important consideration on any project. Workers must take precautions to avoid situations where injury or damage can occur. When installing an electrical system, the state of the system (which parts are energized and which are not) must be known at all times.

The system must operate safely after it is installed. For systems that operate in hazardous areas, the *Code* requires that specific components and methods are used. A ***hazardous area*** is an area where conditions exist that could cause an explosion.

Explosions

There are many factors that can increase the likelihood of explosion. However, for an explosion to occur, three elements must be present:

- Explosive substance.
- Oxygen.
- Source of ignition.

One of the prime sources of ignition is electricity. Devices such as receptacles, power plugs, pushbutton stations, motor starters, and circuit breakers often produce arcs during their normal operation. Other items, such as motors and lamps, produce heat. Uninsulated conductors can also be a source of ignition.

When installing electrical systems in hazardous areas, every effort must be made to prevent accidental ignition of flammable substances. Following the recommendations, rules, and requirements of the *Code* and other authorities is crucial to electrical safety in these areas.

Hazardous Locations

Hazardous locations are divided into classes based on the type of explosive material in the area. There are three classes:

- **Class I**—Areas containing flammable or explosive vapors or gases.
- **Class II**—Areas containing combustible dust.
- **Class III**—Areas containing ignitable fibers.

In addition, the *Code* divides these classes into divisions and groups. The divisions define whether the explosive substance is present in normal operations or under unusual circumstances. The group designation identifies the specific substance that is present. There are two divisions under each class:

- **Division 1**—Areas where a flammable, ignitable, or explosive condition exists during normal operations.
- **Division 2**—Areas where a flammable, ignitable, or explosive condition exists under unusual circumstances.

Class I Locations

A Class I location contains an explosive vapor or gas. If this substance is present under normal conditions, the area is identified as Class I, Division 1. If the gas or vapor is only present in unusual circumstances, the area is identified as Class I, Division 2. Class I locations are further identified by group, depending on the type of gas present:

- **Group A**—Acetylene.
- **Group B**—Hydrogen, fuel gases over 30% hydrogen, butadiene, ethylene oxide, propylene oxide, acrolein, or similar gases.

- **Group C**—Ethylene, ethyl ether, or similar gases.
- **Group D**—Acetone, ammonia, cyclopropane, hexane, methanol, methane, natural gas, naptha, propane, gasoline, alcohol, benzene, butane, ethanol, or similar gases.

NEC NOTE 100

Volatile Flammable Liquid: A flammable liquid having a flash point below 100°F (38°C), or a flammable liquid whose temperature is above its flash point, or a Class II combustible liquid having a vapor pressure not exceeding 40 psia (276 kPa) at 100°F (38°C) whose temperature is above its flash point.

In the *Code*, *Article 501—Class I Locations* contains requirements for the wiring materials and methods used in these areas. The requirements vary for the two divisions.

Lighting fixtures must be approved for use in Class I areas. An unapproved fixture may produce enough heat to ignite gases. The mounting method must also be approved. Normally, the fixture has an explosionproof wiring chamber sealed off from the lamp compartment. This eliminates the need to install a separate seal adjacent to the lighting fixture. See **Figure 16-1.**

NEC NOTE 100

Explosionproof Apparatus: Apparatus enclosed in a case that is capable of withstanding an explosion of a specified gas or vapor that may occur within it and of preventing the ignition of a specified gas or vapor surrounding the enclosure by sparks, flashes, or explosion of the gas or vapor within, and that operates at such an external temperature that a surrounding flammable atmosphere will not be ignited thereby.

Plugs and receptacles must be designed to eliminate arcing outside explosionproof enclosures. The contacts must not create a spark in the explosive atmosphere. Two methods are used to accomplish this:

- Delayed-action plugs and receptacles are designed to prevent rapid withdrawal of the plug from the receptacle. This allows the metal contacts time to cool before being exposed to the surrounding explosive gases.
- Receptacle contacts can be designed with an integral switch. The contacts are not energized until the plug is fully inserted. The switch is placed in the Off position before the plug is removed. See **Figure 16-2.**

Controllers can also be sources of electrical arcs. Therefore, motor control stations, pushbutton switches, circuit breakers, and other controllers must be explosionproof in Class I areas.

Figure 16-1. Low-profile, explosionproof lighting fixture approved for hazardous locations. (Crouse-Hinds Division, Cooper Industries)

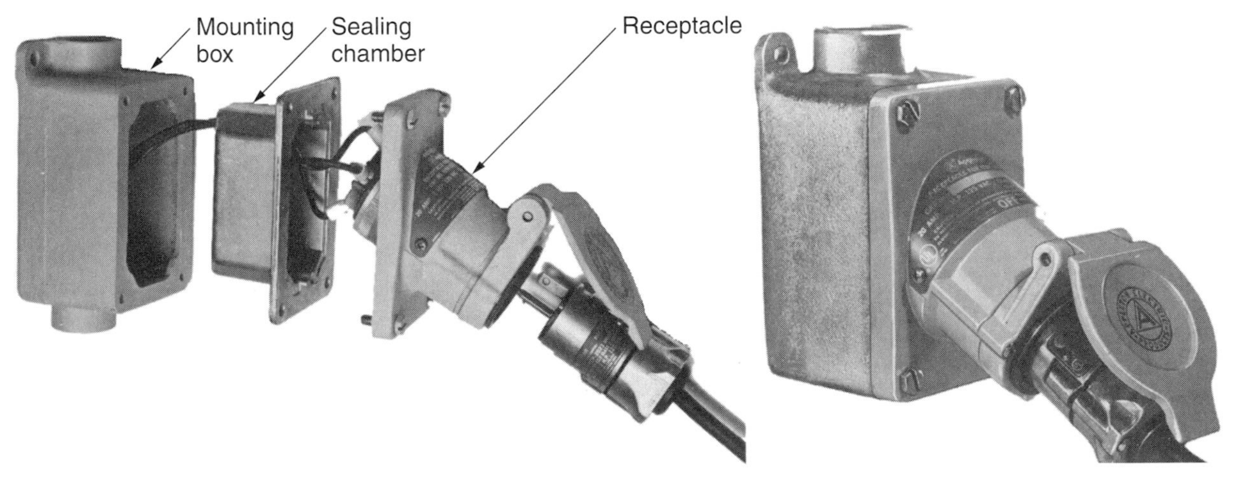

Figure 16-2. Explosionproof receptacle for hazardous areas is factory sealed. Left—Exploded view. Right—Fully assembled. (Appleton Electric Co.)

Sealing conduit fittings (called ***seals***) must be installed in Class I areas. Seals are filled with an approved sealing compound that prevents an explosion from moving through the conduit. See **Figure 16-3.** Seals are installed in two situations:

- Conduit attached to an enclosure where arcing devices, taps, terminals, splices, or high temperature devices are housed must be sealed. The seal must be installed within 18″ of the enclosure it isolates.
- A seal is placed where a conduit passes from a hazardous area to a nonhazardous area.

Factory-sealed devices, where the seal is designed into the unit, eliminate the use of hand-poured seals. This saves time and money. The installer simply connects the pigtails or terminal screws of the factory-sealed unit to the circuit wires.

Class I, Division 1

Rigid metal conduit, intermediate metal conduit, and mineral-insulated cable (type MI) are allowed for use in Class I, Division 1 areas. The conduit must be threaded and cannot be connected by set screw or compression fittings. All enclosures, boxes, and fittings must be explosionproof and have threaded openings or hubs, **Figure 16-4.** If an explosion occurs within the enclosure, the body must be strong enough to withstand it, so that any surrounding gas does not ignite.

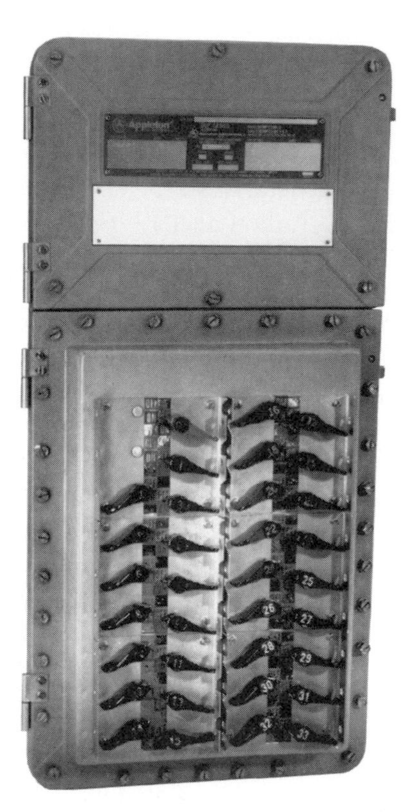

Figure 16-4. Explosionproof enclosure suitable for use in Class I, Division 1 areas. (Appleton Electric Co.)

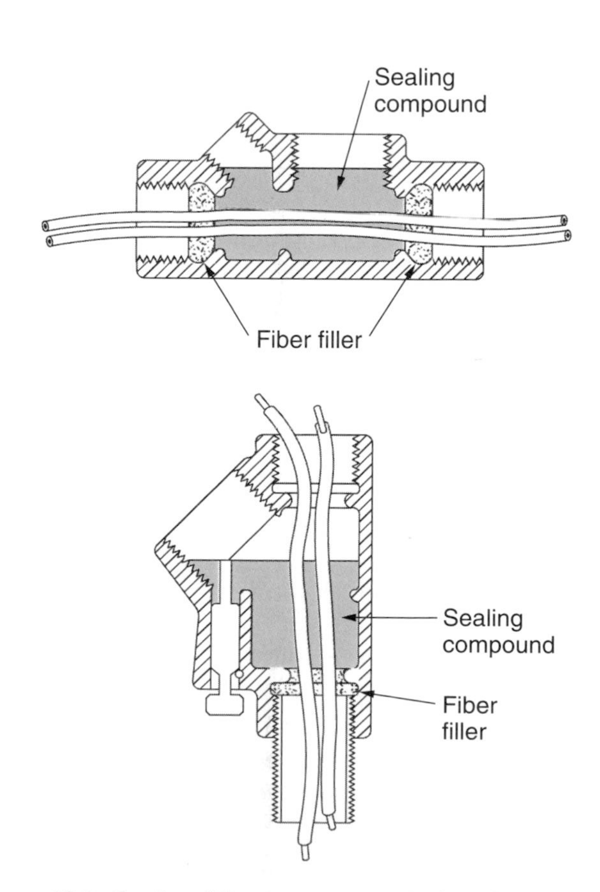

Figure 16-3. Sealing fittings prevent explosions from traveling through conduit. (Appleton Electric Co.)

Mineral-insulated cable consists of copper conductors carefully spaced in magnesium oxide covered with a copper sheath and overall jacketing. It is fireproof and can operate at a temperature range of –40°F to 480°F (–40°C to 250°C) with the appropriate cable fittings and terminals.

Motors used in Class I, Division 1 areas must be explosionproof, pressurized with purified air, and completely enclosed. Open motors cannot be used. When repairs are made, procedures must be followed to ensure that the motor maintains its explosionproof integrity.

Class I, Division 2

Along with the raceway and cable types allowed in Division 1 areas, metal-clad cable (MC) and tray cable (TC) are permitted in Division 2 areas. Cables must be sealed at the boundary between Division 2 and a nonhazardous area if the cable does not have an airtight, continuous sheath.

Class II Locations

Class II areas contain combustible dust. If the dust is present under normal conditions, the area is identified as

Class II, Division 1. If the dust is only present in unusual circumstances, the area is defined as Class II, Division 2. Class II locations are further identified by group, depending on the type of dust present:

- **Group E**—Metal dust.
- **Group F**—Carbon dust.
- **Group G**—Flour dust, wood dust, plastic dust, and other similar combustible substances.

In the *Code, Article 502—Class II Locations* contains the requirements for the wiring methods and materials used in these areas. These requirements differ for the two divisions.

The types of wiring permitted for use in Class II, Division 1 locations are metal conduit and mineral-insulated (MI) cable. Enclosures must have threaded hubs or bosses to accommodate connections. Enclosure covers must be ***dusttight*** and all openings must be completely closed to prevent dust from entering.

NEC NOTE 100

Dusttight: Constructed so that dust will not enter the enclosing case under specified test conditions.

Flexible metal conduit and flexible cord may be used. Both ends of the wiring must be sealed if electrically conductive dust is present (Division 1 areas).

Electrical equipment containing taps, joints, or terminal connections in Class II locations must be ***dust-ignitionproof.*** The surface temperature must remain low so that dust will not carbonize or ignite. Maximum temperatures for different equipment ratings are shown in **Figure 16-5.**

Class II Temperatures

Identification Number	Maximum Temperature (°F)
T1	842
T2	572
T2A	536
T2B	500
T2C	446
T2D	419
T3	392
T3A	356
T3B	329
T3C	320
T4	275
T4A	248
T5	212
T6	185

Figure 16-5. Maximum temperatures for equipment rated for use in Class II locations. This table is based on *Table 500.8(B)* of the *Code*.

NEC NOTE 500.2

Dust-Ignitionproof. Equipment enclosed in a manner that excludes dusts and does not permit arcs, sparks, or heat otherwise generated or liberated inside of the enclosure to cause ignition of exterior accumulations or atmospheric suspensions of a specified dust on or in the vicinity of the enclosure.

Motors used in Class II areas must be enclosed and approved for use in Class II locations. In Division 1 areas, the motors must be pipe-ventilated to meet temperature restrictions. Motors in Class II, Division 2 areas must be enclosed, but do not need to be pipe-ventilated. See *Section 502.125(B)* of the *Code* for further information.

Dry-type transformers should be used in Class II areas. Oil-filled transformers are not permitted in Class II areas. Lighting fixtures for Class II, Division 1 locations must be dusttight, with no openings that allow arcs to exit the unit. Each fixture must be approved for use in Class II and be so marked. The maximum wattage must be indicated.

Class II receptacles and attachment plugs should be designed so no live parts are exposed at the connections. Seals are required at all dust-ignitionproof boxes and fittings.

Class III Locations

Ignitable fibers are extremely dangerous. They ignite easily and spread quickly, causing flash fires similar to an explosion. Equipment used in Class III locations is designed to minimize the entry of fibers and to prevent the escape of arcs and sparks. Enclosures must be dusttight, and the surface temperature is limited to a maximum of 250°F (120°C) during an overload condition. Maximum operating temperature must not exceed 330°F (165°C) under normal conditions.

Article 503—Class III Locations lists the various types of wiring methods permitted in these areas:

- Rigid metal conduit
- Intermediate metal conduit
- Electrical metallic tubing
- Rigid nonmetallic conduit
- Wireway (dusttight)
- Metal-clad cable
- Mineral-insulated cable
- Shielded nonmetallic cable
- Liquidtight flexible metal conduit
- Liquidtight flexible nonmetallic conduit
- Flexible cord (with grounding conductor)

Seals are not required in Class III areas, but all boxes, fittings, and equipment enclosures must be dusttight to prevent fibers from entering and sparks from escaping. Motors must be enclosed.

Dry-type transformers can be used in Class III locations. Control transformers must be enclosed in dusttight enclosures and must comply with temperature restrictions identical to those outlined for Class II.

Repair and Storage Garages

Areas where automobiles (and other similar vehicles having engines that use volatile or flammable fuels) are repaired or stored are covered under *Article 511—Commercial Garages, Repair and Storage.*

NOTE

Article 511 is not applicable to properly ventilated parking garages.

Gasoline vapor is heavier than air. In a garage with little or no air circulation, gasoline vapor concentration can increase near the floor and in service pits. This causes these areas to be considered hazardous. Pit areas below the finished floor are considered Class I, Division 1 areas. The area extending 18″ above the floor is considered Class I, Division 2. See **Figure 16-6.**

Proper ventilation can reduce or remove the hazardous classifications from garage areas. If the pit area has an exhaust fan that performs six air changes per hour, the pit area becomes a Class I, Division 2 area. If the exhaust fan for the entire garage area performs four air exchanges per hour, the garage is no longer considered a hazardous area.

NOTE

The authority having jurisdiction decides if the exhaust system is adequate to allow the more lenient classifications.

Rooms that are separated from the garage area—offices, storerooms, and bathrooms—are not considered hazardous areas. Rooms that are not fully partitioned from the garage area may be considered hazardous. The authority having jurisdiction makes this judgment, usually based on the ventilation system.

Equipment and Wiring Methods

General-purpose electrical equipment can be installed 12′ above the floor of a garage area. Equipment installed within 18″ of the floor must be approved for Class I locations. Equipment installed above 18″ and below 12′ may be either general purpose or "approved for the purpose," depending on the type of equipment. For example, receptacles installed above 18″ level can be of the general purpose type, but lighting fixtures installed between the 18″ and 12′ level must be approved for the purpose—that is, primarily tight and enclosed.

Other items, such as generators and motors, must be totally enclosed. All receptacles within the garage repair area must have GFCI protection. The following methods of wiring are permitted above the 18″ level:

- Rigid nonmetallic conduit
- Rigid metal conduit
- Intermediate metal conduit
- Electrical metallic conduit
- Electrical nonmetallic tubing
- Metal-clad cable (MC)
- Mineral-insulated cable (MI)
- Flexible heavy-duty service cord
- Shielded nonmetallic cable (SNM)
- Tray cable (TC)
- Flexible metal conduit
- Liquidtight flexible metal conduit
- Liquidtight flexible nonmetallic conduit

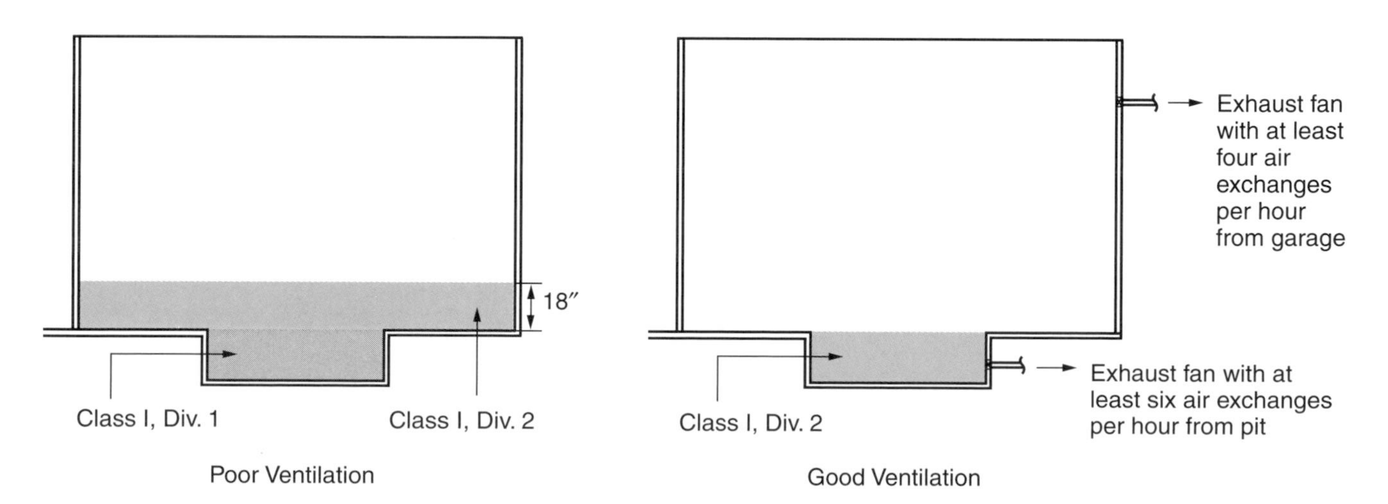

Figure 16-6. Portions of commercial garage facilities are considered hazardous. Wiring methods in those areas must comply with *Article 511.*

Portable lamps (droplights) must be approved for Class I, Division 1 areas. They must have a handle and a guard with a hook. There cannot be a plug-in receptacle in portable lamps, and they must be the unswitched type.

Aircraft Hangars

Areas within an aircraft hangar are classified based on the proximity to the aircraft. The areas where the fuel compartment and other fuel-carrying components of the aircraft are located are of principal concern. See **Figure 16-7.** In the *Code*, the regulations for wiring methods and materials in these areas are discussed in *Article 513—Aircraft Hangars*.

The floor area within the hangar (from grade up to 18″) is considered Class I, Division 2. Service pits and depressions in the floor are classified as Class I, Division 1. All areas within 5′ horizontally and 5′ above the wings and engine enclosures are considered to be Class I, Division 2. Any electrical equipment capable of arcing or

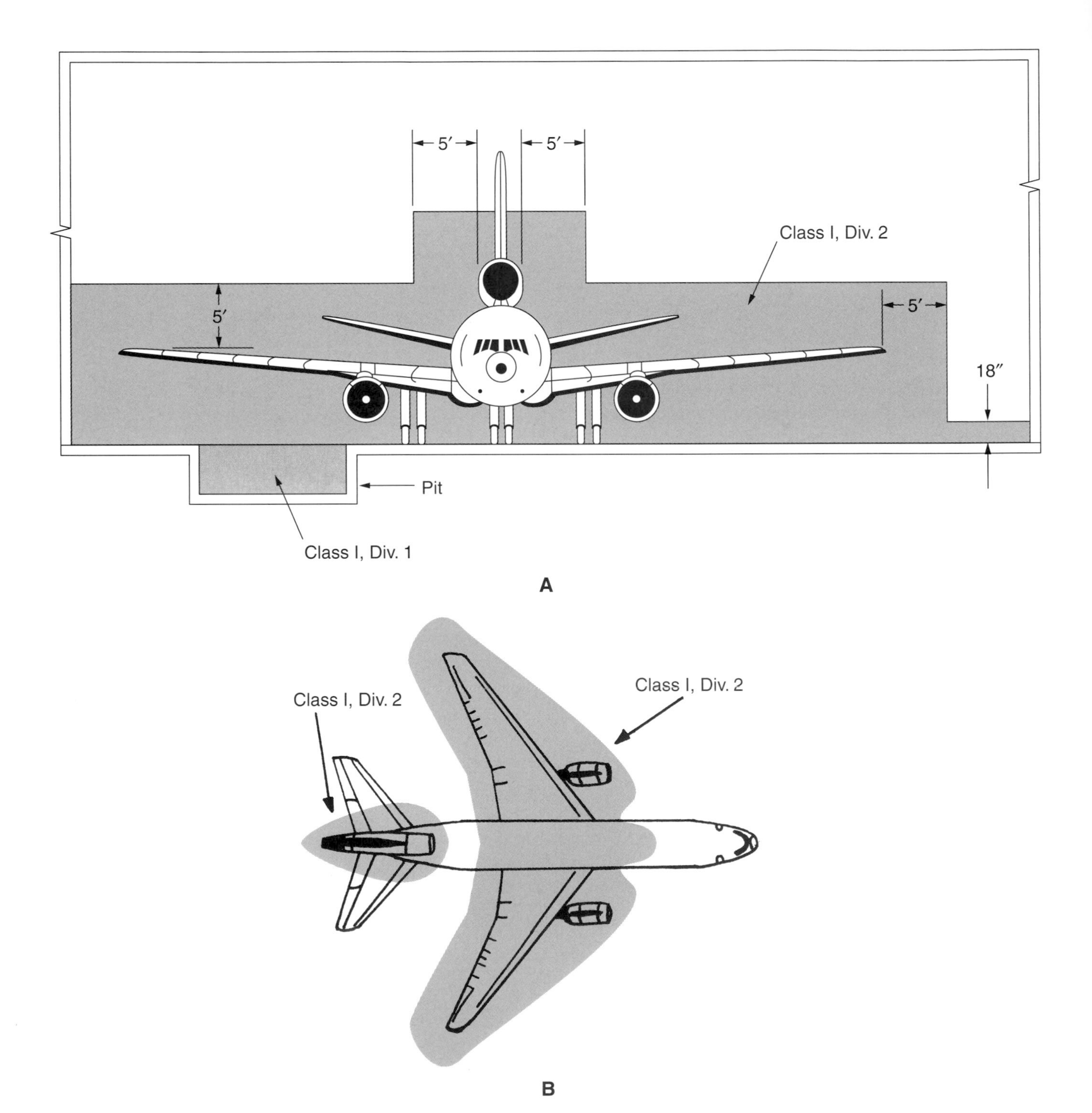

Figure 16-7. Classification of aircraft hangars is based on location of the aircraft.

generating sparks must be totally enclosed if located within 10′ above the wings or engine.

Rooms adjacent to a hangar are considered nonhazardous. General purpose wiring can be used in these areas.

Motor Fuel Dispensing Facilities

Locations where fuel is transferred from a dispensing pump to a fuel tank are discussed in the *Code* in *Article 514—Motor Fuel Dispensing Facilities*. The area where the pump is located is hazardous due to the flammable liquids and vapors present.

NEC NOTE 514.1

Motor fuel dispensing facilities include locations where gasoline or other volatile flammable liquids or liquefied flammable gases are transferred to the fuel tanks (including auxiliary fuel tanks) of self-propelled vehicles or approved containers.

Outdoor areas within 20′ horizontally and 18″ vertically of a fuel-dispensing pump are Class I, Division 2 areas, **Figure 16-8.** In addition, the area around each underground tank pipe 10′ horizontally and 18″ above grade is also considered Class I, Division 2. Further, wiring and equipment located below grade of a Class I, Division 2 location is classified as Class I, Division 1, up to and including the point where the equipment surfaces from beneath grade level. Space within the gasoline dispenser itself is considered Class I, Division 1.

For overhead fuel-dispensing pumps, the space within 18″ horizontally (in all directions) from the dispensing unit is classified as Class I, Division 2. The dispensing unit, nozzle, and hose are considered Class I, Division 1.

When making electrical repairs at the service pump, power to the pump must be disconnected. All controls and circuit breakers must open both the hot and neutral conductors. This eliminates any hazards that would be present if the neutral and hot wire were reversed. Also, junction boxes at dispensers must be explosionproof.

A sealing fitting is needed for every conduit run connected to the dispenser. The sealing fitting must be the first fitting in the conduit after it exits from below grade. An additional sealing fitting is needed when the conduit crosses the boundary of the Class I area. Conduits supplying lighting fixtures, poles, and lighted signs within the hazardous area must be sealed at both ends. See **Figure 16-9.** Lighting fixtures located less than 12′ above grade must be totally enclosed or sealed.

NEC NOTE 514.13

Each dispensing device shall be provided with a means to remove all external voltage sources, including feedback, during periods of maintenance and service of the dispensing equipment. The location of this means shall be permitted to be other than inside or adjacent to the dispensing device. The means shall be capable of being locked in the open position.

Spraying, Dipping, and Coating

Spraying, dipping, and coating processes often involve paints, solvents, thinners, and other flammable substances. In such areas, not only are the vapors and mists ignitable, but

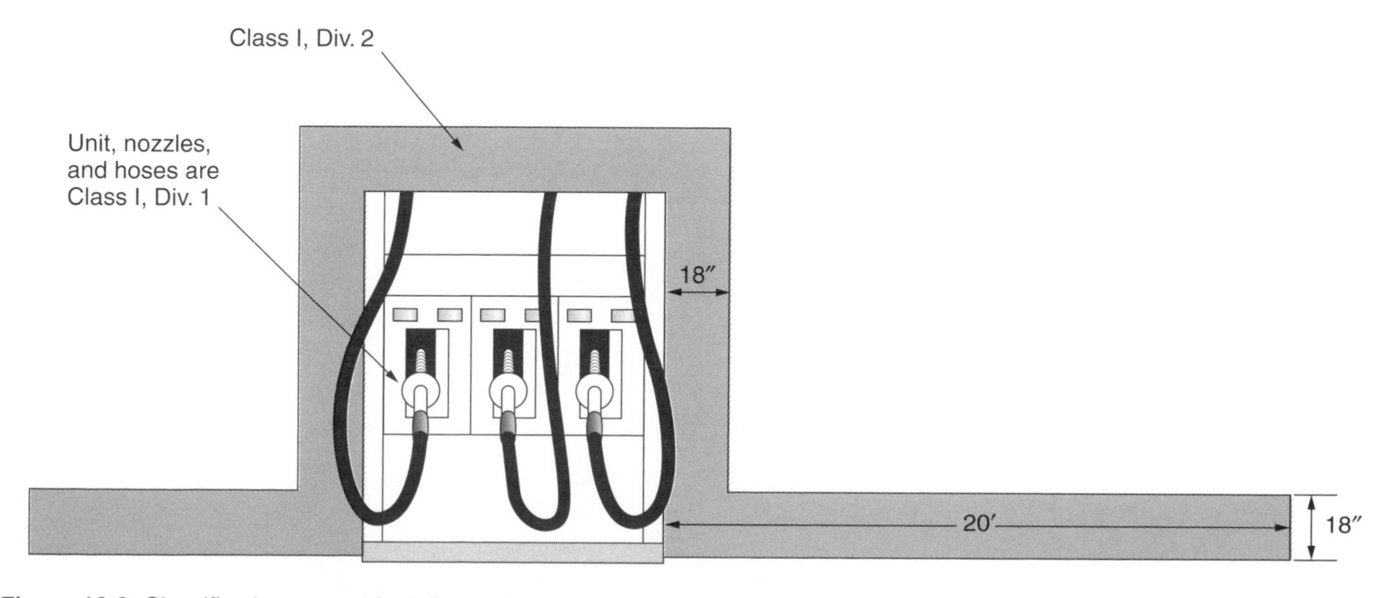

Figure 16-8. Classification around fuel dispensing pumps.

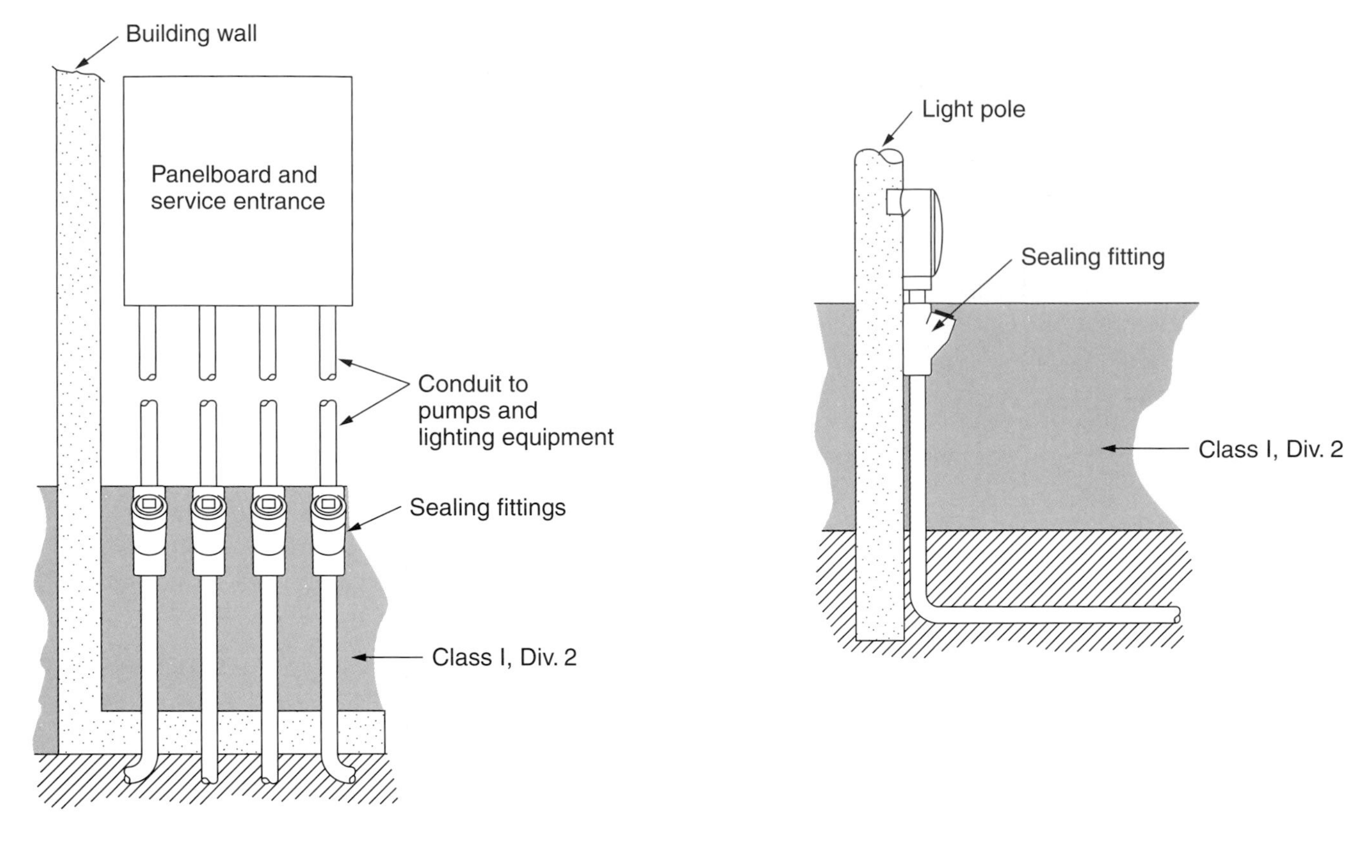

Figure 16-9. Conduit seals and explosionproof equipment must be installed and placed properly to effectively meet the *Code* requirements.

the residues—dusts and similar deposits—are also dangerous fire hazards. *Article 516—Spray Application, Dipping, and Coating Processes* in the *Code* specifies the wiring methods and materials to be used in these areas.

Many factors determine the classification of a process area. The amount of hazardous material present must be considered. The ventilation system also affects the classification, as does the building design. The general discussion in this text serves as an introduction to the wiring requirements in these areas. When working with process applications, consult the *Code* for the specific conditions and consult with the authority having jurisdiction.

For dipping and coating process areas, the area 5′ horizontally around the operation is a Class I, Division 1 area. This classification extends to the floor. Further, the area beyond this is Class I, Division 2, extending to 8′ from the operation. Any low areas, including pits, within 25′ of the dipping or coating origin point are Class I, Division 1 areas. See **Figure 16-10.**

Open spraying operations are classified as Class I, Division 1 for a space extending 20′ horizontally and 10′ vertically from the spray source. If the spray area is enclosed by partitions or walls (without a roof), this distance is reduced accordingly. However, for a distance of 3′ above the open roof and outside entrance openings, the space is considered Class I, Division 2. If the spraying is performed within a fully enclosed booth (with a roof), the Class I, Division 2 space extends 3′ in all directions from any opening. See **Figure 16-11.**

Wiring Equipment

All wiring materials, methods, and equipment installed within spray, dipping, and coating process areas must be suitable and approved for Class I locations. The wiring and equipment must be installed as outlined under the provisions contained in *Article 501—Class I Locations*. Particular attention must be paid to grounding, bonding, and providing conduit sealing fittings wherever wiring enters or exits the hazardous areas. Equipment within the hazardous area must be enclosed to prevent arc or spark escape.

Review Questions

Answer the following questions. Do not write in this book.

1. What three elements must be present for an explosion to occur?
2. Which of the three elements mentioned in Question 1 does the *Code* requirements for hazardous locations attempt to eliminate?

3. What types of materials are present in each of the three hazardous location classes?
4. Explain the difference between a Division 1 area and a Division 2 area.
5. Why are delayed-action receptacles used in hazardous areas?
6. Explain why below-floor pit areas in commercial garages are classified as hazardous locations.
7. How many air exchanges per hour are required to make the general garage area nonhazardous?
8. Who determines if the ventilation system in a garage is sufficient to reduce the hazardous classification?
9. How far horizontally from a fuel-dispensing pump does the hazardous area extend?
10. What is the hazardous classification and division of an area 7′ away from a vat used in a dipping process?

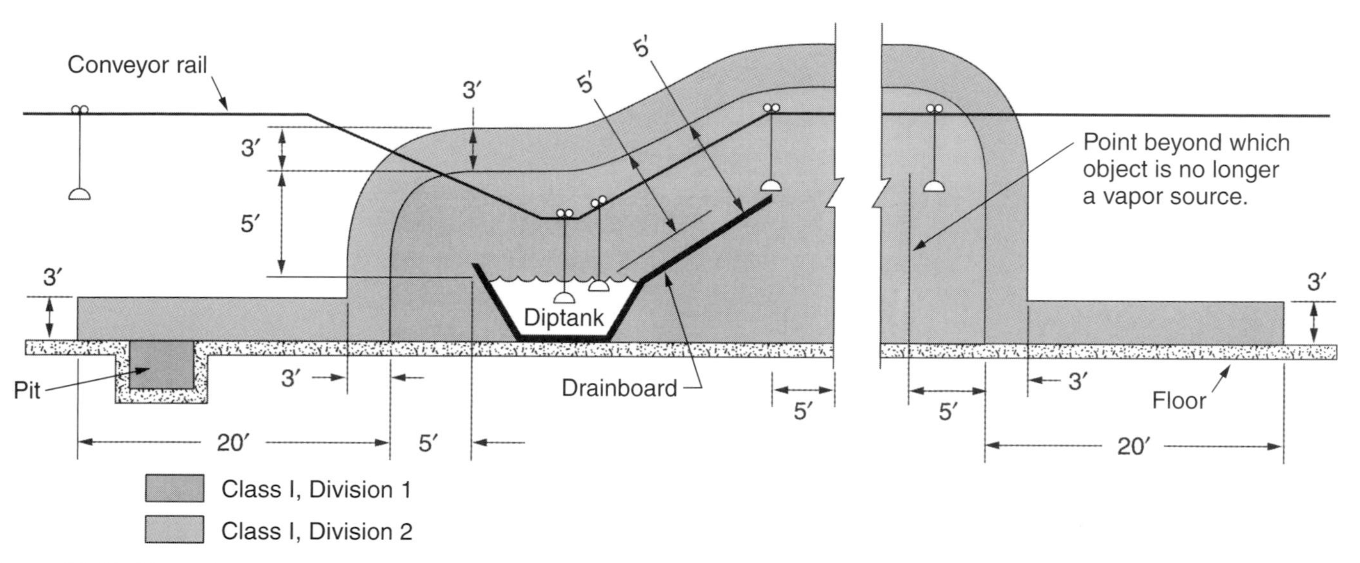

Figure 16-10. Hazardous area classifications for dipping and coating processes. This figure is adapted from *Figure 516.3(B)(5)* in the *Code.*

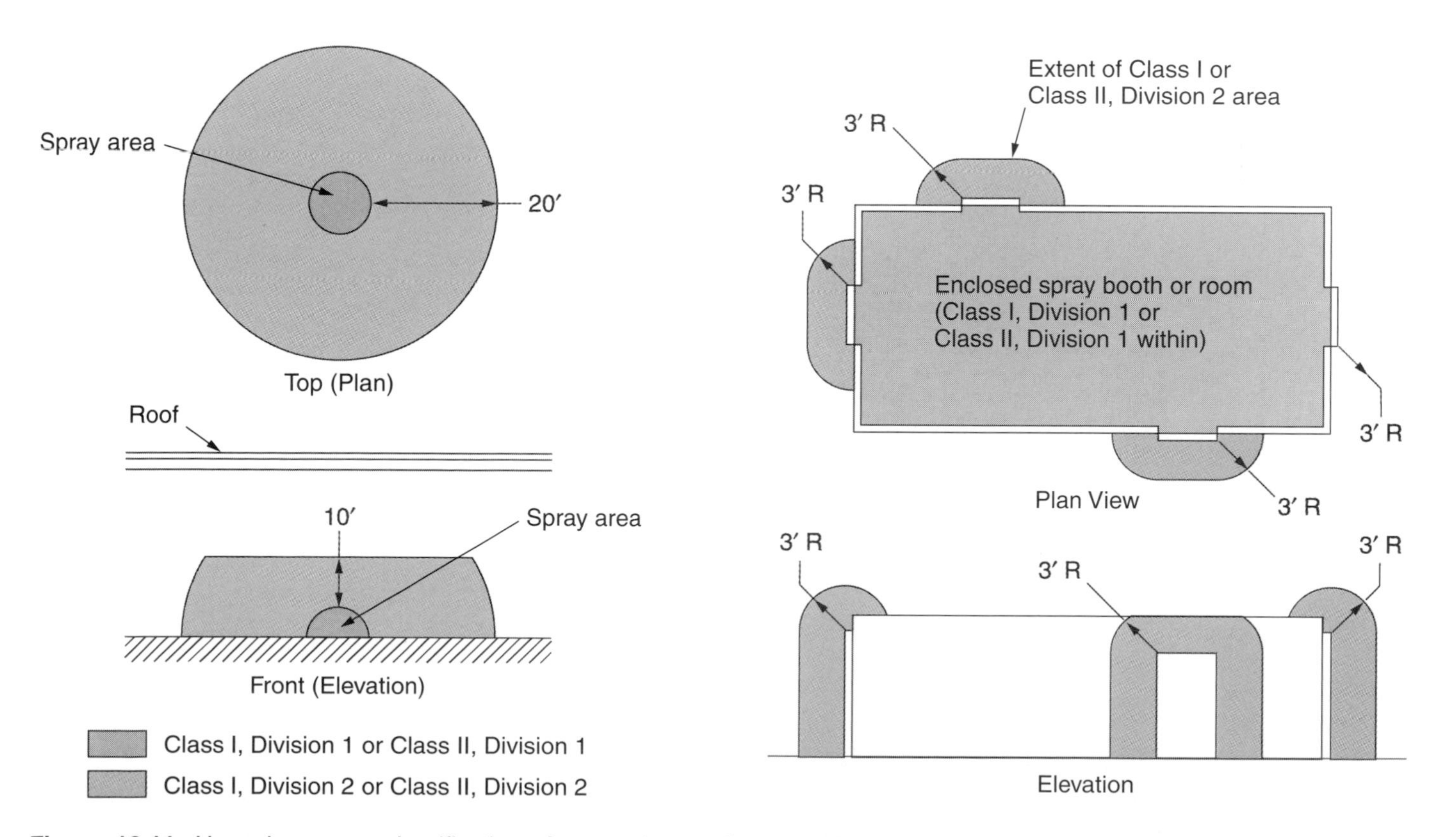

Figure 16-11. Hazardous area classifications for spraying applications. This figure is adapted from *Figures 516.3(B)(1)* and *516.3(B)(4)* in the *Code.*

USING THE NEC

Refer to the National Electrical Code to answer the following questions. Do not write in this book.

1. Which section of *Article 500* describes the classification of a Class II, Division 2 location?
2. In addition to *Article 501,* with which other article must surge arresters in Class I, Division 1 areas comply?
3. List the three requirements for installing askarel-filled transformers rated in excess of 25 kVA in Class II, Division 2 areas.
4. Under what conditions can multiwire branch circuits be installed in Class II, Division 1 areas?
5. List the three types of motors normally allowed in Class III, Division 1 areas.
6. In which hazardous class and group do atmospheres containing wood dusts belong?
7. Are connectors for recharging electric vehicles allowed in Class I areas?
8. Name the three purposes of posting signs in areas used for electrostatic spraying.
9. Which wiring methods can be used for underground wiring at service stations?
10. What are the *Code* requirements for electrical systems of aircraft in hangars?

Chapter 17

Pools and Fountains

Technical Terms

Common bonding grid
Deck box
Dry-niche fixture
Wet-niche fixture

Objectives

After completing this chapter, you will be able to:

- Cite the requirements for receptacles and lighting fixture placement around pools.
- Describe various types of underwater lighting fixtures.
- Identify limitations and clearances for conductors passing over pools.
- List the *Code* rules for bonding and grounding equipment near a pool.
- Identify the *Code* requirements for spa and hot tub installation.

Special precautions must be taken when installing electrical wiring and equipment in the vicinity of water. The *Code* requirements for these types of areas are found in *Article 680—Swimming Pools, Fountains, and Similar Installations.* This article is applicable for pools, fountains, spas, and hot tubs.

To avoid the obvious hazards created by the combination of water and electricity, receptacles and lighting fixtures near water must be located with safety in mind. If the electrical circuit becomes connected to the body of water, an extremely dangerous condition can occur. The *Code* regulations attempt to minimize the possibility of this situation.

WARNING

Due to the ease with which water can cause electrical shorts and faults, you must follow all rules for device placement, bonding, and grounding in and around a pool.

Receptacles

Normally, receptacles are not allowed within 10′ of the inside wall of a pool or fountain. However, for permanently installed pools, a receptacle used to power the water-pump motor is permitted between 5′ and 10′ from the pool. Also, there are no minimum distance requirements for receptacles separated from the pool by a wall or other barrier. See **Figure 17-1.**

All receptacles within 20′ of a pool must be protected by a GFCI. Permanently installed pools must have at least one 125-volt receptacle within 20′ for maintenance use.

Lighting Fixtures

The *Code* allows lighting fixtures to be installed directly above and within 5′ horizontally from the inside wall of a pool as long as the fixture is at least 12′ above the maximum water level. For indoor pools, this height may be reduced to 7′-6″ if two conditions are met:

- The fixture must be on a circuit protected by a GFCI.
- The fixture must be totally enclosed.

Existing lighting fixtures and outlets are allowed to be within 5′ of the pool if they are 5′ above the maximum water level, rigidly attached to an existing structure, and protected by a GFCI. See **Figure 17-2.**

Fixtures installed in an area located 5′ to 10′ horizontally from the inside wall of a pool must be GFCI-protected unless they meet the following requirements:

- Fixtures must be 5′ above the maximum water level of the pool.
- Fixtures must be firmly attached to a structure.

Cord-and-plug devices, such as table lamps, can be used in an area less than 16′ from the inside wall of the pool if they are rated 20 amperes or less. For permanently installed pools, the flexible cord must be 3′ long or

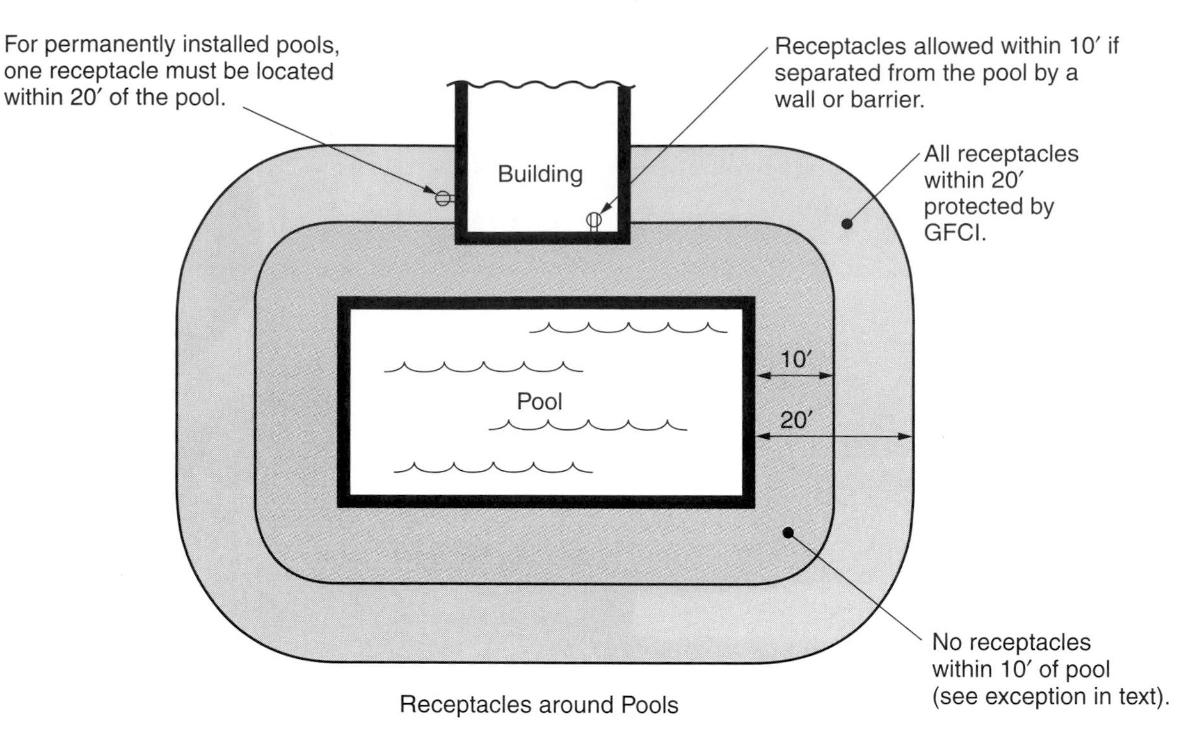

Figure 17-1. Guidelines for receptacle placement around pools.

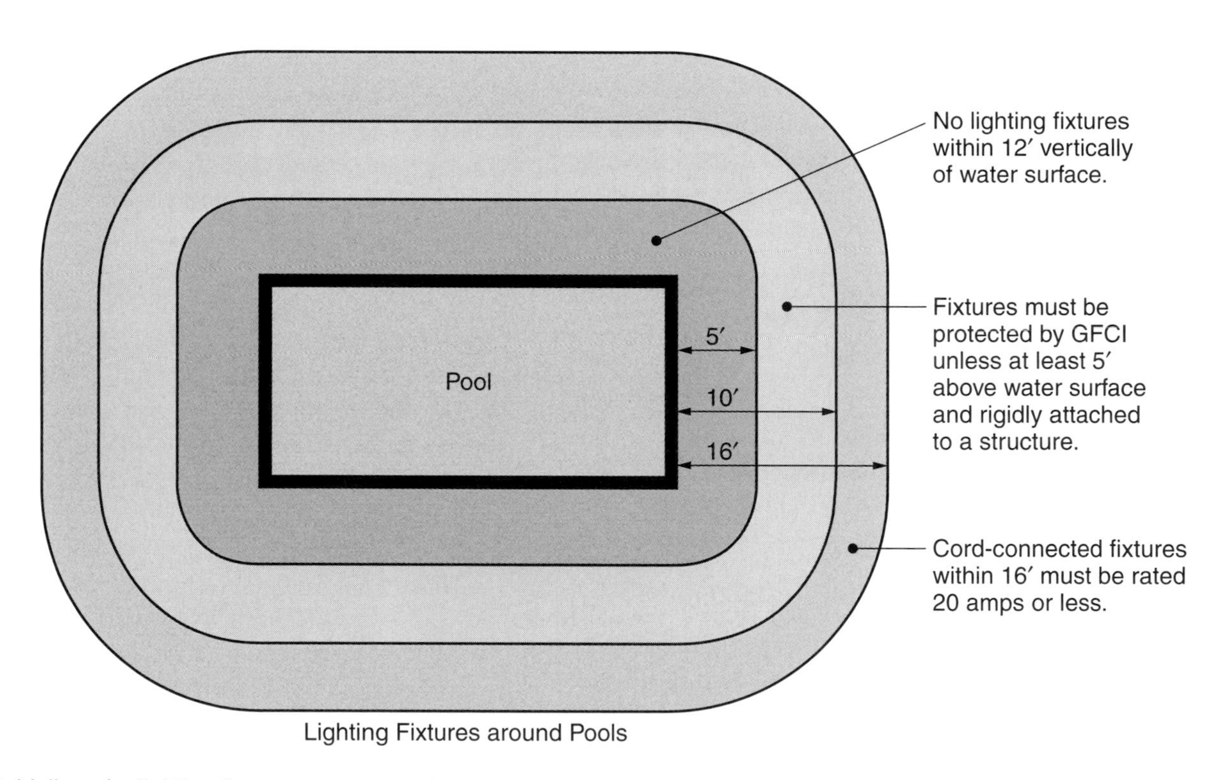

Figure 17-2. Guidelines for lighting fixtures around pools.

shorter, and must include an equipment grounding conductor 12 AWG or larger and a grounding-type attachment plug.

Further guidance regarding the installation of fixtures and receptacles above and around swimming pools is found in *Section 680.22* of the *Code*.

Underwater Lighting

Lighting within permanently installed swimming pools can be attractive and safe if correctly installed. These same fixtures can be deadly if not properly installed. For this reason, the *Code* requirements in

Section 680.22 and *Section 680.23* must be fully understood and strictly followed. There are several general requirements:

- Fixtures must be installed at least 18″ below the normal water level of the pool.
- The circuit supplying the fixtures cannot be over 150 volts.
- Fixtures must be designed and installed so that there is no possibility of shock hazard, even without GFCI protection. However, all lighting fixtures that operate at more than 15 volts must have GFCI protection.
- Fixtures used for underwater lighting must be of an approved type and listed by Underwriters Laboratories or another recognized test laboratory.

NOTE

Lighting fixtures listed and identified for use at a minimum depth of 4″ are permitted.

Wet-niche fixtures

Wet-niche fixtures are installed in a metallic housing or inside the pool wall and submerged in the water. These fixtures can be lifted out of the forming shell for servicing. The forming shell is equipped with threaded entry ports for conduit, which can be rigid metal, IMC, liquidtight flexible nonmetallic, or rigid nonmetallic. Metallic conduit must be brass or other corrosion-resistant material.

NEC NOTE 680.2

Wet-Niche Lighting Fixture: A lighting fixture intended for installation in a forming shell mounted in a pool or fountain structure where the fixture will be completely surrounded by water.

The forming shell has a lug for grounding and bonding. An equipment grounding conductor and bonding jumper to the ground grid are attached to the lug.

Wet-niche fixtures have a flexible cord that is at least 12′ long. This allows the fixture to be removed and lifted to the pool deck for bulb changes. The fixture cord must be potted (sealed) at the end attached to the fixture and at its terminals to prevent water entry. If the cord needs shortening, it must be trimmed at the supply end, not at the fixture end.

When nonmetallic conduit is used, a solid or stranded 8 AWG insulated conductor must be run with the circuit conductors and attached to the inside of the grounding lug.

Figure 17-3 illustrates the requirements for a wet-niche lighting fixture. The *Code* discusses wet-niche fixtures in *Section 680.23(B)*.

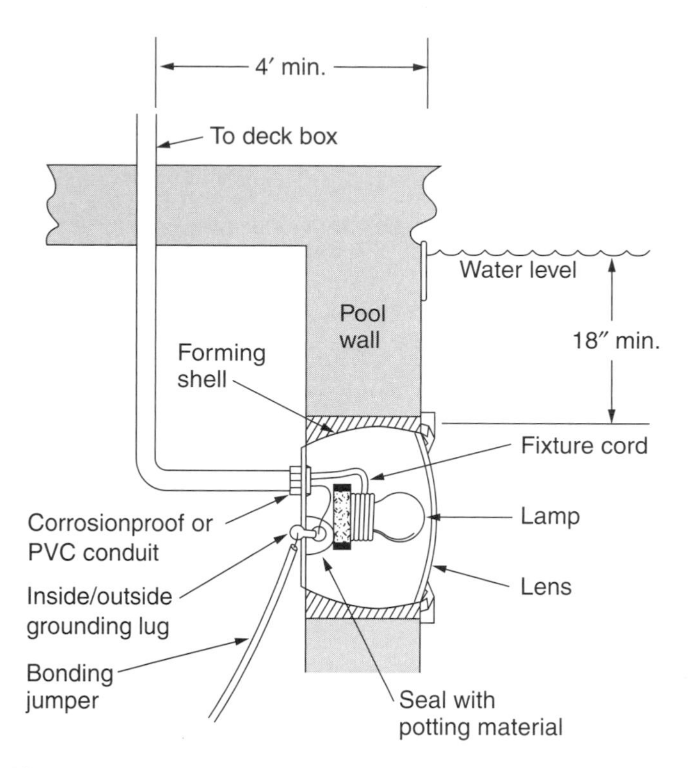

Figure 17-3. Requirements for a wet-niche pool fixture.

Dry-niche fixtures

Dry-niche fixtures are installed in closed recessed formations outside the wall of a pool. A ***deck box*** (junction box) set on or in the concrete around the pool is used to supply the fixture. This deck box can be fed using rigid metal conduit, IMC, or nonmetallic rigid conduit. EMT is permitted for indoor use only.

NEC NOTE 680.2

Dry-Niche Fixture: A lighting fixture intended for installation in the wall of a pool or fountain in a niche that is sealed against the entry of pool water.

The deck box for a dry-niche fixture is not required to be elevated 4″, or set back 4′ from the pool edge. In fact, many styles of dry-niche fixtures have integral flush deck boxes to permit easier maintenance.

The recession that houses the dry-niche fixture must have a drain at its low point. This allows any water that leaks in to drain out, preventing damage to the fixture. **Figure 17-4** shows the requirements for a dry-niche fixture installation.

Deck boxes

In general, deck boxes used for connections to underwater fixtures must be installed in a location where they do not pose a tripping hazard. Common locations include under an observation platform or diving board. The boxes must be elevated at least 4″ and located 4′ from the pools edge—except as noted with dry-niche fixtures.

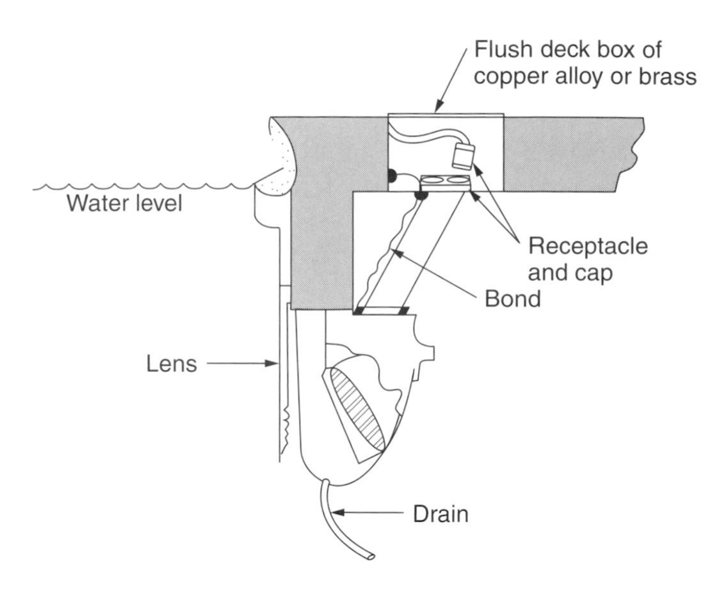

Figure 17-4. Requirements for a dry-niche pool fixture.

Deck boxes must be constructed of corrosion-resistant material. A flush deck box can be used for lighting operation at 15 volts or less if it is sealed or potted. Other provisions and requirements concerning the installation of deck boxes, as well as boxes used to house GFCIs and transformers around pools, are found in *Section 680.24.* Also, further specific information and suggestions should be obtained from the utility company and local inspection authority.

Overhead Conductors

Conductors that are not owned and maintained by the electric utility cannot be strung over a pool or within 10′ horizontally from the pool. See **Figure 17-5A.** This applies to applications such as service drops and conductors for pole-mounted lights.

Conductors that are owned by an electric utility can be located over a pool as long as the clearances specified in *Section 680.8* are maintained. The required clearance depends on the voltage carried in the line. Also, the clearance over the water surface differs from the clearance over a diving board or observation stand. These clearances must be maintained at least 10′ horizontally from the pool. See **Figure 17-5B.**

NOTE

Communication lines can pass over the pool if they maintain a 10′ vertical clearance of diving boards and platforms.

Bonding

Bonding is addressed in *Section 680.26* of the *Code.* All metal parts within 5′ of the inside edge of the pool walls must be bonded together. This may include any of the following items:

- Forming shells of the underwater light fixtures.
- Drains and screens.
- Reinforcing bars in concrete.
- Metal pool framing.
- Ladders and handrails.
- Metal deck boxes.
- Transformers and transformer enclosures.
- Diving board framing and supports.
- Pumps and pump motors.
- Metal conduit.
- Metal pool cover components.

These items must be bonded to a ***common bonding grid.*** See **Figure 17-6.** The connection to the grid must be made with a copper conductor no smaller than 8 AWG. The common bonding grid may be any of the following:

- The tied reinforcement bars in a concrete pool.
- The bolted or welded metal pool wall.
- A solid copper conductor that is no smaller than 8 AWG.

The purpose of connecting all of these parts together is to keep everything at the same electrical potential and eliminate voltage gradients. This is called an equipotential bond. This bonded group of items is then attached to ground.

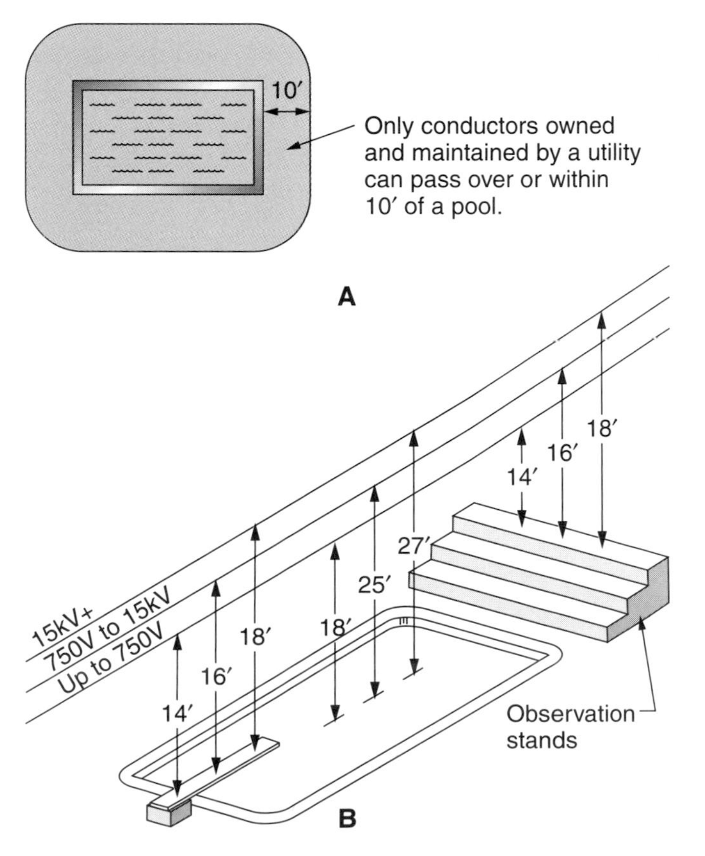

Figure 17-5. Clearances for overhead conductors. A—Only conductors owned and maintained by a utility can pass over or within 10′ of a pool. B—The minimum clearance of overhead conductors varies for different line voltage.

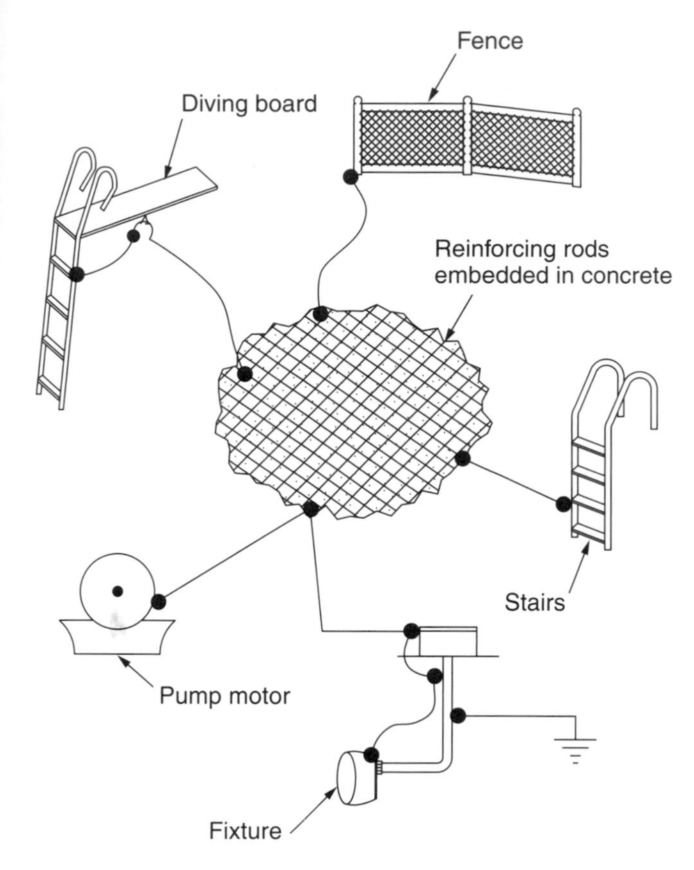

Figure 17-6. All metallic swimming pool parts and accessories must be bonded and grounded to prevent shock hazards in the pool area.

Grounding

All electrical enclosures and equipment near a pool must be grounded. Proper grounding of these items ensures that a short circuit will flow through the ground path and trip the overcurrent protective device. The following items must be grounded:

- Underwater lighting fixtures.
- All electrical equipment within 5′ of the pool.
- All recirculating system equipment.
- Junction boxes.
- Transformer enclosures.
- Panels that supply electrical equipment for the pool but are not part of the service equipment.

The size of the grounding conductor is determined from *Table 250.122* of the *Code*. The grounding conductor for pool-related equipment and devices is normally no smaller than 12 AWG. This conductor cannot be spliced.

The equipment grounding conductor must be installed with the circuit conductors in rigid metal conduit, IMC, or rigid nonmetallic conduit. For indoor installations, EMT can be used. Cord-supplied components must have an equipment grounding conductor as part of the cord assembly.

NOTE

Remember, all grounding conductors must be properly identified. Grounding conductors must be green insulated or green with a yellow stripe, if 8 AWG or smaller. If the grounding conductors are 6 AWG or larger, they can be any color, but must be marked green at all accessible places (pull boxes, terminal boxes, etc.) and terminations.

Storable Pools

Part III of *Article 680* addresses storable pools. A storable pool is defined by the *Code* as a pool constructed aboveground capable of holding water at a maximum depth of 42″, or any size pool with nonmetallic, molded walls or inflatable walls.

A cord-connected pump is often used with storable pools. The pump should be double-insulated and have an equipment grounding conductor in its cord. The receptacle into which the pump is plugged must include an equipment grounding connection.

Cord- and plug-connected lighting fixtures can be installed in the walls of a storable pool under the following conditions:

- The lamp must operate at 15 volts or less.
- The lens, body, and transformer must be impact-resistant.
- The assembly must be listed for its purpose and cannot have exposed metal parts.
- The transformer must be identified for its purpose and must be an isolated winding type with a grounded metal barrier between the primary and secondary windings.

WARNING

All electrical equipment used with storable pools must be protected by a ground-fault circuit-interrupter.

Spas and Hot Tubs

Hotels, motels, condominiums, exercise facilities, and retreats may require spas, hot tubs, and hydromassage equipment. These units require careful installation. *Part IV* of *Article 680* of the *Code* contains regulations regarding these types of pools.

Circuits supplying power to spas, hot tubs, therapeutic pools, and hydromassage units must be protected by a GFCI. This protects those using the unit from shocks.

Hydromassage Tubs

Wiring methods used with hydromassage tubs and therapeutic units must comply with *Chapters 1–4* of the

Code. If the unit is part of a bathroom, the wiring must comply with all electrical requirements related to such rooms. Electrical components for these tubs must be GFCI protected.

Outdoor Units

Outdoor units must comply with all the requirements applicable to pools, as found in *Part I, Part II,* and *Part IV* of *Article 680*. Some of these units are supplied by the manufacturer as assembled units. A cord from an assembled unit can be plugged into a receptacle that is protected by a GFCI. However, the cord cannot exceed 15′ in length.

If the assembly includes a factory-supplied panel, it can be connected to the service or main panelboard. Metal parts should be bonded together. Proper grounding of the system is essential.

Indoor Units

When installed indoors, approved packaged units can be connected with a flexible cord to a 20-amp circuit. These units must be no less than 5′ from any receptacles in the same room. If receptacles are also within 5′ to 10′ of the unit, the receptacles must have GFCI protection. This also applies to wall switches.

Lighting fixtures and paddle fans located within a 5′ perimeter must be at least 7′-6″ above the maximum waterline. See **Figure 17-7.** Also, the fixtures must be protected by a GFCI. Fixtures over 12′ above spas or tubs need no additional protection.

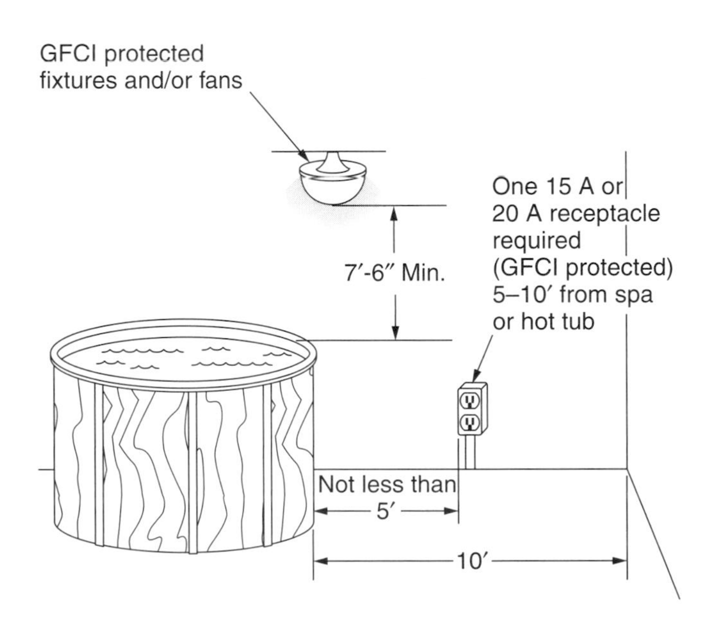

Figure 17-7. Lighting fixtures, paddle fans, and other fixtures located within 5′ of a spa, hot tub, or therapeutic pool must be at least 7′-6″ above the waterline and be GFCI protected.

Review Questions

Answer the following questions. Do not write in this book.

1. What is the minimum horizontal distance from the edge of a pool for overhead conductors that are not owned by an electric utility?
2. What is the maximum voltage allowed for underwater lighting for indoor pools?
3. For what reason may a receptacle be located within 5′ to 10′ of the edge of a pool?
4. What is the minimum distance away from the pool's edge for a deck box to be located?
5. Why is the flexible cord on a wet-niche fixture required to be 12′ long?
6. What is the minimum vertical clearance above indoor hot tubs for GFCI-protected lighting fixtures and ceiling fans?
7. What two conditions must be met in order for a lighting fixture to be installed from 7′-6″ to 12′ above an indoor pool?
8. What is the minimum clearance above a diving board for a 240-volt utility-owned service conductor?

USING THE NEC

Refer to the National Electrical Code to answer the following questions. Do not write in this book.

1. How far from a pool must a motor for an electrically operated pool cover be located?
2. Name three types of buildings in which the installation of a pool or hot tub would be subject to satisfying the requirements found in *Part VI* of *Article 680*.
3. Compare the requirements for receptacle location around an indoor spa with the receptacle location requirements around a hydromassage bathtub.
4. If electrical equipment in a fountain depends on submersion for cooling, what type of protection against overheating is required?
5. According to the *Code,* what is the primary purpose for construction of permanently installed decorative fountains and reflection pools?
6. Are the metal bands or hoops that secure the wooden sides of an outdoor hot tub required to be bonded?
7. When permanently wired radiant heaters are used to heat the deck area of a permanently installed pool, what are the minimum horizontal and vertical clearances from the pool for the heaters?
8. What is the maximum allowed overcurrent protective device for pool water heaters?

Chapter 18

Maintenance and Troubleshooting

Technical Terms

Factory acceptance testing
Field proof testing
Installation acceptance testing
Megohmmeter (megger)
Planned maintenance testing
Preventive maintenance
Single-phasing
Troubleshooting testing

Objectives

After completing this chapter, you will be able to:

- Care for testing equipment.
- Troubleshoot a power circuit, power supply, control circuit, and control transformer.
- Identify major causes of motor breakdown.
- Explain the importance of preventive maintenance.
- Perform several troubleshooting tests.

Today's businesses require high production, cost-effective processes, and minimized downtime to remain competitive. Maintenance is no longer a luxury to be performed whenever time allows. It must be carefully planned, scheduled, and performed on a regular basis to prevent costly shutdowns.

To properly conduct routine preventive maintenance on electrical systems, commercial electricians must understand electrical testing and troubleshooting procedures. This chapter will look at the subjects of maintenance, testing, and troubleshooting techniques. A strong emphasis is placed on motors, which are typically the most expensive individual components, as well as the most critical to operation.

Troubleshooting

Commercial electricians are frequently called to solve electrical malfunctions. Often, these electrical failures are simple and can be easily remedied. Sometimes they are more complex problems and require systematic troubleshooting procedures.

Testing Equipment

In order to determine what is wrong with an electrical system, testing equipment is needed. The better the testing equipment, the easier the job and the more accurate the readings. There are several general rules for the use and care of testing equipment:

- Keep test equipment in good, clean storage. Make sure all components are in proper condition and are well maintained. Be sure internal fuses, power supply cords, and batteries are in good condition.
- Keep the test equipment instruction booklet and refer to it when necessary to ensure the equipment is being used correctly and operating accurately.
- Use the test equipment for its intended purpose.
- Do not "push" the equipment beyond its rated maximums.
- When using instruments with several scales, always begin testing with the highest scale to prevent overloading.
- When making resistance measurements, never work on an energized circuit.
- Make sure clamp-on instrument jaws are completely closed and on one conductor at a time.
- Inspect insulation on test leads before and after testing. Look carefully for any cuts, abrasions, or nicks.
- Observe all safety rules while testing.

Troubleshooting Methods and Procedures

Troubleshooting consists of finding and correcting a problem. The goal is to do this as quickly as possible to avoid costly delays and equipment downtime. Experience, knowledge, and technique are the main ingredients to successful troubleshooting.

The problems that arise are varied, and it would be impossible to cover every possible situation and the methods used to correct them in this text. However, there are some sound general methods that can be applied to most troubleshooting tasks.

When troubleshooting, it is important to understand how the equipment functions and what types of circuits are present. It may be necessary to find a schematic drawing of the circuitry. Sometimes, a few minutes of looking over the wiring may be enough.

Use the process of elimination. There will be certain parts that can be easily checked. For example, power connections to the supply, circuit breakers, and fuses can all be quickly checked.

Check each distinct part of the equipment. Test the line side of the power circuit, then the load side. Move on to the control circuitry. Move logically from one aspect of the electrical system to the next, rather than jumping from part to part.

Troubleshooting power circuits

When checking a power circuit, first the source is considered and then the control circuit is isolated:

1. Use a voltmeter to check each of the incoming lines to confirm that voltage is present. Verify that the voltage is at the required level. See **Figure 18-1.**
2. Isolate the control circuit from the power circuit using a "jumper," as shown in **Figure 18-2.** For manual motor starters, use a screwdriver to close the contacts. Once the contacts are closed, check to verify there is voltage at the T1, T2, and T3 terminals.

Troubleshooting the power source

Troubles are often encountered at the supply of the circuit. A breaker may trip or a fuse may blow, creating a problem. When this is the case, the reason for the failure of the protective device must be found.

The failure is normally a result of a short-circuit or current surge, but it must be investigated and resolved before replacing the breaker or fuse. Even before checking, make sure there is power up to the protective device. See **Figure 18-3.**

If power is present, the breaker or fuse should be checked. Connect one side of the voltmeter to the line side of one breaker and the other lead to the load side of a different breaker. If the voltage is correct, the breaker is good.

CAUTION

Never connect the voltmeter leads across the same breaker.

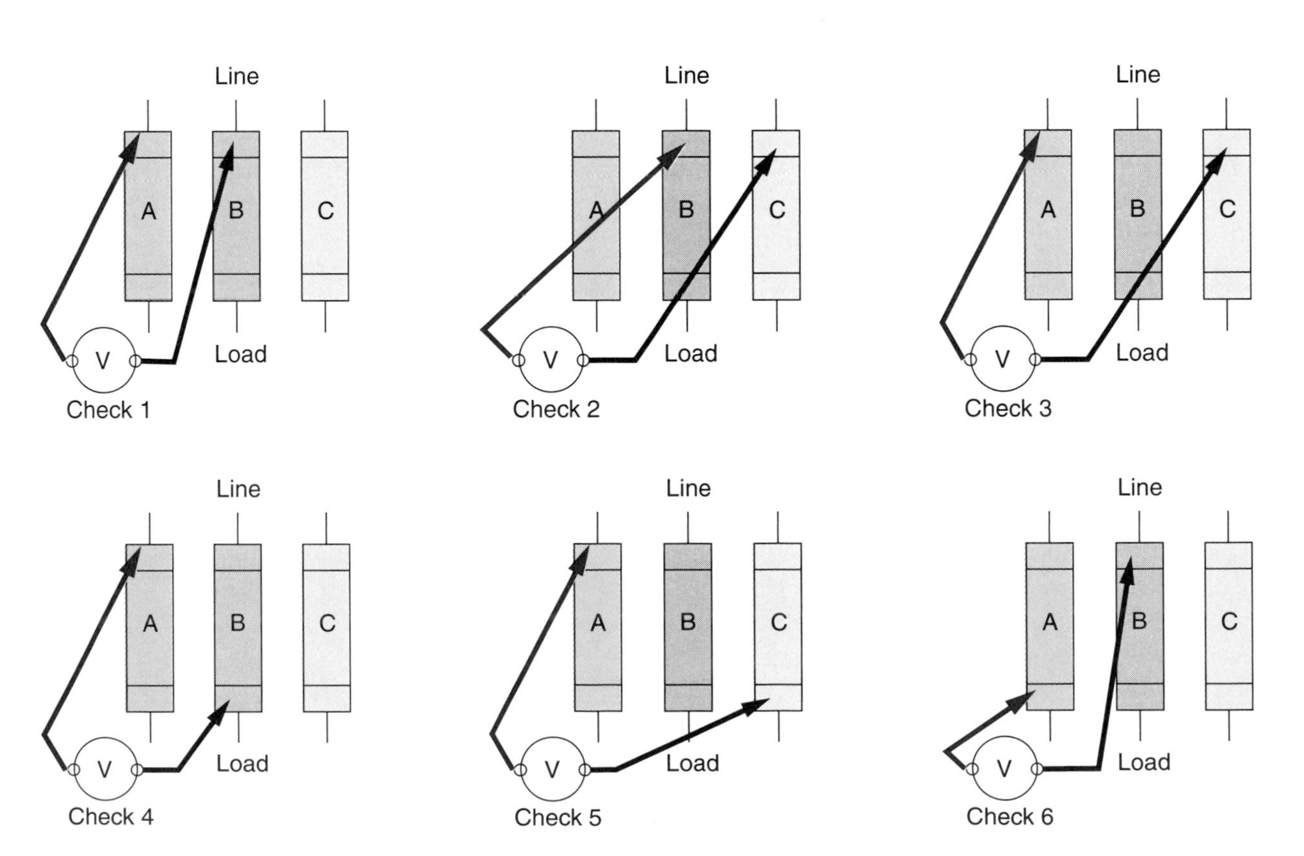

Figure 18-1. When testing a power circuit, use a voltmeter to confirm the voltage at the supply conductors.

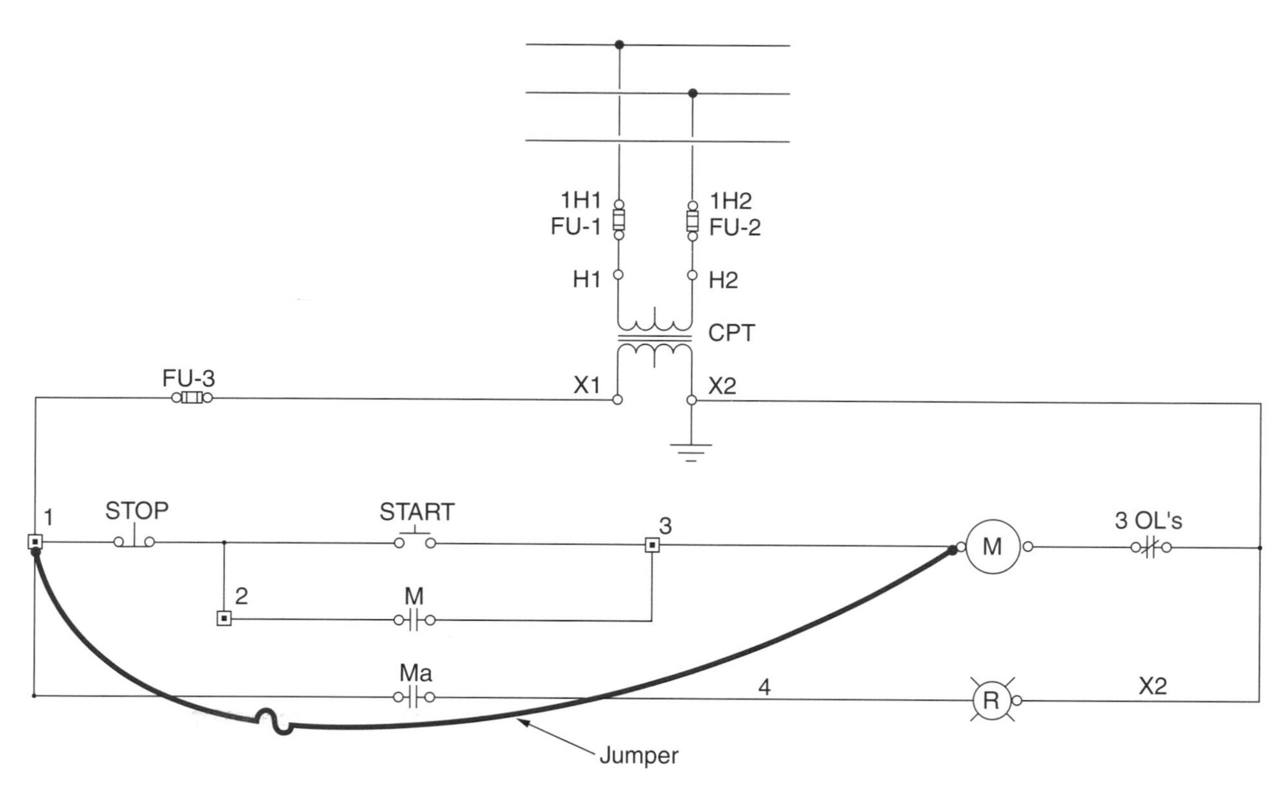

Figure 18-2. Use a jumper to isolate the control circuit. The fuse must be large enough to let enough current through for the starting coil.

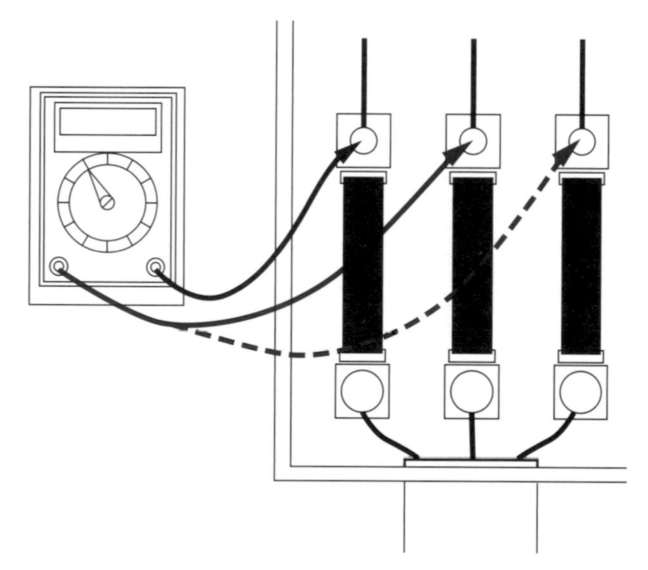

Figure 18-3. When a fuse blows or a breaker trips, make sure power is present up to the overcurrent protective device using a voltmeter.

Troubleshooting control circuits

If the control circuit may be the source of the trouble, follow the steps illustrated in **Figure 18-4:**

1. Check the input voltage to the control transformer.
2. Check the output voltage at the control transformer.
3. Verify the current draw of the control circuit.
4. Isolate the control circuit with a fused jumper. If the motor starts with the jumper installed, the control circuit is faulty.
5. If the motor doesn't start with the jumper installed, check the voltage across the overloads.

Troubleshooting control transformers

Sometimes the control transformer supplying voltage to the control circuit is defective and must be replaced. Check the input and output voltage of the transformer, **Figure 18-5.** The voltage should be within 7.5% of its rating.

Sample Problem 18-1

Problem: A control transformer that steps voltage down from 480 volts to 120 volts is being tested. What are acceptable voltage readings for the input and the output?

Solution: Acceptable voltage readings are within 7.5% of the nominal voltage:

$$480 \text{ V} \times 0.075 = 36 \text{ V}$$
$$\text{Acceptable input range} = 444 \text{ V to } 516 \text{ V}$$

$$120 \text{ V} \times 0.075 = 9 \text{ V}$$
$$\text{Acceptable output range} = 111 \text{ V to } 129 \text{ V}$$

The acceptable input voltage range is 444 volts to 516 volts. The acceptable output voltage range is 111 volts to 129 volts.

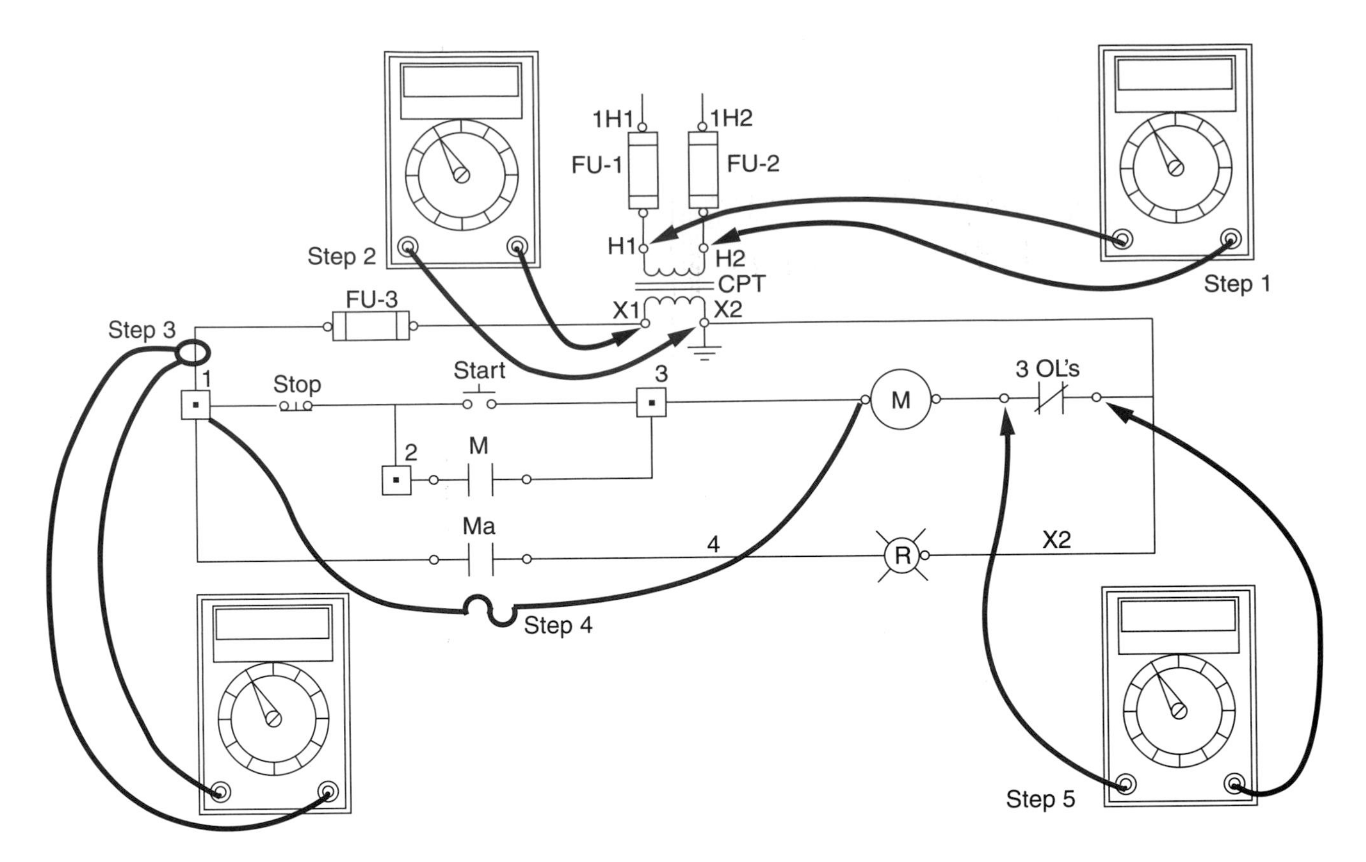

Figure 18-4. Troubleshooting a control circuit.

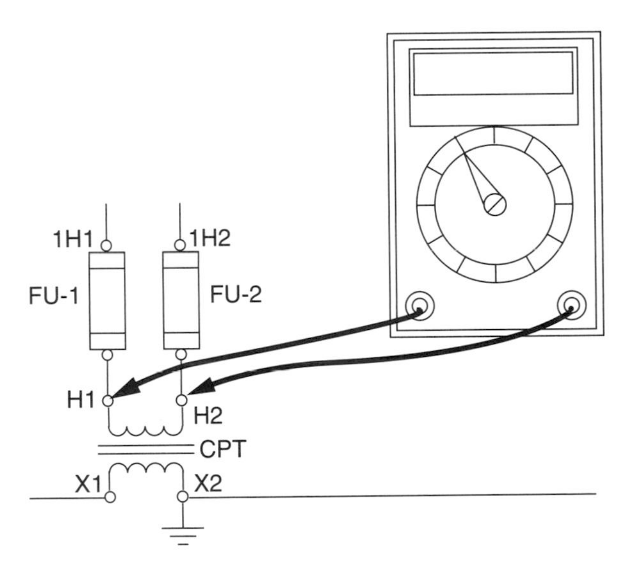

Figure 18-5. When troubleshooting a control transformer, use a voltmeter to check input and output voltage.

Control transformers must also be checked for grounding problems. Performing a ground test on a control transformer can be done easily. See **Figure 18-6.**

Motor Repair

Motors are generally the most expensive electrical equipment within a commercial or light industrial operation. Motors are dependable and require only a minimum amount of maintenance. However, they eventually fail and must be repaired or replaced.

Small defective motors are usually discarded and replaced with new ones. It costs less to replace a small motor than it does to repair it. Large motors are replaced to prevent downtime, and the old motor is sent to the shop for repair. The repaired motor can be placed in service later.

Troubleshooting motors involves two basic operating concepts: determine the cause of failure and reduce the chance of failure occurring again.

In general, a motor may fail for one of several reasons:

- High or low voltage.
- Unbalanced voltage or current at the source.
- Excessive load.
- Overheating.
- Single-phasing.
- Excessive cycling.
- Excessive moisture.
- Motor control center problems.

High or low voltage

Check for high or low voltage at the motor circuit source, **Figure 18-7.** The voltage should be within 7.5% of the motor rating. If the voltage is excessively low or

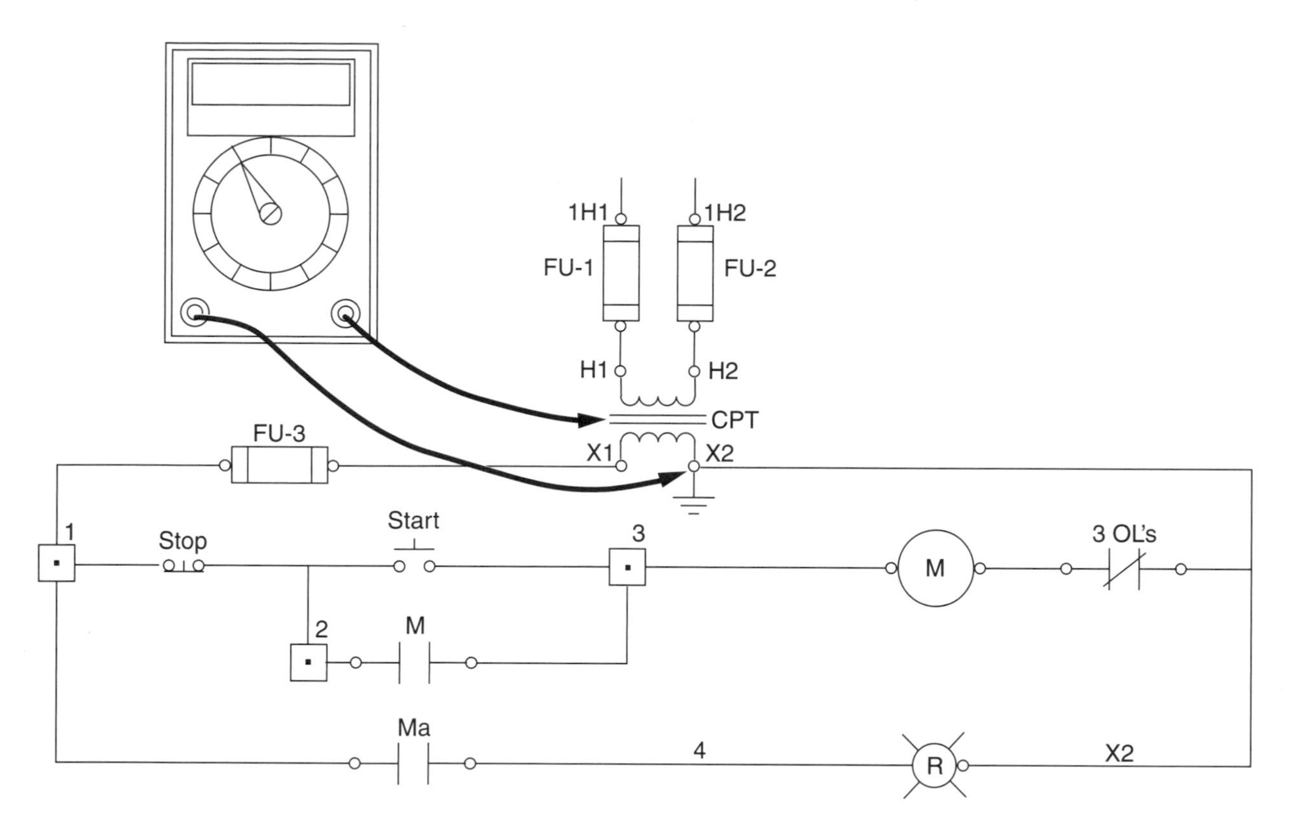

Figure 18-6. Checking for grounding problems in a control transformer.

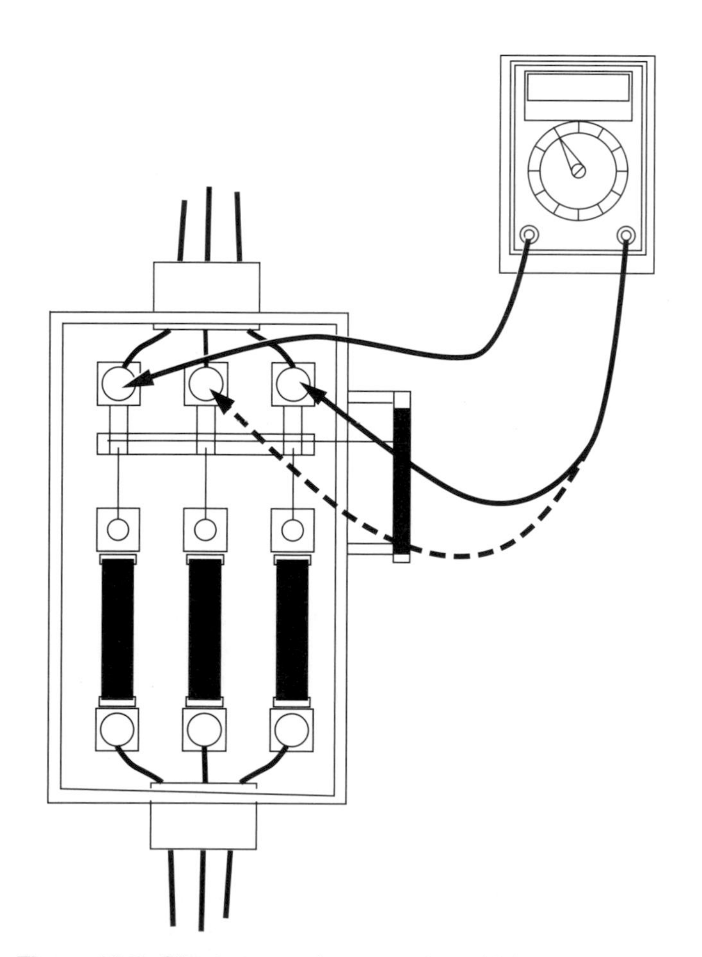

Figure 18-7. Check motors for excessively high or low voltage at the power supply.

high, this must be corrected before placing another motor in service.

Unbalanced voltage or current at the source

An unbalanced voltage at the power source is detrimental to the motor windings. Unequal voltages cause the current to be several times higher (relative to the voltage imbalance) and heat the motor windings. This reduces the torque, causing vibration and, eventually, bearing or winding damage.

The voltage should be checked with a voltmeter to ensure that the feeder conductors are carrying the same voltage to the motor leads. See **Figure 18-8.** Any significant differences in the phase voltages should be corrected.

Excessive load

Excessive loading is one of the major causes of motor breakdown. When motors have excessive loads placed on them, they try to compensate by working harder than they are designed to work. The motor will draw excessive current. This excess current causes heat, and the additional heat imposed in the windings will cause the motor to fail prematurely.

Check the current drawn by the motor to verify that it does not exceed the current rating. The current shown on the nameplate is the maximum rated current for the motor. During normal motor operation, the current draw should be somewhat less. See **Figure 18-9.**

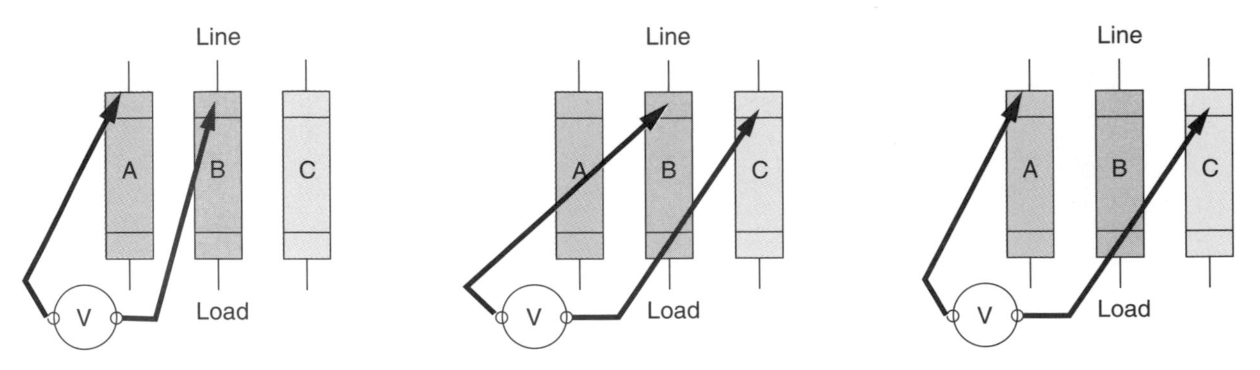

Figure 18-8. Feeder conductors should have equal voltage per phase.

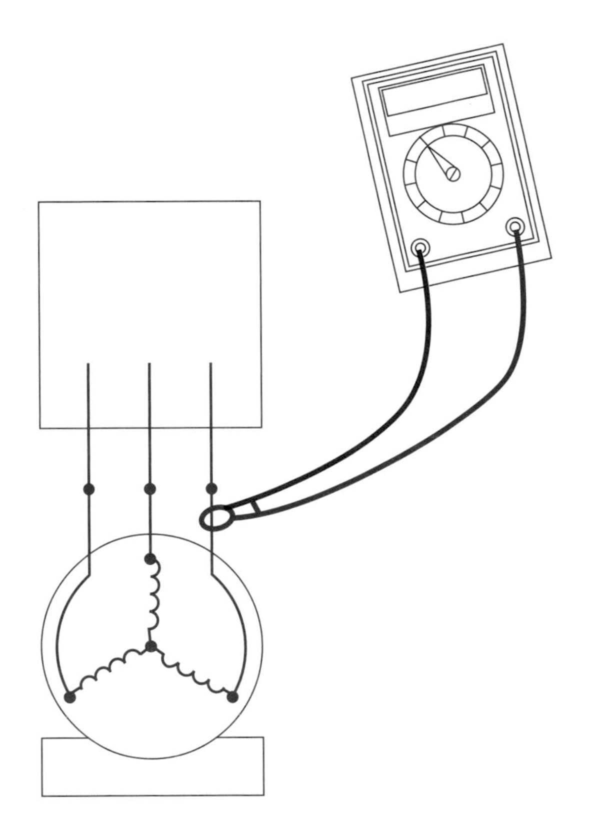

Figure 18-9. Use a clamp-on meter to check the current draw by the motor. If the current is too high, the motor is overloaded.

Overheating

Motors should be checked periodically to make sure air passages are free of dirt and other debris. If the air passages are clogged for long periods, the motor will overheat and the windings will fail.

Motors that operate in areas where the temperatures are unusually high must have additional means of cooling. Venting will reduce heat buildup and help the motor dissipate heat.

Single-phasing

Single-phasing, a common cause of motor failure, occurs when one phase of the three-phase power opens, leaving the motor running on the remaining two lines. This causes the motor to slip, which in turn creates heat and vibration.

Sometimes, single-phasing occurs when the motor is not running. When the motor is eventually started, it will either start in the wrong direction or simply vibrate and hum.

Excessive cycling

When motors are started, they draw much greater current then when running. Repeated starting, jogging, and reversing will not allow the motor sufficient time to cool. Heat buildup is certain to occur, causing the motor to "burn up" after a relatively short service life.

Excessive moisture

Unless a motor is designed to be placed in a wet environment, as in the case of totally enclosed motors, it will not last long in such conditions. Moisture in any form can be severely damaging to open-frame motors. Precautions should be taken to protect motors from moisture in any form.

Motor control center problems

Motor problems often turn out to be motor control center problems. The table in **Figure 18-10** lists common MCC problems and remedies.

Preventive Maintenance

Preventive maintenance means keeping electrical equipment in good condition so that the operation of the facility can continue with little or no interruption. This keeps problems at a minimum and avoids costly plant shutdowns.

Today's business philosophy places preventive maintenance as a high priority. The reasoning is simple—downtime results in production loss and income loss, and preventive maintenance reduces downtime.

All equipment fails eventually. A regularly implemented maintenance program can substantially lengthen service life and minimize potential hazards. As a bare minimum, the following equipment should be regularly inspected:

- Transformers
- Lighting fixtures
- Cables
- Switchgear components
- Circuit breakers
- Relays
- Motors
- Metering equipment

For many of these items, visual inspection may be sufficient. However, motors and transformers require regularly scheduled maintenance.

Motor maintenance

Motor breakdown results in lost production time, which results in lost revenue. The cost of stopping production to repair a motor may be greater than the value of the motor. Therefore, motor maintenance is needed to prevent and reduce breakdowns and increase the life of the motor.

Required maintenance for a motor is specified in the manufacturer's manual. The following are some general procedures normally required for motor maintenance:

- **Lubrication**—Check oil level and change the oil according to the manufacturer's directions. Use the proper type of lubricant.

CAUTION

Do not use too much lubricant on a motor. Excess lubricant can leak onto the motor windings and cause serious damage.

- **Bearings**—Bearing problems have many causes—excessive loading, poor lubrication, and installation of improper type of bearing. These bearing problems will often result in motor failure. Bearings should be checked regularly with a stethoscope to help detect problems before they become damaging. The temperature of bearing surfaces should also be checked periodically.
- **Motor temperature**—Check the motor's running temperature. If the temperature is outside the recommended range, determine what is causing the problem and correct it. Running too hot or too cold will cause premature motor failure.
- **Rotor and stator**—The gap between the rotor and stator should be measured in several locations. If the gaps are uneven, check for bearing wear.
- **Belt**—Belt tension should be checked. Belts that are too loose or too tight can be adjusted. Worn belts should be replaced.
- **Motor mount**—Visually inspect the motor's anchor bolts and concrete pad for signs of wear.
- **Motor controls**—Keep motor controls clean. Test the moving parts and contactors regularly. Overload relay settings should also be checked.

Transformer maintenance

Transformer maintenance is often neglected because transformers have no moving parts and tend to have a long useful life. However, periodic maintenance of transformers can help to detect potential problems.

The transformer temperature, oil level, and gas pressure (where appropriate) should all be checked. Check bushings and accessories such as fans and pumps. For oil-filled transformers, oil samples can be analyzed to check for oil degradation.

Testing

Testing measures the condition of an electrical device or system. There are five broad categories of testing electrical equipment:

- Factory acceptance testing
- Field proof testing
- Installation acceptance testing
- Planned maintenance testing
- Troubleshooting testing

Factory Acceptance Testing

Factory acceptance testing is conducted prior to shipping from the place of manufacture. The tests are performed to verify the equipment will function according to specifications. An independent testing company or quality control personnel perform the tests, which may be witnessed by the customer or his designated representative.

Field Proof Testing

Field proof testing is performed on the equipment after it is installed, but before it is energized. The tests are performed to verify the equipment is ready for service. Measurements taken during field proof testing can serve as data for future reference.

Installation Acceptance Testing

Installation acceptance testing is similar to field proof testing, except these tests are done after the new equipment is energized. Data recorded at this initial stage is saved for future use. Initial test readings establish excellent baseline comparison data.

Planned Maintenance Testing

Planned maintenance testing is performed at set intervals over the service life of the equipment. The record of test data is extremely important in establishing base information to keep the equipment performing properly. This information also gives the maintenance technicians direct indication of whether equipment performance is improving or deteriorating.

Motor Controls Troubleshooting Chart

Trouble	Cause	Solution
Contacts		
Contact chatter	1. Poor contact in control circuit. 2. Low voltage.	1. Replace the contact device or use holding circuit interlock (three-wire control). 2. Check coil terminal voltage and voltage dips during starting.
Welding or freezing	1. Abnormal inrush of current. 2. Rapid jogging. 3. Insufficient tip pressure. 4. Low voltage preventing magnet from sealing. 5. Foreign matter preventing contacts from closing. 6. Short circuit or ground fault.	1. Check for grounds, shorts, or excessive motor load current, or use larger contactor. 2. Install larger device rated for jogging service. 3. Replace contacts and springs, check carrier for deformation or damage. 4. Check coil terminal voltage and voltage dips during starting. 5. Clean contacts, starters, and control accessories used with very small current or low voltage. 6. Remove fault and check to be sure fuse or breaker size is correct.
Short tip life/Overheating of tips	1. Filing or dressing. 2. Interrupting excessively high currents. 3. Excessive jogging. 4. Weak tip pressure. 5. Dirt or foreign matter on contact surface. 6. Short circuits or ground fault. 7. Loose connection in power circuit. 8. Sustained overload.	1. Do not file silver tips. Rough spots or discoloration will not harm tips or impair their efficiency. 2. Install larger device or check for grounds, shorts, or excessive motor currents. 3. Install larger device rated for jogging service. 4. Replace contacts and springs, check contact carrier for deformation or damage. 5. Clean contact. Take steps to reduce entry of foreign matter into enclosure. 6. Remove fault and check to be sure fuse or breaker size is correct. 7. Clean and tighten. 8. Check for excessive motor load current or install larger device.
Coils		
Open circuit	1. Excessive voltage or high ambient temperature. 2. Incorrect coil. 3. Shorted turns caused by mechanical damage or corrosion. 4. Insufficient voltage, failure of magnet to seal in. 5. Dirt or rust on pole faces. 6. Mechanical obstruction.	1. Check coil terminal voltage, which should not exceed 110% of coil rating. 2. Install correct coil. 3. Replace coil. 4. Check coil terminal voltage, which should be at least 85% of coil rating. 5. Clean pole faces. 6. Check for free movement of contact and armature assembly.

(Continued on the following page.)

Figure 18-10. Possible causes of MCC failure and remedies.

Motor Controls Troubleshooting Chart *Continued*

Trouble	Cause	Solution
Overload relays		
Tripping	1. Sustained overload. 2. Loose or corroded connection in power circuit. 3. Incorrect thermal units. 4. Excessive coil voltage.	1. Check for excessive motor currents or current unbalance, and correct cause. 2. Clean and tighten. 3. Thermal units should be replaced with correct size for the application conditions. 4. Voltage should not exceed 110% of coil rating.
Failure to trip	1. Incorrect thermal units. 2. Mechanical binding, dirt, corrosion, etc. 3. Relay previously damaged by short circuit. 4. Relay contact welded or not in series with contactor coil.	1. Check thermal unit selection table. Install proper thermal units. 2. Replace relay and thermal units. 3. Replace relay and thermal units. 4. Check circuit for a fault and correct condition. Replace contact or entire relay as necessary.
Magnetic and mechanical parts		
Noisy magnet	1. Broken shading coil. 2. Dirt or rust on magnet faces. 3. Low voltage.	1. Replace magnet and armature. 2. Clean. 3. Check coil terminal voltage and voltage dips during starting.
Failure to pick-up and seal	1. No control voltage. 2. Low voltage. 3. Mechanical obstruction. 4. Coil open or overheated. 5. Wrong coil.	1. Check and control circuit for loose connection or poor continuity of contacts. 2. Check coil terminal voltage and voltage dips during starting. 3. Check for free movement of contact and armature assembly. 4. Replace. 5. Replace.
Failure to drop-out	1. Gummy substance on pole faces. 2. Voltage not removed. 3. Worn or corroded parts causing binding. 4. Residual magnetism due to lack of air gap in magnet path.	1. Clean pole faces. 2. Check coil terminal voltage and control circuit. 3. Replace parts. 4. Replace magnet and armature.
Pneumatic timers		
Erratic timing	1. Foreign matter in valve.	1. Replace complete timing head or return timer to factory for repair and adjustment.
Contacts do not operate	1. Misadjustment of actuating screw. 2. Worn or broken parts in snap switch.	1. Adjust per instruction in service bulletin. 2. Replace snap switch.
Limit switches		
Broken parts	1. Excessive travel of actuator.	1. Use resilient actuator or operate within tolerances of the device.
Manual starters		
Failure to reset	1. Latching mechanism worn or broken.	1. Replace starter.

Figure 18-10. Possible causes of MCC failure and remedies.

Troubleshooting Testing

Troubleshooting testing varies greatly, depending on the type of equipment being serviced. However, there are certain tests that are common to troubleshooting many different types of equipment.

Insulation resistance test

An insulation resistance test measures the conductor insulation resistance using a ***megohmmeter (megger)***. The megger can be electronic, motorized, or hand-cranked. It applies a dc voltage of 100 to 5000 volts to the conductor. Good insulation will show a steady increase in its resistance as the voltage is applied. Insulation that is poor or contaminated with grease, dirt, or moisture will have a low resistance.

There are several types of insulation resistance tests, each designed to detect certain characteristics of conductor insulation. **Figure 18-11** illustrates one type of megger available.

High-potential test

A high-potential test (or high-pot test) determines the condition and reliability of conductor insulation by applying high ac voltage to the conductor for short, repetitive periods. See **Figure 18-12.**

Protective device test

Protective device tests include a broad range of tests performed on relays, circuit breakers, instrument transformers, and switchgear components. These tests verify that the device is capable of performing its intended function.

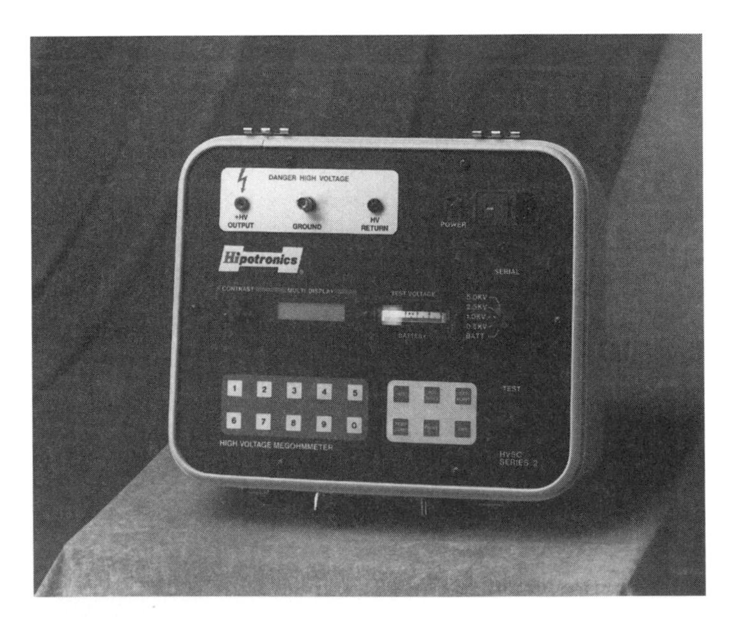

Figure 18-11. A megger (megohmmeter) measures the conductor insulation resistance. (Hipotronics, Inc.)

Figure 18-12. This high-pot tester is used to check the condition of conductor insulation. (AVO International and Biddle Instruments)

Grounding electrode (resistance) test

A grounding electrode test is performed to ensure that the grounding integrity is satisfactory to protect equipment and personnel. A simple ground resistance set is used to perform the test.

Infrared inspection test

This test is used to detect high-temperature areas (hot spots) in switchgear, motor control centers, and other parts of the electrical system. It is an excellent way to determine bad splices, poor terminations, loose connections, damaged insulation, and overloaded circuits. Infrared monitors indicate hot spots on a temperature scale or a screen that displays a bright color spot where the high temperature exists.

The table in **Figure 18-13** lists various tests and the equipment commonly tested. When used in conjunction with periodic maintenance and inspection, these tests will help keep equipment operating efficiently for many years.

Troubleshooting Motors

It is fairly difficult to distinguish dual-voltage motors as wye- or delta-wound when the identifying tags are no longer visible or missing. However, there is an easy way to check and properly identify the leads so the motor can be correctly terminated.

Electrical Equipment Test Guide

	Insulation Resistance	High Potential	Power Factor	Polarity	Winding Resistance	Insulating Liquid
Motors and generators	Yes	Yes	Yes		Yes	
Switchgear	Yes	Yes				
Cables, conductors	Yes	Yes				
Circuit breakers	Yes	Yes				
Power transformers	Yes	Yes	Yes	Yes	Yes	Yes
Instrument transformers	Yes	Yes	Yes	Yes	Yes	
Capacitors	Yes	Yes				
Arresters	Yes	Yes	Yes			

Figure 18-13. This chart shows the type of equipment for which various tests are applicable.

Wye-wound dual-voltage motors have four circuits. One circuit has three leads; the remaining three circuits have two leads each. Thus, there are a total of nine lead wires, **Figure 18-14.**

As can be seen in the illustration, T1-T4 is one two-wire circuit, T2-T5 is another two-wire circuit, T3-T6 accounts for the third two-wire circuit. T7-T8-T9 forms the three-wire circuit. When troubleshooting, you must determine which circuit is which:

1. Using a continuity tester or ohmmeter, connect one probe to any of the T-leads and check for continuity by touching the other probe to each of the remaining eight leads. The lead producing a reading is part of the same circuit. If two leads produce a reading, the three make up the three-wire circuit.
2. Identify the leads for the three-wire circuit as T7, T8, and T9 (in any order). Temporarily mark the leads for the two-wire circuits as T1-T4, T2-T5, and T3-T6.

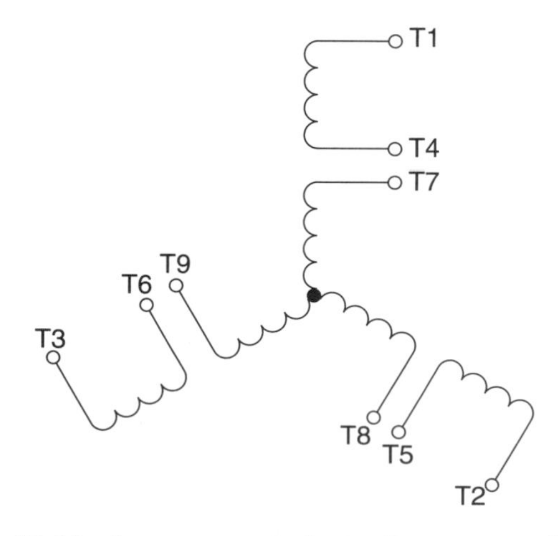

Figure 18-14. A wye-wound dual-voltage motor has four circuits and a total of nine lead wires.

3. Connect the three-wire leads to a 230-volt source and run the motor. Each of the two-wire circuits should have an induced voltage of 120 volts. This can be verified using a voltmeter.
4. With the motor still running, connect T1 to T7. Read the voltage between T4 and T8, and then between T4 and T9. If the voltage readings are 330 volts to 350 volts, T1 and T4 are correctly identified. If the voltage readings are near 120 volts, the labels for T1 and T4 need to be switched.
5. Repeat these steps to determine the correct marking of T2-T5 and T3-T6. Once everything is properly tagged, connect T4, T5, and T6 together and read the voltage between T1, T2, and T3, which should be 225 volts to 235 volts.

A delta-wound three-phase motor has three three-wire circuits, **Figure 18-15.** When troubleshooting, you must determine which leads are associated with each circuit:

1. Using an ohmmeter or continuity tester, identify the three, three-wire circuits in the same manner as described for wye-wound motors.
2. Determine which of the three leads is common between the two coils. This is done by checking the resistance between the leads, as shown in **Figure 18-16.** The resistance between the common lead and another lead is half as much as the resistance between the two "noncommon" leads.
3. Mark the common lead as T1 and the other two leads as T4 and T9. Identify the leads for the other two circuits in the same manner, labeling the common leads as T2 and T3 and the other leads as T5-T7 and T6-T8, respectively.
4. Connect T1, T4, and T9 to a 230-volt power source. Place a jumper between T7 to T4 and start the motor.

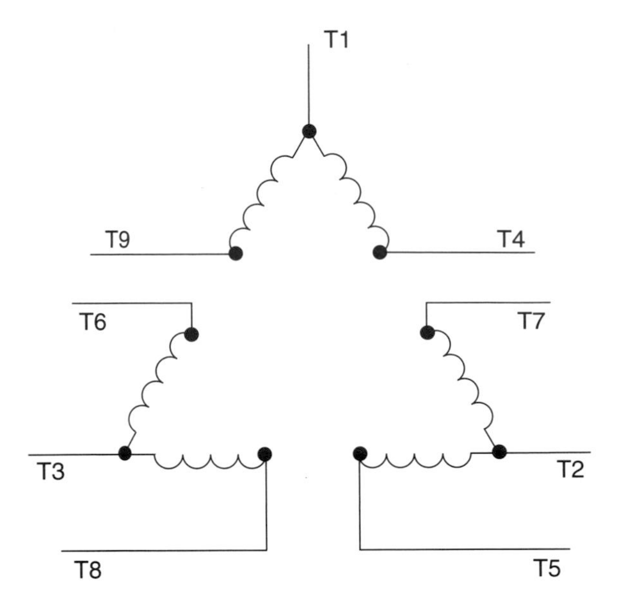

Figure 18-15. A delta-wound, three-phase motor has three three-wire circuits.

5. Read the voltage between T1 and T2. If the voltage is 440 volts to 480 volts, the leads are correctly marked. If the reading is below 400 volts, switch T4 and T9 and repeat the voltage reading. If the voltage is still low, switch T7 and T5 (as well as T9 and T4) and there should be a higher voltage reading.
6. Connect T6 or T8 to T9 and look for a voltage reading of approximately 460 volts between T1 and T3. If the voltage is too low (below 400 volts) switch T6 and T8 and repeat the voltage reading. When everything is correct, permanently mark the leads and the motor is ready to run.

Review Questions

Answer the following questions. Do not write in this book.

1. Why is regularly scheduled maintenance of electrical equipment needed?
2. What are the two basic steps used when troubleshooting a power circuit?
3. When troubleshooting a control circuit, which device should you check first?
4. What is the acceptable voltage range for a control transformer?
5. List eight possible causes of motor failure.
6. Name two causes of motor overheating.
7. When does *single-phasing* occur?
8. Explain how excessive cycling can lead to motor breakdown.
9. List two motor bearing checks that should be regularly performed.
10. What is the purpose of having oil samples from transformers analyzed?
11. Explain the difference between factory acceptance testing and field proof testing.
12. What does an infrared inspection test detect?

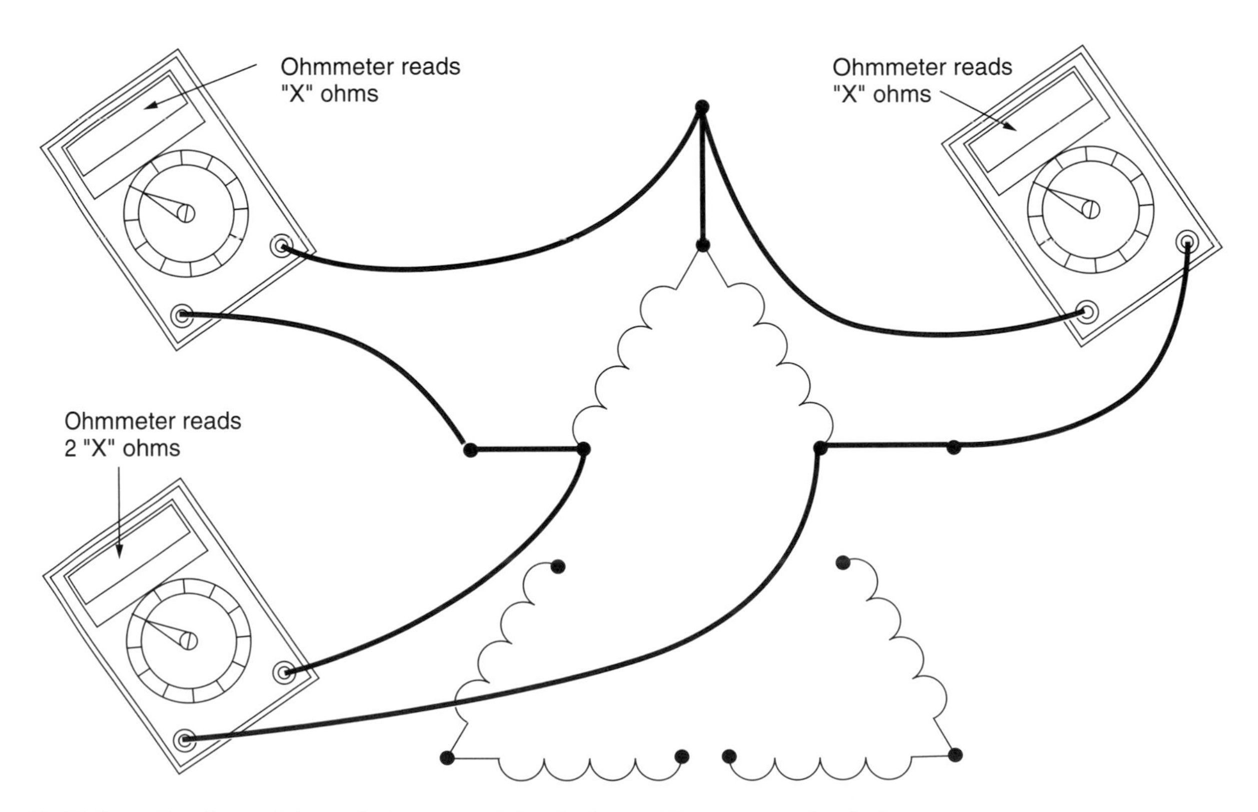

Figure 18-16. Checking the resistance between each lead wire and the common lead wire.

Code Reference Section

The following tables from the *Code* are provided for your reference. These tables are in no way a substitute for the *Code*. Before using any *Code* tables, be sure you have become familiar with all applicable *Code* sections regarding the specific topic.

Motor Full-Load Currents

Table 430.247 Full-Load Current in Amperes, Direct-Current Motors

The following values of full-load currents* are for motors running at base speed.

	Armature Voltage Rating*					
Horse-power	90 Volts	120 Volts	180 Volts	240 Volts	500 Volts	550 Volts
¼	4.0	3.1	2.0	1.6	—	—
⅓	5.2	4.1	2.6	2.0	—	—
½	6.8	5.4	3.4	2.7	—	—
¾	9.6	7.6	4.8	3.8	—	—
1	12.2	9.5	6.1	4.7	—	—
1 ½	—	13.2	8.3	6.6	—	—
2	—	17	10.8	8.5	—	—
3	—	25	16	12.2	—	—
5	—	40	27	20	—	—
7 ½	—	58	—	29	13.6	12.2
10	—	76	—	38	18	16
15	—	—	—	55	27	24
20	—	—	—	72	34	31
25	—	—	—	89	43	38
30	—	—	—	106	51	46
40	—	—	—	140	67	61
50	—	—	—	173	83	75
60	—	—	—	206	99	90
75	—	—	—	255	123	111
100	—	—	—	341	164	148
125	—	—	—	425	205	185
150	—	—	—	506	246	222
200	—	—	—	675	330	294

*These are average dc quantities.

Table 430.247 contains full-load currents for dc motors. The horsepower rating and voltage rating of the motor must be known to determine the voltage rating.

Table 430.248 Full-Load Currents in Amperes, Single-Phase Alternating-Current Motors

The following values of full-load currents are for motors running at usual speeds and motors with normal torque characteristics. The voltages listed are rated motor voltages. The currents listed shall be permitted for system voltage ranges of 110 to 120 and 220 to 240 volts.

Horsepower	115 Volts	200 Volts	208 Volts	230 Volts
1/6	4.4	2.5	2.4	2.2
1/4	5.8	3.3	3.2	2.9
1/3	7.2	4.1	4.0	3.6
1/2	9.8	5.6	5.4	4.9
3/4	13.8	7.9	7.6	6.9
1	16	9.2	8.8	8.0
1 1/2	20	11.5	11.0	10
2	24	13.8	13.2	12
3	34	19.6	18.7	17
5	56	32.2	30.8	28
7 1/2	80	46.0	44.0	40
10	100	57.5	55.0	50

Table 430.248 contains full-load currents for single-phase, ac motors. The horsepower rating and voltage rating of the motor must be known to determine the voltage rating.

Table 430.249 Full-Load Current, Two-Phase Alternating-Current Motors (4-Wire)

The following values of full-load current are for motors running at speeds usual for belted motors and motors with normal torque characteristics. Current in the common conductor of a 2-phase, 3-wire system will be 1.41 times the value given.

The voltages listed are rated motor voltages. The currents listed shall be permitted for system voltage ranges of 110 to 120, 220 to 240, 440 to 480, and 550 to 600 volts.

	Induction Type Squirrel Cage and Wound Rotor (Amperes)				
Horsepower	115 Volts	230 Volts	460 Volts	575 Volts	2300 Volts
½	4.0	2.0	1.0	0.8	—
¾	4.8	2.4	1.2	1.0	—
1	6.4	3.2	1.6	1.3	—
1 ½	9.0	4.5	2.3	1.8	—
2	11.8	5.9	3.0	2.4	—
3	—	8.3	4.2	3.3	—
5	—	13.2	6.6	5.3	—
7 ½	—	19	9.0	8.0	—
10	—	24	12	10	—
15	—	36	18	14	—
20	—	47	23	19	—
25	—	59	29	24	—
30	—	69	35	28	—
40	—	90	45	36	—
50	—	113	56	45	—
60	—	133	67	53	14
75	—	166	83	66	18
100	—	218	109	87	23
125	—	270	135	108	28
150	—	312	156	125	32
200	—	416	208	167	43

Table 430.249 contains full-load currents for two-phase, ac motors. The horsepower rating and voltage rating of the motor must be known to determine the voltage rating.

Table 430.250 Full-Load Current Three-Phase Alternating-Current Motors

The following values of full-load currents are typical for motors running at speeds usual for belted motors and motors with normal torque characteristics.

The voltages listed are rated motor voltages. The currents listed shall be permitted for system voltage ranges of 110 to 120, 220 to 240, 440 to 480, and 550 to 600 volts.

	Induction Type Squirrel Cage and Wound Rotor (Amperes)							Synchronous-Type Unity Power Factor* (Amperes)			
Horsepower	115 Volts	200 Volts	208 Volts	230 Volts	460 Volts	575 Volts	2300 Volts	230 Volts	460 Volts	575 Volts	2300 Volts
½	4.4	2.5	2.4	2.2	1.1	0.9	—	—	—	—	—
¾	6.4	3.7	3.5	3.2	1.6	1.3	—	—	—	—	—
1	8.4	4.8	4.6	4.2	2.1	1.7	—	—	—	—	—
1 ½	12.0	6.9	6.6	6.0	3.0	2.4	—	—	—	—	—
2	13.6	7.8	7.5	6.8	3.4	2.7	—	—	—	—	—
3	—	11.0	10.6	9.6	4.8	3.9	—	—	—	—	—
5	—	17.5	16.7	15.2	7.6	6.1	—	—	—	—	—
7 ½	—	25.3	24.2	22	11	9	—	—	—	—	—
10	—	32.2	30.8	28	14	11	—	—	—	—	—
15	—	48.3	46.2	42	21	17	—	—	—	—	—
20	—	62.1	59.4	54	27	22	—	—	—	—	—
25	—	78.2	74.8	68	34	27	—	53	26	21	—
30	—	92	88	80	40	32	—	63	32	26	—
40	—	120	114	104	52	41	—	83	41	33	—
50	—	150	143	130	65	52	—	104	52	42	—
60	—	177	169	154	77	62	16	123	61	49	12
75	—	221	211	192	96	77	20	155	78	62	15
100	—	285	273	248	124	99	26	202	101	81	20
125	—	359	343	312	156	125	31	253	126	101	25
150	—	414	396	360	180	144	37	302	151	121	30
200	—	552	528	480	240	192	49	400	201	161	40
250	—	—	—	—	302	242	60	—	—	—	—
300	—	—	—	—	361	289	72	—	—	—	—
350	—	—	—	—	414	336	83	—	—	—	—
400	—	—	—	—	477	382	95	—	—	—	—
450	—	—	—	—	515	412	103	—	—	—	—
500	—	—	—	—	590	472	118	—	—	—	—

*For 90 and 80 percent power factor, the figures shall be multiplied by 1.1 and 1.25, respectively.

Table 430.250 contains full-load currents for three-phase, ac motors. The horsepower rating and voltage rating of the motor must be known to determine the voltage rating. Note the separate columns for induction type and synchronous type motors.

Table 430.251(A) Conversion Table of Single-Phase Locked-Rotor Currents for Selection of Disconnecting Means and Controllers as Determined from Horsepower and Voltage Rating

For use only with Sections 430.110, 440.12, 440.41, and 455.8(C).

	Maximum Locked-Rotor Current in Amperes, Single Phase		
Rated Horsepower	115 Volts	208 Volts	230 Volts
½	58.8	32.5	29.4
¾	82.8	45.8	41.4
1	96	53	48
1 ½	120	66	60
2	144	80	72
3	204	113	102
5	336	186	168
7 ½	480	265	240
10	600	332	300

Table 430.251(A) is used to select ratings of disconnecting means and controllers for single-phase motors. Note the applicable sections listed in the table.

Table 430.251(B) Conversion Table of Polyphase Design B, C, and D Maximum Locked-Rotor Currents for Selection of Disconnecting Means and Controllers as Determined from Horsepower and Voltage Rating and Design Letter
For use only with Sections 430.110, 440.12,* 440.41,* and 455.8(C).

Rated Horsepower	Maximum Motor Locked-Rotor Current in Amperes Two- and Three-Phase, Design B, C, and D					
	115 Volts	200 Volts	208 Volts	230 Volts	460 Volts	575 Volts
	B, C, D	B, C, D	B, C, D	B, C, D	B, C, D	B, C, D
½	40	23	22.1	20	10	8
¾	50	28.8	27.6	25	12.5	10
1	60	34.5	33	30	15	12
1 ½	80	46	44	40	20	16
2	100	57.5	55	50	25	20
3	—	73.6	71	64	32	25.6
5	—	105.8	102	92	46	36.8
7 ½	—	146	140	127	63.5	50.8
10	—	186.3	179	162	81	64.8
15	—	267	257	232	116	93
20	—	334	321	290	145	116
25	—	420	404	365	183	146
30	—	500	481	435	218	174
40	—	667	641	580	290	232
50	—	834	802	725	363	290
60	—	1001	962	870	435	348
75	—	1248	1200	1085	543	434
100	—	1668	1603	1450	725	580
125	—	2087	2007	1815	908	726
150	—	2496	2400	2170	1085	868
200	—	3335	3207	2900	1450	1160
250	—	—	—	—	1825	1460
300	—	—	—	—	2200	1760
350	—	—	—	—	2550	2040
400	—	—	—	—	2900	2320
450	—	—	—	—	3250	2600
500	—	—	—	—	3625	2900

*Design A motors are not limited to a maximum starting current or locked rotor current.

Table 430.251(B) is used to select ratings of disconnecting means and controllers for two-phase and three-phase motors. Note the applicable sections listed in the table and the motor design letters.

Table 430.52 Maximum Rating or Setting of Motor Branch-Circuit Short-Circuit and Ground-Fault Protective Devices

	Percentage of Full-Load Current			
Type of Motor	Nontime Delay Fuse[1]	Dual Element (Time-Delay) Fuse[1]	Instantaneous Trip Breaker	Inverse Time Breaker[2]
Single-phase motors	300	175	800	250
AC polyphase motors other than wound-rotor				
Squirrel cage — other than Design B energy effecient	300	175	800	250
Design B energy effecient	300	175	1100	250
Synchronous[3]	300	175	800	250
Wound rotor	150	150	800	150
Direct current (constant voltage)	150	150	250	150

Note: For certain exceptions to the values specified, see 430.54.
[1]The values in the Nontime Delay Fuse column apply to Time-Delay Class CC fuses.
[2]The values given in the last column also cover the ratings of nonadjustable inverse time types of circuit breakers that may be modified as in Section 430.52(C).
[3]Synchronous motors of the low-torque, low-speed type (usually 450 rpm or lower), such as are used to drive reciprocating compressors, pumps, etc., that start unloaded, do not require a fuse rating or circuit-breaker

Table 430.52 is used to select ratings of overcurrent protective devices for motors. Note that the type of protective device and the type of motor determine the device rating.

Allowable Ampacities for Conductors

Table 310.16 Allowable Ampacities of Insulated Conductors Rated 0 Through 2000 Volts, 60°C Through 90°C (140°F Through 194°F), Not More than Three Current-Carrying Conductors in Raceway, Cable, or Earth (Directly Buried), Based on Ambient Temperature of 30°C (86°F)

Size AWG or kcmil	Temperature Rating of Conductor (See Table 310.13.)						Size AWG or kcmil
	60°C (140°F)	75°C (167°F)	90°C (194°F)	60°C (140°F)	75°C (167°F)	90°C (194°F)	
	Types TW, UF	Types RHW, THHW, THW, THWN, XHHW, USE, ZW	Types TBS, SA, SIS, FEP, FEPB, MI, RHH, RHW-2, THHN, THHW, THW-2, THWN-2, USE-2, XHH, XHHW, XHHW-2, ZW-2	Types TW, UF	Types RHW, THHW, THW, THWN, XHHW, USE	Types TBS, SA, SIS, THHN, THHW, THW-2, THWN-2, RHH, RHW-2, USE-2, XHH, XHHW, XHHW-2, ZW-2	
	COPPER			ALUMINUM OR COPPER-CLAD ALUMINUM			
18	—	—	14	—	—	—	—
16	—	—	18	—	—	—	—
14*	20	20	25	—	—	—	—
12*	25	25	30	20	20	25	12*
10*	30	35	40	25	30	35	10*
8	40	50	55	30	40	45	8
6	55	65	75	40	50	60	6
4	70	85	95	55	65	75	4
3	85	100	110	65	75	85	3
2	95	115	130	75	90	100	2
1	110	130	150	85	100	115	1
1/0	125	150	170	100	120	135	1/0
2/0	145	175	195	115	135	150	2/0
3/0	165	200	225	130	155	175	3/0
4/0	195	230	260	150	180	205	4/0
250	215	255	290	170	205	230	250
300	240	285	320	190	230	255	300
350	260	310	350	210	250	280	350
400	280	335	380	225	270	305	400
500	320	380	430	260	310	350	500
600	355	420	475	285	340	385	600
700	385	460	520	310	375	420	700
750	400	475	535	320	385	435	750
800	410	490	555	330	395	450	800
900	435	520	585	355	425	480	900
1000	455	545	615	375	445	500	1000
1250	495	590	665	405	485	545	1250
1500	520	625	705	435	520	585	1500
1750	545	650	735	455	545	615	1750
2000	560	665	750	470	560	630	2000

CORRECTION FACTORS

Ambient Temp. (°C)	For ambient temperatures other than 30°C (86°F), multiply the allowable ampacities shown above by the appropriate factor shown below.						Ambient Temp. (°F)
21–25	1.08	1.05	1.04	1.08	1.05	1.04	70–77
26–30	1.00	1.00	1.00	1.00	1.00	1.00	78–86
31–35	0.91	0.94	0.96	0.91	0.94	0.96	87–95
36–40	0.82	0.88	0.91	0.82	0.88	0.91	96–104
41–45	0.71	0.82	0.87	0.71	0.82	0.87	105–113
46–50	0.58	0.75	0.82	0.58	0.75	0.82	114–122
51–55	0.41	0.67	0.76	0.41	0.67	0.76	123–131
56–60	—	0.58	0.71	—	0.58	0.71	132–140
61–70	—	0.33	0.58	—	0.33	0.58	141–158
71–80	—	—	0.41	—	—	0.41	159–176

* See 240.4(D).

Table 310.16 contains allowable conductor ampacity for an ambient temperature of 86°F when three or less conductors are contained in a raceway. Correction factors are used to adjust the values when there is a different ambient temperature or if there are more than three conductors in the raceway.

Table 310.17 Allowable Ampacities of Single-Insulated Conductors Rated 0 Through 2000 Volts in Free Air, Based on Ambient Air Temperature of 30°C (86°F)

Size AWG or kcmil	Temperature Rating of Conductor (See Table 310.13.)						Size AWG or kcmil
	60°C (140°F)	75°C (167°F)	90°C (194°F)	60°C (140°F)	75°C (167°F)	90°C (194°F)	
	Types TW, UF	Types RHW, THHW, THW, THWN, XHHW, ZW	Types TBS, SA, SIS, FEP, FEPB, MI, RHH, RHW-2, THHN, THHW, THW-2, THWN-2, USE-2, XHH, XHHW, XHHW-2, ZW-2	Types TW, UF	Types RHW, THHW, THW, THWN, XHHW	Types TBS, SA, SIS, THHN, THHW, THW-2, THWN-2, RHH, RHW-2, USE-2, XHH, XHHW, XHHW-2, ZW-2	
	COPPER			ALUMINUM OR COPPER-CLAD ALUMINUM			
18	—	—	18	—	—	—	—
16	—	—	24	—	—	—	—
14*	25	30	35	—	—	—	—
12*	30	35	40	25	30	35	12*
10*	40	50	55	35	40	40	10*
8	60	70	80	45	55	60	8
6	80	95	105	60	75	80	6
4	105	125	140	80	100	110	4
3	120	145	165	95	115	130	3
2	140	170	190	110	135	150	2
1	165	195	220	130	155	175	1
1/0	195	230	260	150	180	205	1/0
2/0	225	265	300	175	210	235	2/0
3/0	260	310	350	200	240	275	3/0
4/0	300	360	405	235	280	315	4/0
250	340	405	455	265	315	355	250
300	375	445	505	290	350	395	300
350	420	505	570	330	395	445	350
400	455	545	615	355	425	480	400
500	515	620	700	405	485	545	500
600	575	690	780	455	540	615	600
700	630	755	855	500	595	675	700
750	655	785	885	515	620	700	750
800	680	815	920	535	645	725	800
900	730	870	985	580	700	785	900
1000	780	935	1055	625	750	845	1000
1250	890	1065	1200	710	855	960	1250
1500	980	1175	1325	795	950	1075	1500
1750	1070	1280	1445	875	1050	1185	1750
2000	1155	1385	1560	960	1150	1335	2000
	CORRECTION FACTORS						
Ambient Temp. (°C)	For ambient temperatures other than 30°C (86°F), multiply the allowable ampacities shown above by the appropriate factor shown below.						Ambient Temp. (°F)
21–25	1.08	1.05	1.04	1.08	1.05	1.04	70–77
26–30	1.00	1.00	1.00	1.00	1.00	1.00	78–86
31–35	0.91	0.94	0.96	0.91	0.94	0.96	87–95
36–40	0.82	0.88	0.91	0.82	0.88	0.91	96–104
41–45	0.71	0.82	0.87	0.71	0.82	0.87	105–113
46–50	0.58	0.75	0.82	0.58	0.75	0.82	114–122
51–55	0.41	0.67	0.76	0.41	0.67	0.76	123–131
56–60	—	0.58	0.71	—	0.58	0.71	132–140
61–70	—	0.33	0.58	—	0.33	0.58	141–158
71–80	—	—	0.41	—	—	0.41	159–176

* See 240.4(D).

Table 310.17 contains allowable conductor ampacity for an ambient temperature of 86°F when conductors are not contained in a raceway. Correction factors are used to adjust the values when there is a different ambient temperature.

Table 310.18 Allowable Ampacities of Three Single-Insulated Conductors Rated 0 Through 2000 Volts, 150°C Through 250°C (302°F Through 482°F), in Raceway or Cable, Based on Ambient Air Temperature of 40°C (104°F)

Size	Temperature Rating of Conductor (see Table 310-13)				Size
	150°C (302°F)	200°C (392°F)	250°C (482°F)	150°C (302°F)	
AWG or kcmil	Type Z	Types FEP, FEPB, PFA	Types PFAH, TFE	Type Z	AWG or kcmil
	COPPER		NICKEL OR NICKEL-COATED COPPER	ALUMINUM OR COPPER-CLAD ALUMINUM	
14	34	36	39	—	14
12	43	45	54	30	12
10	55	60	73	44	10
8	76	83	93	57	8
6	96	110	117	75	6
4	120	125	148	94	4
3	143	152	166	109	3
2	160	171	191	124	2
1	186	197	215	145	1
1/0	215	229	244	169	1/0
2/0	251	260	273	198	2/0
3/0	288	297	308	227	3/0
4/0	332	346	361	260	4/0
			CORRECTION FACTORS		
Ambient Temp. (°C)	For ambient temperatures other than 40°C (104°F), multiply the allowable ampacities shown above by the appropriate factor shown below.				Ambient Temp. (°F)
41–50	0.95	0.97	0.98	0.95	105–122
51–60	0.90	0.94	0.95	0.90	123–140
61–70	0.85	0.90	0.93	0.85	141–158
71–80	0.80	0.87	0.90	0.80	159–176
81–90	0.74	0.83	0.87	0.74	177–194
91–100	0.67	0.79	0.85	0.67	195–212
101–120	0.52	0.71	0.79	0.52	213–248
121–140	0.30	0.61	0.72	0.30	249–284
141–160	—	0.50	0.65	—	285–320
161–180	—	0.35	0.58	—	321–356
181–200	—	—	0.49	—	357–392
201–225	—	—	0.35	—	393–437

Table 310.18 contains allowable conductor ampacity for an ambient temperature of 104°F when three or less conductors are contained in a raceway. Correction factors are used to adjust the values when there is a different ambient temperature or if there are more than three conductors in the raceway.

Table 310.19 Allowable Ampacities of Single-Insulated Conductors Rated 0 Through 2000 Volts, 150°C Through 250°C (302°F Through 482°F), in Free Air, Based on Ambient Air Temperature of 40°C (104°F)

Size	Temperature Rating of Conductor (see Table 310.13)				Size
	150°C (302°F)	200°C (392°F)	250°C (482°F)	150°C (302°F)	
AWG or kcmil	Type Z	Types FEP, FEPB, PFA	Types PFAH, TFE	Type Z	AWG or kcmil
	COPPER		NICKEL OR NICKEL-COATED COPPER	ALUMINUM OR COPPER-CLAD ALUMINUM	
14	46	54	59	—	14
12	60	68	78	47	12
10	80	90	107	63	10
8	106	124	142	83	8
6	155	165	205	112	6
4	190	220	278	148	4
3	214	252	327	170	3
2	255	293	381	198	2
1	293	344	440	228	1
1/0	339	399	532	263	1/0
2/0	390	467	591	305	2/0
3/0	451	546	708	351	3/0
4/0	529	629	830	411	4/0
			CORRECTION FACTORS		
Ambient Temp. (°C)	For ambient temperatures other than 40°C (104°F), multiply the allowable ampacities shown above by the appropriate factor shown below.				Ambient Temp. (°F)
41–50	0.95	0.97	0.98	0.95	105–122
51–60	0.90	0.94	0.95	0.90	123–140
61–70	0.85	0.90	0.93	0.85	141–158
71–80	0.80	0.87	0.90	0.80	159–176
81–90	0.74	0.83	0.87	0.74	177–194
91–100	0.67	0.79	0.85	0.67	195–212
101–120	0.52	0.71	0.79	0.52	213–248
121–140	0.30	0.61	0.72	0.30	249–284
141–160	—	0.50	0.65	—	285–320
161–180	—	0.35	0.58	—	321–356
181–200	—	—	0.49	—	357–392
201–225	—	—	0.35	—	393–437

Table 310.19 contains allowable conductor ampacity for an ambient temperature of 104°F when three or less conductors are contained in a raceway. Correction factors are used to adjust the values when there is a different ambient temperature.

Index of Code References

The following *Code* articles, sections, and tables are discussed in this text. Items are arranged by article and section number. An index of topics begins on page 233.

MasterFormat Reference Section

Division Numbers and Titles

PROCUREMENT AND CONTRACTING REQUIREMENTS GROUP

Division 00 Procurement and Contracting Requirements

SPECIFICATIONS GROUP

GENERAL REQUIREMENTS SUBGROUP

Division 01 General Requirements

FACILITY CONSTRUCTION SUBGROUP

Division 02 Existing Conditions
Division 03 Concrete
Division 04 Masonry
Division 05 Metals
Division 06 Wood, Plastics, and Composites
Division 07 Thermal and Moisture Protection
Division 08 Openings
Division 09 Finishes
Division 10 Specialties
Division 11 Equipment
Division 12 Furnishings
Division 13 Special Construction
Division 14 Conveying Equipment
Division 15 Reserved
Division 16 Reserved
Division 17 Reserved
Division 18 Reserved
Division 19 Reserved

FACILITY SERVICES SUBGROUP

Division 20 Reserved
Division 21 Fire Suppression
Division 22 Plumbing
Division 23 Heating, Ventilating, and Air Conditioning
Division 24 Reserved
Division 25 Integrated Automation
Division 26 Electrical
Division 27 Communications
Division 28 Electronic Safety and Security
Division 29 Reserved

SITE AND INFRASTRUCTURE SUBGROUP

Division 30 Reserved
Division 31 Earthwork
Division 32 Exterior Improvements
Division 33 Utilities
Division 34 Transportation
Division 35 Waterway and Marine Construction
Division 36 Reserved
Division 37 Reserved
Division 38 Reserved
Division 39 Reserved

PROCESS EQUIPMENT SUBGROUP

Division 40 Process Integration
Division 41 Material Processing and Handling Equipment
Division 42 Process Heating, Cooling, and Drying Equipment
Division 43 Process Gas and Liquid Handling, Purification, and Storage Equipment
Division 44 Pollution Control Equipment
Division 45 Industry-Specific Manufacturing Equipment
Division 46 Reserved
Division 47 Reserved
Division 48 Electrical Power Generation
Division 49 Reserved

Note: Contact the Construction Specifications Institute at www.csi.org for a complete index

DIVISION 26 ELECTRICAL

26 00 00 ELECTRICAL
- **26 01 00 Operation and Maintenance of Electrical Systems**
  - 26 01 10 Operation and Maintenance of Medium-Voltage Electrical Distribution
  - 26 01 20 Operation and Maintenance of Low-Voltage Electrical Distribution
  - 26 01 26 Maintenance Testing of Electrical Systems
  - 26 01 30 Operation and Maintenance of Facility Electrical Power Generating and Storing Equipment
  - 26 01 40 Operation and Maintenance of Electrical and Cathodic Protection Systems
  - 26 01 50 Operation and Maintenance of Lighting
- **26 05 00 Common Work Results for Electrical**
  - 26 05 13 Medium-Voltage Cables
  - 26 05 19 Low-Voltage Electrical Power Conductors and Cables
  - 26 05 23 Control-Voltage Electrical Power Cables
  - 26 05 26 Grounding and Bonding for Electrical Systems
  - 26 05 29 Hangers and Supports for Electrical Systems
  - 26 05 33 Raceway and Boxes for Electrical Systems
  - 26 05 36 Cable Trays for Electrical Systems
  - 26 05 39 Underfloor Raceways for Electrical Systems
  - 26 05 43 Underground Ducts and Raceways for Electrical Systems
  - 26 05 46 Utility Poles for Electrical Systems
  - 26 05 48 Vibration and Seismic Controls for Electrical Systems
  - 26 05 53 Identification for Electrical Systems
  - 26 05 73 Overcurrent Protective Device Coordination Study
- **26 06 00 Schedules for Electrical**
  - 26 06 10 Schedules for Medium-Voltage Electrical Distribution
  - 26 06 20 Schedules for Low-Voltage Electrical Distribution
  - 26 06 30 Schedules for Facility Electrical Power Generating and Storing Equipment
  - 26 06 40 Schedules for Electrical and Cathodic Protection Systems
  - 26 06 50 Schedules for Lighting
- **26 08 00 Commissioning of Electrical Systems**
- **26 09 00 Instrumentation and Control for Electrical Systems**
  - 26 09 13 Electrical Power Monitoring and Control
  - 26 09 23 Lighting Control Devices
  - 26 09 26 Lighting Control Panelboards
  - 26 09 33 Central Dimming Controls
  - 26 09 36 Modular Dimming Controls
  - 26 09 43 Network Lighting Controls
  - 26 09 61 Theatrical Lighting Controls

26 10 00 MEDIUM-VOLTAGE ELECTRICAL DISTRIBUTION
- **26 11 00 Substations**
- **26 12 00 Medium-Voltage Transformers**
  - 26 12 13 Liquid-Filled, Medium -Voltage Transformers
  - 26 12 16 Dry-Type, Medium -Voltage Transformers
  - 26 12 19 Pad-Mounted, Liquid-Filled, Medium -Voltage Transformers
- **26 13 00 Medium-Voltage Switchgear**
  - 26 13 13 Medium-Voltage Circuit Breaker Switchgear
  - 26 13 16 Medium-Voltage Fusible Interrupter Switchgear
  - 26 13 19 Medium-Voltage Vacuum Interrupter Switchgear
- **26 18 00 Medium-Voltage Circuit Protection Devices**
  - 26 18 13 Medium-Voltage Cutouts
  - 26 18 16 Medium-Voltage Fuses
  - 26 18 19 Medium-Voltage Lightning Arresters
  - 26 18 23 Medium-Voltage Surge Arresters
  - 26 18 26 Medium-Voltage Reclosers
  - 26 18 29 Medium-Voltage Enclosed Bus
  - 26 18 33 Medium-Voltage Enclosed Fuse Cutouts
  - 26 18 36 Medium-Voltage Enclosed Fuses
  - 26 18 39 Medium-Voltage Motor Controllers

26 20 00 LOW-VOLTAGE ELECTRICAL TRANSMISSION
- **26 21 00 Low-Voltage Overhead Electrical Power Systems**
- **26 22 00 Low-Voltage Transformers**
- **26 23 00 Low-Voltage Switchgear**
- **26 24 00 Switchboards and Panelboards**
  - 26 24 13 Switchboards
  - 26 24 16 Panelboards
  - 26 24 19 Motor-Control Centers
- **26 25 00 Enclosed Bus Assemblies**
- **26 26 00 Power Distribution Units**
- **26 27 00 Low-Voltage Distribution Equipment**
- **26 28 00 Low-Voltage Circuit Protective Devices**
- **26 29 00 Low-Voltage Controllers**
  - 26 29 13 Enclosed Controllers
  - 26 29 23 Variable-Frequency Motor Controllers

26 30 00 FACILITY ELECTRICAL POWER GENERATING AND STORING EQUIPMENT
- **26 31 00 Photovoltaic Collectors**
- **26 32 00 Packaged Generator Assemblies**
  - 26 32 13 Engine Generators
  - 26 32 16 Steam-Turbine Generators
  - 26 32 19 Hydro-Turbine Generators
  - 26 32 23 Wind Energy Equipment
  - 26 32 26 Frequency Changers
  - 26 32 29 Rotary Converters
  - 26 32 33 Rotary Uninterruptible Power Units
- **26 33 00 Battery Equipment**
  - 26 33 13 Batteries
  - 26 33 16 Battery Racks
  - 26 33 19 Battery Units
  - 26 33 23 Central Battery Equipment
  - 26 33 33 Static Power Converters
  - 26 33 43 Battery Chargers
  - 26 33 46 Battery Monitoring
  - 26 33 53 Static Uninterruptible Power Supply
- **26 35 00 Power Filters and Conditioners**
  - 26 35 13 Capacitors
  - 26 35 16 Chokes and Inductors
  - 26 35 23 Electromagnetic -Interference Filters
  - 26 35 26 Harmonic Filters
  - 26 35 33 Power Factor Correction Equipment
  - 26 35 36 Slip Controllers
  - 26 35 43 Static-Frequency Converters
  - 26 35 46 Radio-Frequency-Interference Filters
  - 26 35 53 Voltage Regulators
- **26 36 00 Transfer Switches**

26 40 00 ELECTRICAL AND CATHODIC PROTECTION
- **26 41 00 Facility Lightning Protection**
  - 26 41 13 Lightning Protection for Structures
  - 26 41 16 Lightning Prevention and Dissipation
  - 26 41 19 Early Streamer Emission Lightning Protection
  - 26 41 23 Lightning Protection Surge Arresters and Suppressors
- **26 42 00 Cathodic Protection**
  - 26 42 13 Passive Cathodic Protection for Underground and Submerged Piping
  - 26 42 16 Passive Cathodic Protection for Underground Storage Tank
- **26 43 00 Transient Voltage Suppression**

26 50 00 LIGHTING
- **26 51 00 Interior Lighting**
- **26 52 00 Emergency Lighting**
- **26 53 00 Exit Signs**
- **26 54 00 Classified Location Lighting**
- **26 55 00 Special Purpose Lighting**
  - 26 55 23 Outline Lighting
  - 26 55 29 Underwater Lighting
  - 26 55 33 Hazard Warning Lighting
  - 26 55 36 Obstruction Lighting
  - 26 55 53 Security Lighting
  - 26 55 59 Display Lighting
  - 26 55 61 Theatrical Lighting
  - 26 55 63 Detention Lighting
  - 26 55 70 Healthcare Lighting
- **26 56 00 Exterior Lighting**
  - 26 56 13 Lighting Poles and Standards
  - 26 56 16 Parking Lighting
  - 26 56 19 Roadway Lighting
  - 26 56 23 Area Lighting
  - 26 56 26 Landscape Lighting
  - 26 56 29 Site Lighting
  - 26 56 33 Walkway Lighting
  - 26 56 36 Flood Lighting
  - 26 56 68 Exterior Athletic Lighting

Index

This is an index of topics discussed within this text. For listings of specific *Code* Articles, Sections, and tables, refer to the Index of Code References on page 229.

A

B

C

D

E

F

G

H

I

J

K

L

M

N

O

P

R

S

T

U